NUCLEUS

The History of Atomic Energy of Canada Limited

Robert Bothwell

NUCLEUS

The History of Atomic Energy of Canada Limited

UNIVERSITY OF TORONTO PRESS

Toronto Buffalo London

Toronto Buffalo London
Printed in Canada
ISBN 0-8020-2670-2

Printed on acid-free paper

Canadian Cataloguing in Publication Data

Bothwell, Robert, 1944–
Nucleus: the history of Atomic Energy of Canada Limited

Includes bibliographical references and index.
ISBN 0-8020-2670-2

1. Atomic Energy of Canada Limited – History.
2. Nuclear industry – Canada – History.
I. Title.

HD9698.C34A76 1988 621.48'0971 C88-093265-1

Contents

Foreword vii
Acknowledgements ix
Introduction xi
Abbreviations xix

1 Early Days 3
2 Errand in the Wilderness 54
3 Atomic Energy? 106
4 A Canadian Reactor 148
5 Nuclear Power Decisions 177
6 A Competitive World 212
7 The Flight of the Arrow 247
8 The Burden of Proof: Douglas Point 279
9 Pickering 300
10 Experience Abroad 346
11 Big Science 393
12 Epilogue; or the Price of Success 422

Notes 453
A Note on Sources 503
Index 510

Foreword

Nothing is as perishable, or as forgotten, as the recent past. This phenomenon is as true of companies as it is of individuals. As I discovered when I became president of Atomic Energy of Canada Limited, the usable memory bank of a large company may go back five years, or ten, but in the nature of things it is confined to the recollections of the employees who have had the opportunity, and the time, to dwell on the past and its meaning. As retirements and death took their toll, AECL's corporate memory was dwindling and our company faced the prospect of losing its history.

In my view it is important that this should not happen. AECL and its predecessor organizations have played a major part in the development of Canadian science and engineering over the past forty-five years. We have a unique history as one of the principal scientific arms of the government of Canada, but AECL is also the centre of a significant and highly skilled branch of Canadian industry. AECL has fostered the development of a distinctly Canadian nuclear reactor system, the CANDU, and a flourishing radiochemical company; and AECL's employees are proud of their company's accomplishment.

AECL is a publicly owned company. As such it is regularly poked, prodded, and scrutinized by the regulatory and supervisory organs of the Government of Canada. But such examinations can only tell part of the story: they answer the question 'what?' but not 'why?' or even 'how?' As a crown company, AECL has the responsibility to offer an accounting of its activities, in as clear and comprehensible a form as possible. With the passing of our founding generation, it became clear that only a history could do that.

To that end we engaged a prominent Canadian historian, Professor Robert Bothwell of the History Department of the University of Toronto. Professor Bothwell is a specialist in the history of twentieth-century Canada, and he is well known for his earlier history of Eldorado Nuclear and his biography of C.D. Howe. We gave Professor Bothwell access to all our files, except personnel. Some of our veterans read his drafts and offered criticism, but their role throughout was advisory rather than supervisory. Professor Bothwell was free to find his own evidence, and to draw his own conclusions.

Much of the evidence lies in documents, but much of it came from the people who shaped AECL's history. Consequently, although this is a corporate history, it is also essentially the history of the generation of Canadians who took up the challenge of atomic energy – of, for, and in Canada. Dr Bothwell has captured their story in the pages that follow. He has also brought to life their spirit. Our founder, C.D. Howe, caught that spirit in 1942 when he set Canadian atomic research in motion: 'Okay,' he said 'let's go!'

JAMES DONNELLY
President, Atomic Energy of Canada Limited
January 1988

Acknowledgements

Any book of this kind, depending as it does on frequent and lengthy travel, makes heavy demands on the organization and tolerance of the author's family; to my wife Heather and daughter Eleanor I can only offer thanks and gratitude.

Thanks and gratitude are also due to my research assistants, especially Paul Marsden, who for two years coped valiantly with disorganized and dusty papers and managed this project's book-keeping; for that, and for good advice throughout, I am most grateful. Linda Goldthorp and Ruth Fawcett also worked long hours through several summers to retrieve lost and doubtful documents, sometimes in unusual circumstances, while Anne Smith did yeoman service in the Public Archives of Canada. Liz Evans in England was extraordinarily thorough.

Lorna Arnold, John Clarke, and Robin Nicholson in Great Britain and Bertrand Goldschmidt in France were both encouraging and helpful throughout. The US Atomic Energy Commission historians received their Canadian colleagues hospitably, but their procedures are slow and their staff (doubtless) few. Documents from Washington were received too late; they do not modify any of the conclusions in this book, although they shed some light and some colour on American support for heavy-water research in the 1960s and on US assistance in the Chalk River clean up of 1952–3. Ontario Hydro opened its archives to research, and its veterans devoted much time and effort to helping this history. The same cannot be said for the custodians of the Ontario premiers' papers.

Although other historians have received access to these rare documents, my own access was long delayed by the Ontario cabinet secretary; grudging permission came too late to be of help in this volume. By contrast, the federal government's various agencies, especially the Privy Council Office and the Department of External Affairs, were extremely prompt and helpful.

Two former presidents of AECL, W.J. Bennett and Lorne Gray, devoted long hours to answering questions about the company; both also made documents available. Les Cook, an AECL veteran, generously contributed his own draft memoirs and a great deal of time to answering questions about his time at AECL; his lively interest is much appreciated. David Kirkbride of CIL gave me a morning, and useful photographs and documents.

At AECL, Ernie Siddall, Ray Burge, Gene Critoph, and Ted Smale read drafts of the manuscript and offered comments; they are not, however, responsible for the final result. Ray Burge, in addition, procured photos and illustrations from the AECL collection. When not otherwise stated, that is their source. Ron Veilleux, in addition to reading the manuscript, acted as liaison to the rest of the company; Arlene Brunke very capably managed the network of computers, computer specialists, couriers, phone calls, and paper mounds that were necessary to sustain this history. At CRNL, Ruth Everson and Carol Lachapelle greatly assisted the digging out of files. In addition, I am indebted to my friend Norman Hillmer of the Historical Division of the Department of National Defence, who read the manuscript and offered valuable comments.

Finally, I wish to thank Rosemary Shipton and Mary McDougall Maude, who furnished editorial counsel on the structure of the book; Rosemary Shipton alone confronted the many glitches that then occurred as this book, with its attendant new technology, lurched towards publication. Her intrepid response to computer disasters several times kept the book on track and on time.

Introduction

The appearance of another book on Canadian nuclear policy is bound to evoke a highly mixed response. Those already interested in the subject will apply tests of origin and attitude. Having ascertained where the book stands they will alternatively read and profit, or abhor and condemn. They will then await the next book or article, and start over.

This is a pity. Any debate frozen in time and attitude necessarily contributes little to public understanding, either of issues or, more modestly, of the circumstances or context of a large public question. And atomic energy is a very large public question. It has accounted, over time, for a high proportion of this country's capital investment. The expenditure has created a large and highly skilled workforce; but it has caused critics to wonder whether 'the atomic industry' is not a leviathan racing out of control. It has been an important economic factor; but it has also raised eyebrows and stimulated questions as to whether, after all, the money poured into atomic energy has been worth it. Yet it contributes, at the time of writing, 46.5 per cent of the electric supply in Canada's largest province, Ontario, 42.9 per cent in New Brunswick, as well as a minuscule proportion of Quebec's. It poses undoubted hazards to health and even life, should something go awry; but the number of people harmed by atomic energy is likely to be fewer than the number harmed by sulphur dioxide emissions from coal-burning alternative power stations.

The agenda of paradoxes could be extended. Even a truncated list indicates the importance of atomic power as a public issue in Canada in the 1980s and 1990s. Nor would the question vanish if, like the genie in Aladdin's fable, atomic power could be wished back into some elemental bottle and pitched into the sea. The knowledge of atomic technology will persist and the potential for its use will continue, if not in Canada then in other of the twenty-six countries that currently produce atomic power – or possibly in the much larger number of states that have sufficient knowledge to dabble in atomic reactions, for good or ill.

The possibility of good or evil has been present in atomic energy from the start, as it is in almost any form of human activity. When Alfred Nobel, the Swedish inventor, devised a new form of explosive in the nineteenth century, he believed that he had added to humanity's store of productivity and knowledge, and, not coincidentally, safety; but the features that made dynamite easier and safer to use in a mine shaft also made high explosive bombs safer and easier to handle.

The fact that atomic energy has up until now been a relatively arcane and expensive field was responsible for its translation from university laboratories to factories owned, operated, and directed by governments. In all countries with a nuclear capability, the state exercises a more or less direct authority over the atomic arts. So it does, and has done, in Canada.

The history of atomic energy in Canada begins, as is often the case in the history of Canadian enterprise, outside the country. Nuclear fission, nuclear weapons, and nuclear energy are an international phenomenon, and Canada's atomic energy project grew up in a context that stretched far beyond its borders. The nature of the international connection did not remain constant, of course. It began with a transfer of technology from Great Britain to Canada, continued with the development of co-operation with the United States, and moved on into a situation where Canada competed in reactor design and sales with its two erstwhile partners.

Canada's acquisition of nuclear technology cannot be described as the fruit of anything resembling an 'industrial strategy'; it was, rather, military in its intention. But it was scientific in nature, and the scientific elements gradually came to predominate in the years after 1945. Thus, while the world was buzzing with the news of Hiroshima and the military potential of the atomic bomb, Canada was moving away from the direct military use of nuclear weapons and towards nuclear power. That Canada could do so was a luxury afforded by what came to be called the 'American nuclear umbrella.' Nor did Canada's abstention from producing its own weapons inhibit it from co-operating where it could in the development of American or British nuclear weaponry. The fact remains that the Canadian atomic laboratory at Chalk River had no specific weapons component, and that fact alone helped to shape the nature of the Canadian nuclear program.

If Canada's nuclear technology was imported, so were many of the technicians and scientists who made it work. The progenitor of the original Canadian program, Hans von Halban, was a cosmopolitan refugee; and his two successors as heads of the Canadian laboratory, John Cockcroft and W.B. Lewis, were British. So too were many of the staff both before and after 1945, from scientists down to laboratory assistants and shop hands. For what Canada did not have in the way of skilled labour in 1945, it imported. Atomic energy would have been very different, and probably less successful, had it not been so.

Most of Canada's atomic activity was concentrated at its Chalk River laboratory, a hundred miles northwest of the city of Ottawa. To house the staff, the National Research Council built a suburb in the wilderness and poured money into the creation of a model village, Deep River. It was a place that its staff either loved or left; its existence contributed substantially to the esprit de corps of Canada's atomic scientists. It may, however, have contributed to a sense of isolation and possibly of elitism that detracted from the objectives that the laboratory and especially its leader, W.B. Lewis, wished to accomplish.

While Chalk River looked inward, secure in its consciousness of its own rectitude and proud of its international reputation, its political bases were shifting. Once unique, by 1970 it was no longer so; once buoyed by a postwar popular enthuasism for science and scientific marvels, it was ill-equipped to deal with the drift of public opinion into a suspicion of science, and especially of things nuclear.

Nuclear technology was a large investment for Canada at the time, and it remained so through the 1950s and 1960s. In the late 1940s it was the biggest division of the federal government's National Research Council; it grew so large that it was separated into a crown company of its own, Atomic Energy of Canada Limited (AECL), the principal subject of this book. It seems evident that the responsible minister, C.D. Howe (jokingly but not inaccurately known as the 'Minister of Everything'), saw AECL as a hybrid balanced between being something of a defence project and something of an industrial laboratory. When the balance tipped a few years later towards nuclear power, Howe and AECL's then president, W.J. Bennett, gave the company a mandate to develop nuclear energy.

In Howe's view, AECL's real purpose was research and development. Granted, it was research and development towards an appointed end, nuclear power. But nuclear power, once achieved, should be turned over to private industry for manufacture, and to the utilities that would be expected to buy it. To this end, Bennett and Howe organized the company to ensure maximum co-operation between designer (AECL), manufacturer (Canadian General Electric, CGE), and customer. Here, then, was a clear industrial strategy of a kind often sought but not usually found in Canada.

Once the customer was reached, AECL and its product moved across Canada's delicate jurisdictional boundaries, from 'federal Canada' into 'provincial Canada.' In Canada, electric utilities fall under provincial jurisdiction, and by the 1950s the several provinces directly owned most of them. The story of atomic energy therefore becomes a tale not merely of industrial strategy as directed by Ottawa, but of the smaller and sometimes different policies pursued by Canada's provinces: in this case, the province mostly concerned was Ontario, which had the greatest need of a new source of electricity as well as a number of recently discovered uranium mines.

The relationship between AECL and Ontario Hydro is therefore central to the development of Canada's nuclear industry. That Howe was wise to insist on a close connection between designer and customer – that the supplier adapt a product to the customer rather than force the customer to adapt his needs to what the supplier unveils – soon became evident, and is even more evident when industry-utility quarrels in other countries are scrutinized. The relative, though not total, absence of infighting in Canada is a contrast to the experience of Britain, France, the United States, and a number of other nuclear powers.

Howe would probably have said that Canada, as a small country, could not afford the luxury of jurisdictional squabbling over reactors. Bennett, who was well aware of the financial and technological limitations of Canadian industry, also took the position that AECL could afford to develop only one kind of reactor. Advised by W.B. Lewis, and influenced by the success of Canada's first reactor, NRX, Bennett chose the heavy-water-moderated, natural-uranium-fuelled variety, a kind that was not being strongly pushed in any other country. AECL's past technology, and its fortuitous strength in one particular area, therefore determined the direction that Canada's nuclear technology would take.

These points have made the Canadian experience of nuclear reactors unusual, but there are many other developments or incidents worth attention. Different governments have different strategies, and any plan needs to be adapted to changing circumstances. Thus, AECL removed CGE from Canada's nuclear triumvirate in 1958 and entered into a bilateral partnership with Ontario Hydro. Weakened and ultimately undermined by this development, CGE was unable to make a success of the nuclear business and dropped out as a supplier of reactors a decade later. It remains a lively question whether the right decision was made and whether, as a consequence, AECL was right to move into CGE's place as a purveyor of a reactor package in the domestic and the international marketplace.

In the years after 1958 AECL, under Bennett's successor Lorne Gray, steadily expanded as it strove to create a reliable engineering capacity for the construction of a growing number of reactors. Its earliest effort, the first CANDU built at Douglas Point on Lake Huron, was for some time its least successful reactor; but its second, at Pickering east of Toronto, worked very much better. So well, in fact, did Pickering work that it became a substantial money-maker for the company and a retrospective justification of its daring decision to place itself, and by extension the government of Canada, in business as a consulting engineer.

Two competing and contrasting visions of AECL have therefore overlapped. The first, of a basic research and development company, in which the government's role remains strictly limited and in which success is achieved when public investment is wound up, gave way to a second in which public investment at more and more advanced stages was thought to be necessary. This increasing involvement was justified initially to protect a large public investment, and then to recoup some of that investment through domestic power production and reactor exports.

Exports return us to the international marketplace, and to the performance of AECL as a seller of CANDUS and CANDU concepts. There have been successes: India, where a second generation of Indian-made heavy-water reactors succeeded the first, Canadian-designed ones; Taiwan, where the Nationalist government paid cash for a Canadian-designed research reactor; South Korea, where a CANDU reactor set performance records. But India went ahead and detonated a nuclear device ('peaceful,' so they said, and certainly peacefully), thereby provoking a severing of nuclear relations by Canada. Taiwan, the refuge of a geriatric political party, was excluded as a customer when Canada recognized mainland China. And reactor sales to Argentina and South Korea provoked a political storm in Canada during the mid-1970s, when details of their financing were made public.

As an arm of the Canadian government AECL enjoyed many tangible benefits, but as an actor in the international marketplace it was only one company among many. It was, however, one of the few handicapped by standards of political accountability. The role and performance of crown corporations must also be a serious factor in assessing AECL's record, for while a crown company may look like an ordinary company, AECL's experience shows that it is anything but.

The history of Canada's nuclear technology, and of the company that more than any other institution has embodied that technology, is variously the history of an episode in Canada's economic development, of Canada's international relations, of Canadian federalism, and of the interplay of science and politics. Canadian science has been only infrequently studied by historians, and Canadian history only infrequently by scientists and engineers, even when both groups deeply if unconsciously affect one another. It is useful, and perhaps instructive, to bring the two together.

The history of AECL is also and most importantly the history of a generation. It is the story of a group of men (and a few, a very few, women) in their twenties and thirties who accepted a challenge to grapple with the most advanced aspect of contemporary science. In their laboratory in the Ontario bush they accomplished most of what they set out to do. But not, inevitably, all, and not without a great deal of trouble and travail. It is their story, above all, that is told in the pages that follow.

Abbreviations

AEA	United Kingdom Atomic Energy Authority
AEC	United States Atomic Energy Commission
AECB	Atomic Energy Control Board
AECL	Atomic Energy of Canada Limited
AGR	Advanced Gas-Cooled Reactor
BLW	Boiling Light-Water [Reactor]
CANDU	Canadian Deuterium Uranium [Reactor]
CAPD	Civilian Atomic Power Division
CEGB	Central Electricity Generating Board
CGE	Canadian General Electric
CIL	Canadian Industries Limited
CIR	Canada-India Reactor
CIRUS	Canada India Reactor Uranium System
CP/CPD	Commercial Products Division
CPC	Combined Policy Committee
CRNL	Chalk River Nuclear Laboratory
DIL	Defence Industries Limited
FRUS	*Foreign Relations of the United States*
IAEA	International Atomic Energy Agency
ICI	Imperial Chemical Industries
IEL	Industrial Estates Limited
ING	Intense Neutron Generator
KANUPP	Karachi Nuclear Power Plant
NA	National Archives of the United States

NPD	Nuclear Power Demonstration
NPG	Nuclear Power Group
NPPD	Nuclear Power Plant Division
NRC	National Research Council of Canada
NRU	National Research Universal
NRX	National Research Experimental
OCDRE	Organic-Cooled Deuterium-Moderated Reactor – Experimental
OCR	Organic-Cooled Reactor
OH	Ontario Hydro
OSRD	Office of Scientific Research and Development (US)
OTR	Organic Test Reactor
PAC	Public Archives of Canada
PHW	Pressurized Heavy-Water [Reactor]
PRDPEC	Power Reactor Development Program Evaluation Committee
PRG	Power Reactor Group
PRO	Public Record Office
PWR	Pressurized Light-Water Reactor
RAPP	Rajasthan Power Projects
REL	Research Enterprises Limited
RG	Record Group
SGHWR	Steam-Generating Heavy-Water Reactor
TRE	Telecommunications Research Establishment
UNAEC	United Nations Atomic Energy Commission
WNRE	Whiteshell Nuclear Research Establishment
ZEEP	Zero Energy Experimental Pile

NUCLEUS

The History of Atomic Energy of Canada Limited

1

Early Days

The news during the winter of 1941–2 was dominated by distant battles. In Russia the German army was losing ground, but not enough. In Africa the British were stalled at the bottom of the Gulf of Sirte, on the Mediterranean Sea; in front of them the German army under General Erwin Rommel was starting to regroup. In the Far East the news was unremittingly bad: the British were retreating in Burma and had lost Singapore to the Japanese; to the south the Dutch East Indies had collapsed. The Americans, who had entered the war the previous December, were in the process of losing the Philippines; many doubted that the Japanese could be stopped short of Australia. Many doubted, too, that the Russians could hold out against the inevitable German spring offensive, for behind the German army was the economic power of occupied Europe, dominated from Crete to Brittany by Adolf Hitler.

The expansion of Hitler's power in the 1930s had set in motion a tide of refugees, first from Germany to Austria or Czechoslovakia; then on to France and Britain; and finally, as Germany conquered France and menaced the British Isles, from Britain to far-off America, safe, for the moment, from the reach of Hitler's bombers. But not, perhaps, very safe. It would not take long for the Germans, with their industrial plant and technology, to develop intercontinental bombers or missiles. As to bombs, they might soon develop the capacity to make an explosive more powerful than anything hitherto seen on Earth.

The name for such a bomb had already been devised. It would be an 'atomic bomb.' It would harness the incalculable power of the atom, according to the novelist H.G. Wells, who had written on the subject as long ago as 1913. Wells gave his novel the name *The World Set Free*; it envisaged the discovery of artificial radioactivity in 1933, the subsequent perfection of atomic energy, and, later still, the perversion of that energy into a weapon of mass destruction. 'This book made a very great impression on me,' wrote the physicist Leo Szilard, when he first read it in Berlin in 1932.[1]

The idea evidently made an impression elsewhere. That same year the British novelist Harold Nicolson used it as a theme in his novel *Public Faces*. As a former diplomat and serving politician, Nicolson was actually more pessimistic than Wells; where Wells had predicted an atomic war in 1956, Nicolson situated it in the summer of 1939.[2]

Nicolson was only out by a few months. The war that he forecast turned out in practice to be close to some of Wells's direst fantasies. For the Allies, the first years of the war brought almost unremitting defeat. The English Channel gave Britain a breathing space; behind it, a British and Canadian army was built up to repel an expected German invasion. Only after Hitler attacked the Soviet Union and the United States entered the war did it become possible to contemplate crossing back over the Channel to the continent of Europe; and such an invasion would take a great deal of time to prepare. In the meantime Hitler enjoyed uninterrupted possession of most of Europe, with its skilled population, its scientific traditions, its industrial base. It was too much to hope that he would not combine some of these advantages and proceed to make a bomb. This prospect led to some curious migrations, both of people and of ideas; and when they were over, some of them had lodged in Canada. To discover how, and why, it is necessary to turn to one scientist in particular, Hans von Halban.

Hans von Halban

Hans von Halban was, as his name might suggest, a refugee. Like so many other European anti-Nazis, he was a refugee several times over. Born in Leipzig of Austrian parents in 1908, he had followed his father to the University of Zurich. There he trained as a physicist, receiving his doctorate in 1935. By education Halban was a physicist, but by inclination he was a man of the world, passionately interested in sports and the lively arts and determined to enjoy what life had to offer. His talents and his ambition directed him into nuclear physics; where better to pursue the subject than the Radium Institute in Paris, Madame Curie's laboratory?[3]

At the Radium Institute, Halban spent his days under the overall direction of Frédéric and Irène Joliot-Curie, son-in-law and daughter to the great Marie Curie and joint recipients of the Nobel Prize in 1935 for their discovery of 'artificial radioactivity.' (Despite the hyphenated name, Joliot was still, usually, Joliot.) Their laboratory was, plainly, the place to be; and so was Paris. Paris was still the city of light in a Europe increasingly darkened by totalitarian regimes and increasingly fretted by the imminence of war; Halban enjoyed Paris to the full.

He could not ignore what was happening beyond France's frontiers. His pursuit of nuclear physics took him to Niels Bohr's laboratory in Copenhagen in 1936. What he accomplished there and in Paris would later prove to be of considerable importance; but for the moment Halban's Danish sojourn was notable because he fell ill. The illness left him with a weakened heart and an understandable preoccupation with the fragility of his health. His observation of central Europe made him apply for French citizenship as soon as he returned to Paris in 1937; it was granted in 1939, just in time for the outbreak of war. Life was still good in Paris; it was enhanced for Halban by marriage to a strikingly attractive and well-to-do wife. He was again working with Joliot, in a laboratory at the Collège de France.

By then, Halban had achieved a great deal of recognition. He was, plainly, a rising star and, as such, well placed to take on protégés; one such was Lew Kowarski, a refugee of a rather older vintage. Kowarski, Russian by birth, had emigrated from St Petersburg to Poland and then on to France, where he worked as an engineer while struggling to break into the more prestigious world of science. This he achieved by working as Joliot's research assistant and personal secretary. Physically large, with immense hands, he was a very improbable secretary, but his good temper and sociability made him acceptable; when he approached Halban in 1938 for training in nuclear physics, he was not turned away.[4]

The elegant Halban and the gigantic Kowarski were a disparate pair, the more so because the assistant was slightly older than the master. The impression lingers that to Halban, Kowarski was always his personal Caliban; but like the Shakespearean original, Kowarski eventually found something to complain of in his neo-Prospero's attitude. At first, however, gratitude and the established routine of the laboratory obviated clashes.

The two men joined Joliot and another researcher at the Collège, Francis Perrin, in patenting their theories in the spring of 1939. One patent covered a possible nuclear power reactor; another a possible control; and the last an atomic bomb. At the time the four men patented their discoveries they were far from being in exclusive possession of the various aspects of the means of achieving controlled or uncontrolled chain reactions; but it is their work that would eventually pass into Canadian hands, and on that score it has a certain importance to our story.[5]

It was through the Collège de France team that a link between heavy water and uranium was established. The question was, how could a sustained but controlled chain reaction be achieved? How could the behaviour of neutrons be modified so as to ensure that the proper number of neutrons struck the waiting nuclei, to produce in turn enough ongoing neutrons to fission more nuclei? It was understood that for the process to work, the 'fast' neutrons initially expelled from a fissioning nucleus must be slowed to afford a better chance of striking another nucleus; and this would be accomplished by the interposition of another material to moderate their speed; such a material would, obviously, be called a *moderator*. The likelihood of a neutron hitting a nucleus is called a *cross-section*, a term that calls up a very different image. A cross-section may best be understood in terms of an analogy, as the measurement of the probability of a missile – say a bullet – hitting a target. A good moderator will encourage a better cross-section. And in the opinion of the scientists at the Collège de France, the most promising moderator was heavy water.[6] In the spring of 1939 the Collège de France team appeared to be in the forefront of the search for a chain reaction. Convinced they were on the threshold of scientific immortality, they conceived of a mechanism in which neutrons could be made to react – a *reactor*. It would, first of all, consist of uranium in some form. There would be a large amount of uranium, to be measured in tons, and shaped so as to minimize the escape of neutrons. It would have a moderator, though just what kind of moderator took some consideration. The solution, almost literally, came to Kowarski in the bath, 'which was definitely conducive to the idea of neutrons spreading around in a liquid medium.' A liquid moderator, instead of solid blocks holding pieces of uranium apart, might be the ticket, and if it were, then a *homogeneous* liquid, mixing uranium and moderator together in a slurry, seemed the best bet.

The summer was consumed in experiments with water. The war began, but for various reasons the Joliot team were no further advanced. There was an overestimate of the number of neutrons produced per fission, a crucial error, and an unconcern about the number of neutrons that might be diffused from a possible reactor. And water was not working. Only late in the year did the team's attention turn to another possibility, a *heterogeneous* reactor, one in which uranium and moderator are separate. It remained to find a moderator. It had to be low on the periodic table (that is, with a low atomic number). Hydrogen seemed promising, and could be obtained in a pure form, but hydrogen absorbed neutrons – too many neutrons. Attention gradually came to focus on heavy hydrogen, deuterium, which did not.

Deuterium was a relative newcomer to scientific knowledge. It was produced in quantity only in Norway, where a French purchasing mission rounded it up and brought it back to France after a harrowing series of adventures. Now virtually the entire world supply of heavy water was on French soil – 185.5 kilograms of it. Perhaps the most important effect of this escapade was that the heavy water was denied to the Germans. The scientists intended to use it in a prototype power reactor; but their government intended to have them build a bomb.[7]

The invasion of France on 10 May 1940 soon imperilled Joliot's Paris laboratory. The scientists took flight, at first for central France and then for the port city of Bordeaux, where the French

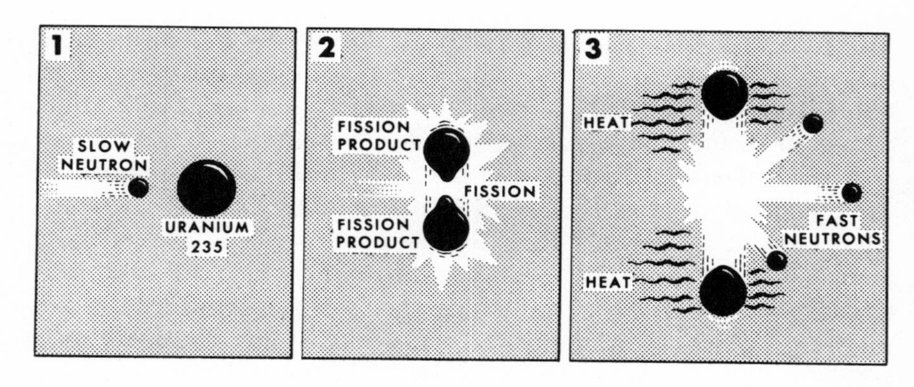

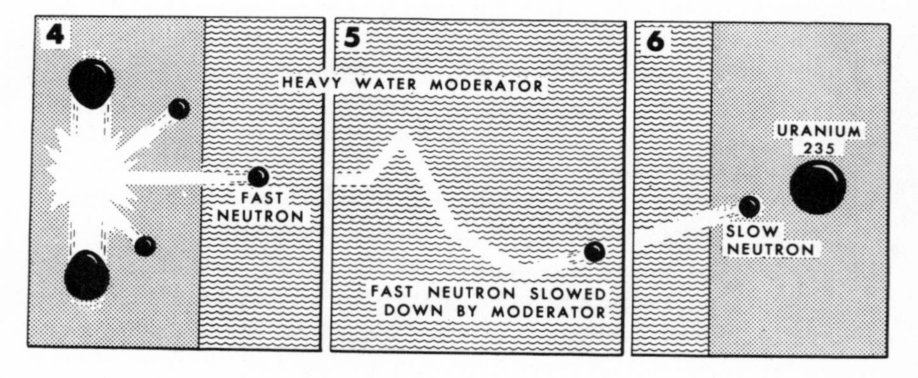

The fission process

government had taken refuge. By the time they reached the city, carrying their precious cargo of heavy water, their mission, as far as wartime France was concerned, was over. The position of the allied armies was hopeless, and it was only a matter of days before the French armies were forced to surrender. Some individuals chose to stay in France; others preferred to struggle on in exile. Joliot was among those who stayed; Halban, Kowarski, and their families joined the exodus.

They were loaded, with their heavy water, onto a British collier, under the unlikely command of the twentieth earl of Suffolk. After an uneventful voyage, they disembarked at Falmouth and were promptly put on a train for London. There most of the Frenchmen were placed in a railway hotel under official protection. Halban, who had never stayed anywhere but the fashionable Mayfair Hotel, hailed a taxi, and proceeded there, with his family in tow and his worldly possessions strapped to the roof. Kowarski, his junior, humbly accepted what the British had to offer.[8]

The Frenchmen had their knowledge, their heavy water, and their patents; the rest the British would have to supply. The heavy water was their ticket of admission, for nobody had anything like it anywhere else in the world; the patents were a guarantee, in a civilized society that respected private property, both of their own standing and of their country's interest in atomic energy. And though official France capitulated to the Germans, 'Free France,' under General Charles de Gaulle, fought on. As 'free Frenchmen,' Halban and Kowarski were still allies, even if they no longer had a government to protect them.

It was natural for the British to integrate their guests into their own atomic research. Before the end of June, Halban and Kowarski were sent to the University of Cambridge, where the Cavendish Laboratory had long been a centre of atomic physics. They continued their work on heavy water and uranium reactors as part of a larger British program, which also included investigation of the possibilities of separating uranium-235 from uranium-238.

One of their earliest encounters was with the distinguished British physicist James Chadwick. Chadwick appreciated the importance of heavy water and speculated whether more could not be manufactured in Canada; indeed, in Chadwick's view, the whole project might be better located in Canada in more favourable circumstances for long-term research.[9] Halban thought well of the prospect, but a few days later, on 7 July, it was firmly squelched. There was no reason, at least for the present, to move atomic research out of England. Halban filed away the idea at the back of his mind; when it recurred, it would not be unfamiliar or unwelcome.

The investigation expanded, to include workers from Britain's major chemical firm, Imperial Chemical Industries. ICI was soon so committed to the possibilities of atomic research, including its industrial potential, that on learning there was some similar investigation proceeding at the National Research Council in

Canada it caused its subsidiary, Canadian Industries Limited (CIL), to donate money for the purpose.[10]

There was no shortage of distinguished physicists in Britain. Many had been engaged in war work for years, first to establish Britain's radar defences, a priority item with Hitler's air force established on nearby French airfields. Some were sent abroad, to the United States and Canada, to renew contacts between scientific workers in Britain and their North American counterparts. Some, however, continued to brood over the feasibility of an atomic weapon in time to affect the outcome of the current war. It was an urgent matter, for as we have seen the possibility of a German bomb was a spur to activity.

Formed into a committee with a misleading title – MAUD – some of the country's leading scientists attempted to reach a conclusion on the practical question of how long it would take, and whether it was therefore a priority, to make an atomic explosion happen. To the committee it seemed that Halban and Kowarski's heavy-water–slow-neutron research was a sideline, even though in Halban's opinion there was a possibility that their reactor might create some other substance with explosive possibilities. The substance, another Cambridge scientist suggested, was the newly discovered plutonium (element 94); but there was not the equipment to pursue the suggestion to any conclusion.

In December 1940 Halban and Kowarski produced experimental proof that a combination of uranium oxide and heavy water could produce a *divergent* chain reaction – where output of neutrons is greater than any input from a neutron source. Such a condition means that a reactor has gone *critical*, and has become self-sustaining. It was an important conclusion, and the foundation of all that would follow. But at the time it appeared that the experiments were not beyond question; and questions did arise, both then and later. The discovery had to compete for attention with the work of others. Halban and Kowarski were not alone; besides their British colleagues there were platoons of refugee scientists from continental Europe, including Germany. They were happy to put their talents to work in the cause of defeating Hitler, especially if there was any possibility that Hitler could make an atomic bomb.[11]

At the end of 1940 it appeared that Britain was in the lead in atomic research, an observation confirmed by the physicist, Professor John Cockcroft of Cambridge, who was touring North America. Cockcroft nevertheless found much ongoing research, including the Canadian National Research Council's investigation of a chain reaction in a uranium and carbon (coke) arrangement. 'News from England,' the Canadians were told, made it very important that this research continue; there were even helpful suggestions as to additional Canadian scientists who might be interested.[12]

The news from England was indeed interesting. The British scientific representative in Ottawa, R.H. Fowler, was informed at the end of December by Cockcroft that 'Halban has obtained strong evidence that the D_2O_1U slow reaction will *go*.... The present suggestion,' he continued, 'is therefore to establish an intense slow neutron chain and to use this to convert 238, 92 [uranium-238] into element 94 [plutonium] and to hope that that will fissure.' Under the circumstances, Cockcroft hoped that 'the uranium investigation' would get a 'more vigorous attack in North America' than it was then receiving.[13]

CHAIN REACTION
The process in which neutrons, released during the splitting of one atom, go on to split other atoms, creating a self-sustaining reaction.

It was by this time the winter of 1941. Although it seemed that there was promise in Halban and Kowarski's work, the result was as far as war purposes were concerned questionable. Even if their reactor produced plutonium, what advantage would plutonium confer on its makers? Heavy water was a difficult proposition, and an expensive one. It might be within the Americans' power, however, and an approach was made to them. Reassured as to the importance of the question, the US scientific agency, OSRD (Office of Scientific Research and Development), started an inquiry that would result, later in the year, in an American emissary paying a call on the Consolidated Mining and Smelting Company (Cominco, a CPR subsidiary) in Trail, British Columbia. It was possible that heavy water could be had as a by-product of Cominco's smelter; and eventually, after much trial and error, it was – in 1943.[14]

Back in England, the heavy-water component of British atomic physics was becoming more and more enmeshed in an industrial complication. We have noted ICI's interest in the potential of atomic research. That interest was now more tangibly expressed. ICI proposed that it should take over heavy-water research, as well as Halban and Kowarski, who had recently reinforced their patent position by taking out a British 'master patent' on their theoretical calculations. This audacious proposal was placed on the British government's table just as the MAUD committee produced its report. Both the proposal and the report were to have momentous, unforeseen, consequences.[15]

The MAUD committee did not conclude that the Halban-Kowarski research was likely to have military significance during the course of the current war. It did, however, conclude that a nuclear explosion could be produced, that a bomb of feasible size (25 pounds) could be built, and that the most promising avenue was through the construction of a separation plant for uranium-235. It recommended that the construction of a bomb be given the highest priority, and it warned 'that the lines on which we are now working are such as would be likely to suggest themselves to any capable physicist.' As for Halban and Kowarski's conception of a uranium reactor, or 'uranium boiler,' it probably could heat a boiler and furnish reliable energy, once certain technical problems, such as controlling the reaction, were out of the way. Given the newly stimulated American interest in heavy water, it would be logical to transfer Halban and Kowarski and their work to the United States, so that they could pursue it close to the sources of heavy water and uranium. The report mentioned in passing that the research on plutonium might alter its perspective, based as it was on the belief that plutonium's fissile capabilities were unproven. As the British official historian has observed, the MAUD committee laboured in ignorance of work simultaneously taking place in the United States that did just that.[16]

The MAUD report secured immediate attention, not merely in Britain but in the United States, where it was transmitted in an early draft in July 1941. It stimulated the directors of American atomic research to reorganize and re-fund their project, which took on a more purposeful, not to say more urgent, air. In Britain the MAUD report was placed before several high-level official committees. To Sir John Anderson, Lord President of the Council, and by an odd coincidence the possessor of a PhD dealing with uranium, it seemed that the MAUD committee in no way exaggerated the significance of their subject. It was too important and too vulnerable to German air attack to build in Britain. Rather, it should be transferred to the United States. As for Canadian uranium, it should be speedily brought under control to prevent speculation.[17]

It was not the last mention of Canada in the aftermath of the MAUD report. A panel of the government's Scientific Advisory Committee, with three Nobel prizewinners among its members, met during September 1941 to consider what it would be best to do. On 3 September it concluded that 'the co-operation of Canada, where the necessary raw material and water power were available, might be enlisted through the National Research Council of Canada.' Professor James Chadwick, a Nobel laureate from Liverpool University, expressed his doubts. Canada, he told the committee on the 16th, might not be able to support such a large project, which required vast quantities of power and which probably should draw on the resources of heavy industry. Three days later the committee returned to the subject. Canada had uranium, and it was believed (wrongly as it turned out) that the Belgians had sent their ore reserves there as well. A joint enterprise should be considered, and the committee should bear in mind that Canada's wide open spaces offered the opportunity to test the product, possibly in Alberta.[18]

The panel's conclusion was that the atomic research program was a 'long term project.' It was too important to leave to private interests – presumably such as ICI – but should involve not only the British government but the American and Canadian governments as well. It would be a good idea to assemble a pilot plant and a separation plant in Canada, using American-made components. Last but not least, it asked for 'adequate arrangements' to secure uranium supplies for the governments concerned.[19]

Two points should be made. First, the panel's report was submitted while the United States was still a neutral in the war, and before its finances and industry were fully committed to war uses.

Second, uranium was held to be in exceedingly short supply, so short as to make its availability a crucial consideration. Whatever misgivings the British authorities might have concerning the state of Canadian science and technology, which they viewed with an affectionate but condescending eye, they would not ignore Canada's possession of uranium.

The American entry into the war soon followed. With it came an acceleration in the tempo of scientific research, which now was organized around three principal centres, the universities of California (Berkeley) and Chicago, and Columbia University in New York. It was at Berkeley that plutonium research was originally concentrated, and at Columbia that crucial early work took place regarding the piling up of uranium in order to stimulate a reaction. But it would be at Chicago that these subjects were pursued, under the general direction of Arthur Compton. Over Compton sat Vannevar Bush, director of the OSRD, and his assistant for the duration, James B. Conant, the president of Harvard University.

In the winter of 1941–2 Conant, Bush, and Compton were enthusiastic supporters of co-operation with the British atomic project; the combination of American science and industry with British experience promised well. So thought Halban when he came to America that winter. He easily found what he had come to see, in abundance, and was duly impressed.

The United States, he reported on 16 March, had completely surpassed the United Kingdom in the resources it was devoting to atomic research. Man for man his research team at Cambridge was outnumbered, and to some extent outclassed. Where in Cambridge it was difficult to unearth a mechanic, the Americans had them by platoons. In Cambridge he had not even been able to procure a lathe, while in America it was a matter of going shopping.[20]

What should be done? Arthur Compton had made two suggestions. Perhaps the British could send over an independent team – Halban's – to work with the Americans, a scheme which would allow them to keep their patents intact while permitting co-operation in the common cause. Alternatively, he would welcome Halban and one or two others of his team to Chicago, where they would join the Americans as specialists in heavy water research. 'In this case,' Halban noted, 'all patents arising would belong to the Americans.'

Halban found the second alternative unattractive. The first had a certain appeal, he believed, and it was the dispatch of an independent team that he recommended in April to Wallace Akers, an ICI official who had become the director of the British 'Tube Alloys' (another meaningless title) project. Akers, of all British officials, was the closest to Halban; he symbolized the relations that had grown up between Halban's brand of industrial atomic energy and private industry, specifically ICI. The British advantage, and hence the British ability to bargain equally with the Americans, was nearly gone, Halban told Akers, and it was essential to make a deal that would allow the harvesting of American know-how. Ideally the proposed British research team would be located in the United States, but if that proved unfeasible then it might, just possibly, be located across the border in Canada.[21]

Akers was not particularly impressed. Where, he asked, would this proposed team be located? What facilities would be available? 'Unless such facilities [in the United States] are negligible we cannot see any advantage in having [an] independent team in Canada where facilities believed practically non-existent and certainly inferior to those in England. Have you any reasons for Canadian suggestion other than advantages of geographical [proximity] British and American workers?'[22]

The Americans would indeed offer 'negligible' facilities. In fact they offered no facilities at all. Halban and one or two others could come, and if they did they would no longer be British agents. Akers liked that prospect no more than Halban liked the idea of becoming a subordinate in somebody else's laboratory. From the conjunction of their dislikes there now flowed a commitment to a new idea: Canada. The Americans would go along with a Canadian-based team, Compton assured Halban. There would be plenty of supplies for all, 'something a man of Compton's power can arrange,' or so Halban, with an inbred acceptance of hierarchy and influence, believed. There was a basis for Canadian research in the National Research Council experiments under Dr George Laurence, a Cambridge-trained physicist. Halban found Laurence 'a very likeable and brilliant man,' and inquiries with council staff revealed that the Halban team, as Halban increasingly thought of it, would be 'very well received.' Besides, Bush and Conant seemed to favour some such scheme, which might therefore cement the Anglo-American atomic relationship.[23]

As letters and cables criss-crossed from London to Chicago and Ottawa, another British mission arrived in the Canadian capital. Its mission was to corner Canada's uranium supplies; its quarry, in the first instance, was the prime minister, Mackenzie King.

Mackenzie King was not a close follower of scientific discoveries, though like many of his generation he believed in the beneficent if not miraculous power of science to assist mankind. Now he learned from Michael Perrin, Akers's assistant, that science had wrought a weapon of infinite destructiveness, that possession of such a weapon might win or lose the war, depending on who got it first, and that Canada's assistance was desired in this desperate enterprise. The British asked for 'the acquisition of some property in Canada,' as King put it in his diary, 'so as to prevent competition in price on a mineral much needed in the manufacture of explosives.' That was Eldorado Gold Mines, a radium producer, with a uranium mine in the Northwest Territories and a radium refinery in Port Hope, Ontario. Perrin explained that its product also involved a 'military weapon of immense destructive force.' He later remembered that 'a look of absolute horror and panic' stole over King's face as he described what such a weapon could do.

The prime minister quickly recovered. 'That's a job for C.D. Howe,' he told his visitors, and he sent them in search of Canada's minister of munitions and supply. Howe in turn sent for C.J. Mackenzie, president of the National Research Council, and in concert with the British they worked out a plan first to nationalize Eldorado in secret and then to transfer its ownership to a governmental syndicate to consist of the British, the Americans, and the Canadians.[24]

Howe's scheme never came to fruition. The importance of his project for our purposes is that it reinforced an existing British belief in the availability of a scarce Canadian resource, uranium. To that could be added British knowledge of the negotiation of a contract between the US government and the BC mining company, Consolidated Mining and Smelting, for the production of heavy water. Now with Halban's own desire to move made plain, and with pleas from Perrin and G.P. Thomson, the former chairman of the MAUD committee, influencing his verdict, it was up to Sir John Anderson to decide what to do.[25]

Anderson delayed, but by the end of July he realized that he could wait no longer. He advised the British prime minister, Winston Churchill, that the health of the British atomic program depended on co-operation with the richer Americans, and that the transfer of the Cambridge heavy-water team to Canada offered the best way to forward British research. Churchill agreed. After one last check to make sure that the Americans agreed, Sir John early in August told Malcolm MacDonald, the British high commissioner in Ottawa, to approach the Canadians. In his explanation to MacDonald, Anderson pointed out that the Halban team would be 'very much out of the picture' if they stayed in Cambridge. The British had reason to think that the Canadians would welcome the move; Mackenzie already knew of the British project, and had indicated in June that he would be happy to receive it.[26]

The British now asked Mackenzie for confirmation. At a meeting with R. Gordon Munro of the British high commission on 17 August Mackenzie did that and more, by taking Munro on to see the inevitable C.D. Howe, his minister. What followed stuck in Mackenzie's memory. 'I remember,' he told a reporter in 1961, 'he sat there and listened to the whole thing, then he turned to me and said: "What do you think?" I told him I thought it was a sound idea, then he nodded a couple of times and said: "Okay, let's go."'[27]

The National Research Council

Howe's decision was important, but now it had to be implemented. How much money would it cost, where would it come from, and who would administer it? The answer to the first question is apparent: nobody really knew, which did not mean that they were unwilling or unable to come up with an estimate – half a million dollars, as Howe was told at the beginning of September 1942. The second question was easier: the British would pay the salaries of the scientists and technicians sent out from England, and the Canadians would pay the rest. As for the third, there was only one possible answer: the National Research Council of Canada (NRC).[28]

The NRC was the Canadian government's official arm for most varieties of science. Founded in 1916, the council was intended to encourage the development of science and applied science in Canada. It promoted science principally through a program of grants to universities; its own activities were almost entirely directed towards industrial research. NRC was, in structure, a crown corporation. Its organs and purposes were defined by an act of parliament in 1924, which gave it a permanent president and an advisory council composed of distinguished citizens drawn from across the country; members of council were usually, but not invariably, scientists or engineers. It was to them that the president submitted his annual reports, which would then be forwarded to higher authority. The higher authority in this case was the minister of trade and commerce, who acted as chairman of a statutory committee of cabinet, formally styled the Privy Council Committee on Scientific and Industrial Research. The fact that NRC was associated with the minister underlined its practical purpose: it existed to assist industry and to enhance Canadian trade. Any purely scientific activity came a long way after.

Some confusion inevitably arises from NRC's nomenclature. It was a council within a council, a laboratory, a granting agency, and a scientific panel all rolled into one and all entitled to be called 'the National Research Council.' But confusing or not, the titles and functions were treated as interchangeable, and the reader must bear with established usage.[29]

NRC's freedom of action was greater than that of an ordinary government department, but it was still limited. It reported to a minister, from whom it received direction. Its funds came from an annual vote of parliament, and its financial procedures were subject to the scrutiny of the Treasury Board, the government's financial watchdog. We shall meet the Treasury Board again; for the moment it is sufficient to know that it had the power to scrutinize all government expenditures, whether on goods or on persons; to the Treasury Board were submitted the annual estimates (budget proposals) of all government departments and agencies. Then, in the words of Canada's wartime finance minister, J.L. Ilsley, 'the staff of the Treasury Board ... go at those estimates and try to have them reduced.' In the case of NRC, the estimates were pared down to $900,000 in 1939, still enough to support a staff of some three hundred, but hardly enough to conduct a first-class research program in any area.[30]

NRC was not in as bad a case as its budgetary figures would suggest. It had had the good fortune to build a large, modern laboratory just as the depression set in, and no amount of subsequent cheese-paring could alter that fact. Its president in the late 1930s, General A.G.L. McNaughton, was active and influential. McNaughton gave as much money to university research as he dared, with the result that some Canadian laboratories were surprisingly up-to-date in their equipment and research programs. Although the general departed in 1939 to head Canada's overseas army, he dedicated the council's activities entirely to the war effort before he left. His hand-picked successor, C.J. Mackenzie (who was ostensibly McNaughton's *locum tenens* until 1944), was determined to carry out the general's intentions to the letter.[31] For the next thirteen years, Mackenzie directed the National Research Council. In that capacity he was called on to advise the government on the appropriateness of locating Halban's laboratory in Canada; and it was on Mackenzie's advice that Howe agreed to go ahead.

It was especially easy for Howe to rely on Jack Mackenzie. Years before, when Howe was a fledgling professor of engineering at Dalhousie University in Halifax, Mackenzie had been one of his first students, only three years younger (born in 1888) than his teacher. Mackenzie had gone off to the prairies after that, to be dean of engineering at the University of Saskatchewan, but Howe's career had taken him there first, to become Canada's foremost builder of grain elevators. Chance brought the two men to Ottawa, Howe as one of Mackenzie King's ministers, and Mackenzie as McNaughton's wartime surrogate.

He was in many ways a good choice for the post he now occupied as the executive head of Canada's national laboratory in Ottawa. All council presidents were aware of its recent establishment, of how much it depended on the favour of a government and civil service whose interest in science was at best abstract and sentimental. Mackenzie's political instincts were particularly well developed, and they had been honed by his experience as a university dean in the cash-starved 1930s. He cultivated the civil service as well as the politicians, and prided himself, then and later, on his status as a bureaucratic mandarin, a 'senior' as he put it, with influence to match. The influence was magnified by a large portion of natural charm, and by an apparently down-to-earth and unassuming disposition.[32]

By his own account, Mackenzie was a cool customer, who deliberately set out to cultivate imperturbability. He had, he wrote, come out of the First World War – he had served in France – 'with a firm resolution that in the future I would allow nothing in the way of misfortune, violent confrontations or frustrations to upset me in any emotional way. Keeping this resolution,' he continued, 'has, I believe, made for efficiency, objectivity and peace of mind under conditions of temporary confusion.'[33]

Temporary confusion was a fact of life in wartime government. It was a situation that was well adapted to Mackenzie's talents. Very early on he demonstrated that he could, when he wanted, secure exemption from ordinary government rules, using the NRC's status as a war research institute as well as its theoretically autonomous standing as a crown corporation. Mackenzie proved, in the early months of the war, that he could act quickly and decisively. His technique was to inject himself into critical situations and personally manage their transformation. Under his direction NRC quickly expanded, but in expanding it also tended to centralize. It was part of Mackenzie's technique that he liked to have his responsibilities as far as possible under his own eye. In discussing the wartime history of NRC it is impossible to ignore Mackenzie's touch; but the magic could not be everywhere nor was it always effective.

As Mackenzie's biographer R.F. Legget put it, '"administration" was a word never heard around NRC in those days. ... Business matters throughout the entire duration of the war were handled by two worthy men who had "grown up" with the Council from its earliest days, aided by a group of clerical assistants and devoted secretaries.' Although this suggests a commendable modesty in limiting overhead, it also suggests a sclerotic and overburdened administrative structure, with a high propensity to rigidity and unimaginativeness. To the outsider looking in, however, NRC's informality and its president's policy of 'flexibility' must have seemed attractive.[34]

That was not the only attraction. Mackenzie enjoyed a wide circle of friends and acquaintances, not only in Canada but in the United States. He had early on made contact with his American counterparts; through them, and through various British official visits, he was aware of the progress that Britain and the United States were making in atomic research; and he arranged for one of his junior physicists, George Laurence, to visit Columbia University and meet some of the scientists there. Mackenzie's personality, and especially his easy informality, made him a welcome guest and encouraged the sharing of confidences; like many Canadians Mackenzie believed that a common North American approach to problems made possible the sharing of confidences across the forty-ninth parallel. Mackenzie's prior knowledge, and his appreciation of the importance of atomic research to both the British and the Americans, helped him come to a quick decision on the British request for assistance to Halban. The importance of the project matched the NRC president's view that Canada needed and should get a scientific boost, and that he should be the instrument of Canadian science's deliverance. Most important of all, Mackenzie believed that he could easily take on Halban's laboratory, and he told Howe so.[35]

The earliest stage of a project, especially a large one, often seems in retrospect to represent a triumph of optimism over reality. Similarly the clarity of an agreement always seems greatest at the beginning. But what happened in the summer of 1942 is remarkably opaque. The Canadian government agreed in August 1942 to accept and support a laboratory of British scientists. They were to work on atomic research. That much is clear. What is not clear is what the laboratory and its director proposed to do when they got to Canada, how long they expected to stay there, and what the object of their work was to be. The Canadian government for its part – through the agency of Howe and Mackenzie – expected to pay many if not most of the lab's expenses, and to send Canadian staff, both scientific and support, to join Halban's team. It seems reasonable to accept that both sides considered the laboratory to be primarily a war project: its existence depended on conditions peculiar to the great conflict and it would terminate at war's end. But what would it leave behind?

Speaking in 1961, C.J. Mackenzie argued that his object in welcoming the British project to Canada was to get in 'on the ground floor of a great technological process for the first time' in Canadian history.[36] This is not surprising. Any administrator, especially in science, was bound to consider the longer term. Mackenzie was anxious to extend Canada's scientific prowess. He was aware that the country was viewed as a scientific backwater. Yet that does not clarify what Mackenzie considered the 'great technological process' to be. He was not especially learned in atomic physics; by training he was a civil engineer who had chosen to practise his profession through university teaching. He kept in distant touch with developments in science, among other things reading the 1939 articles in which atomic physicists revealed to the world the splitting of the uranium nucleus.[37]

It was not impossible, therefore, for Mackenzie to grasp what the British had to say to him, and he had his staff in physics at NRC to bring him up to date on the significance of developments since 1939. He understood the difference between the uranium separation project and the heavy-water project, and he also understood that the latter was not likely to bear warlike fruit for the duration of the current war – as the MAUD committee had argued. So at least he contended in the following year, 1943: the heavy-water reactor, he argued, was 'a peace-time project, and not for this war.' Continuing in the face of objections, Mackenzie summed up: 'Well, when I first heard of it, this section was definitely not for military application.'

There is no reason to doubt Mackenzie's account, which was the standard version of the prospects of Halban and Kowarski's 'boiler' as understood in Britain in 1941. The possibility of producing plutonium from the boiler through the irradiation of uranium struck Mackenzie as remote. Plutonium, as he understood it, was a laboratory creation available in only the minutest quantities. This too was true – in 1941 or earlier. And so when he claimed that 'Nobody considers using 94 [plutonium] for military purposes,' he was doubtless right. The only problem was, as we shall see, that by 1943 his argument was a few years out of date.[38]

It seems reasonable to conclude, then, that the relevant authorities in Canada imperfectly understood what they were committed to accomplish. This limitation would prove a handicap in the future; in 1942 nobody seems to have suspected that some further explanation might be in order.

The explanation would not come from Howe. The minister was busy, and he trusted his executives to handle their own affairs. If Mackenzie told him he would deal with atomic energy, Howe saw no reason not to believe him. While Mackenzie managed the transfer of Halban's laboratory, Howe could worry about tank production, or the shortage of electricity, or about politics. As an American diplomat observed, Howe was the ringmaster 'not of a three ring circus, but of a thirty ring circus.' As of August 1942 he had added a thirty-first ring.[39]

The Thirty-First Ring

The National Research Council was chosen to hold the purse-strings for the Anglo-Canadian atomic project. This brought the NRC into contact with the director of Tube Alloys, Wallace Akers, who had been research director of ICI. Akers had a fine record in industrial research. He had a well-deserved reputation, within his company, of getting his own way. He was not especially accustomed to politics, nor well-adapted to compromise. Because of his connection with ICI, he shared in the odium that attached during the 1930s and 1940s to large business enterprises, reaping their large profits at public expense. His ICI connection was no particular handicap in Canada, where its subsidiary CIL was a mainstay of the war effort, but it was positively harmful in the United States, where it was not just business, but *British* business, come to pasture on green American dollars.

Akers was trusted by Sir John Anderson, whose own abrasive personality mirrored his subordinate's. His relations with Halban were close; having plumped for Halban as director of the Anglo-Canadian project, he backed his man against any and all opposition. As a result, Akers harvested a host of resentments from those who found themselves disagreeing with Halban; they would not be few.

The two men should have been at arm's length, separated by an Anglo-Canadian co-ordinating committee. But though that was true in theory, there was too much practical business afoot. Halban, the begetter of the project, was the inevitable director. This pleased the Canadians. He had charmed Mackenzie, to whom he may have brought a whiff of a distant and glamorous world of prizewinning science, the world of the Curies and Lord Rutherford and their renowned laboratories. Halban presumably knew how to run such a laboratory, a trusting view since no Canadian visitors had crossed the threshold of his Cambridge sanctum.

To manage the new relationship, Mackenzie proposed a series of committees. On top would be Howe and Malcolm MacDonald. Most people in official Ottawa liked MacDonald, and Howe was no exception, but it was also notorious that Howe hated committees. Although Mackenzie was later added to this 'board of directors,' there is no sign that its formal responsibilities weighed on its members, taken as a group. Below the 'directors' came a technical committee (scientific policy committee), chaired by Halban, and consisting of the division heads from his team: chemistry, physics, theoretical and applied, and engineering. As of August 1942 these heads were still to be appointed, but there was little doubt that all would come from Great Britain. To round out the committee there was George Laurence, the NRC's senior man in atomic physics, representing the Canadian interest.[40]

Mackenzie's plan was confirmed in October 1942 by Howe and Anderson. Meeting in Anderson's office in London, the two men discussed the ownership of the Eldorado company. They reiterated what had already been decided, which implied to the British that their uranium supply was airtight. They also ratified financial arrangements for the laboratory. The British were to pay the salaries of those they sent, while the Canadians were expected to pick up the tab on the balance. That, the British estimated, would be around $450,000 a year, as the Canadians had already been told. The British would send some thirty scientists and twenty-five technicians. The rest would have to come from Canada.[41]

So far, apparently, so good. But the relative ease with which arrangements were made belied misgivings on Mackenzie's part, some of which he expressed in a letter to Vannevar Bush, President

Roosevelt's principal adviser on the development of the atomic bomb. 'Personally and unofficially,' Mackenzie wrote in mid-September, 'I view the approach of this 'league of nations' which is coming from England with some mild apprehension as I find that in a general as well as in the most literal sense, I do not easily speak the language of our European colleagues.' This did not augur well; it is of interest that he did not share his feelings with his British partners, an indication that even at this beginning stage there were problems of communication, if not actual lack of trust.[42]

The British team remained blissfully unaware of Mackenzie's sentiments; in a letter to England Halban praised his 'good spirit' and his desire 'to embrace [the project] completely.'[43] The organization of the laboratory was well under way by mid-October and, characteristically, Halban turned his attention first to the selection of his staff; details of administration he left to time and goodwill to sort out.

As these matters proceeded, the Americans were reorganizing their own project. In the summer of 1942 the administration of the American atomic project was handed over to the US Army Corps of Engineers, which designated it the Manhattan Engineering District. To head it they appointed General Leslie Groves, a tough and capable engineer, and they gave Groves the assignment of making sense of the competing ideas and designs for the future bomb. Bush and Conant continued to make policy; but executive authority now passed to Groves.

Halban was for the moment unaware and unconcerned. He had a laboratory, and he could enjoy the pleasures of office. One of these was sorting out the scientific sheep from the subordinate goats. To head experimental physics, he wanted Pierre Auger of the Sorbonne. For chemistry he chose F.A. Paneth, a German refugee teaching at Durham in the north of England; while Paneth was not himself a nuclear chemist, he was senior and very distinguished, and he knew the junior men in the field. G. Placzek, who would direct nuclear physics, came from Czechoslovakia. The only native Englishman in the lot was R.E. Newell, the engineering head, who transferred from ICI. It was on the whole a good list, but there was a striking omission. Kowarski was not on it.

Kowarski's omission was no accident. The two men had drifted apart. While Halban consorted with ICI and travelled around North America, Kowarski stayed in Cambridge, carrying on experiments. Where Halban was little more than an occasional presence to junior members of the Cambridge team, Kowarski, huge, jolly, energetic, and capable, was close to its heart. He was no longer a mere lab assistant but a responsible scientist. Halban would not treat him as such. His reservations about Kowarski were not entirely capricious; nor were they unique to Halban. Seven years later, when Kowarski applied for a position with the British atomic program, its director, J.D. Cockcroft, rejected the offer. The man he had, he wrote, 'is a much better pile physicist than Kowarski will ever be.' Halban's feelings, presumably, were stronger, but they were not his alone. If Kowarski went to Canada, it would be under Auger. Kowarski rebelled. He would not go under such conditions. When other members of the team heard the news, they decided they would not go either. Halban mobilized Akers, and in the end the weight of authority and the appeal to compromise – on Halban's terms – won out. Kowarski stayed in Cambridge with a peripheral research group; the rest, as selected, departed.[44]

The departure was not easy. Canadian immigration formalities had to be gone through, and teams of approved doctors were dispatched from London to Cambridge to tap shins and gaze into ear-drums. Then families had to pack, and stay packed until appropriate convoy arrangements were ready, for this was the height of the German submarine campaign in the Atlantic. Some saw the dark hand of favouritism in the order of their families' passage; others were forbidden to travel on the same ship in case all the available talent in a given area was sunk in the ocean.[45]

Equipment was another headache. The Canadians were not expected to have much, and so the British brought what they could. The Cambridge laboratory was effectively dismantled in November; the vagaries of sea travel kept it closed until it could be unloaded and cleared through Canadian customs the following February or March. The mere fact of transfer had lost the British three or four precious months.

By early November administrative arrangements had been sorted out. J.F. Jackson, a British civil servant designated as chief of administration, established that the new laboratory would do its procurement through the existing NRC purchasing office. 'This,' he reported home, 'has the advantage of simplicity of accounting,' and C.D. Howe himself promised that requests from the atomic scientists would receive the highest priority. Jackson, gratified, was moved to write that 'there will of course be no difficulty about priorities' as a result.[46]

Mackenzie had at first intended to keep the new laboratory in Ottawa, under his eye. A new NRC site was under construction at the city's eastern extremity, and it may have seemed logical to extend the building program there. The prospect did not commend itself to the British. Ottawa was a small (215,000 in the city and suburbs), overcrowded, and underequipped city. It was a civil service town, not an industrial or transportation centre. There was nowhere to buy equipment, and nowhere to place staff and their families. Halban preferred Montreal and, after some persuasion, Montreal it was.

Akers emphatically approved. In Montreal there were many mansions, empty of inhabitants and available for research.[47] There were two universities, one of which, McGill, had a good scientific library. There were colleges and research institutes. The city was mostly French, which should offer an enticement to Halban and his colleagues from Paris. For the English, there was also a large and anglophile English-speaking population. November was spent deciding which of the many mansions would be best. In the end one was chosen, on Simpson Street, but only for temporary accommodation. The top floor of a warehouse was rejected, on the sensible grounds that radioactive emissions would endanger the unsuspecting workers beneath. The best bet turned out to be the Université de Montréal on the north slope of the mountain that dominates the city and gives it its name.

The Université de Montréal was less known than its English counterpart McGill. But because it was a recent foundation, as universities go, it had the newer campus, and its buildings were designed by the famous Montreal architect Ernest Cormier, an exponent of the Art Deco style much favoured in public monuments in eastern Canadian cities. Cormier's constructions at the university included a medical school, recently finished but unoccupied for the duration of the war. Typically, the Anglo-Canadian project on the other side of Mont Royal did not know that there was space to be had; only the intervention of a French refugee scientist, Henri Laugier, excluded from contact with the project by Halban but teaching at the université, brought the vacant space to NRC's notice.[48] The université was agreeable, provided Cormier supervised the work and approved any alterations. The building was to be returned in its original condition, a reasonable precaution considering the nature of the work contemplated. Soon Cormier was once again a familiar figure in his building, where he stalked about carrying his trademark, a goldheaded cane.[49]

After four years of war, Montreal appeared to the British to be the land of milk and honey. Here was life without blackout curtains, but with neon, steaks, and well-stocked stores. Dionne's, the fashionable grocery on St Catherine Street, became a particular favourite. They could afford it, since the British-paid staff drew not merely their salaries (the top, Halban's, being $7500 Canadian; the bottom, that of a laboratory assistant, just under $900) but an overseas allowance, untaxed. It followed that evenings and weekends afforded them a style of life that even in peacetime would have been enviable.[50]

Simpson Street, and later the Université de Montréal, enjoyed an armed guard. Memories of the guard differ. Some scientists recall a superannuated veteran with an obsolete rifle; others have a red-coated Mountie fixed firmly in their minds. On Simpson Street every room, including the bathrooms, was put to use. Files were stacked in the bathtubs, and calls of nature therefore involved the disruption of work in a very real sense. Because it was a secret operation, security rules had to be strictly followed. Secret files could not be left around, but had to be locked up in a safe when the personnel authorized to see them were not in the room. The trouble was that there was a shortage of safes, and NRC procedures demanded that new safes must be purchased only on tender, even though there were plenty of stores in Montreal with safes for sale. Months were to pass before the tenders cast upon the commercial market returned with the requisite number of safes.

When Mackenzie had cheerfully offered the services of NRC's existing staff he had not spelled out the consequences; perhaps, after all, he did not know what they would be. By April 1943 the

British team were contemplating ways of circumventing NRC procedures and ordering equipment to be shipped by air from familiar suppliers in England.[51]

Then there were staff to be procured. From NRC came George Laurence, whose participation had been cited as one of the reasons for moving to Canada. From the University of British Columbia there came George Volkoff, to work in physics. Volkoff did not leave teaching entirely, as it turned out, because his lecture skills were highly prized and were used to introduce recruits to the mysteries of nuclear physics.[52] From Queen's University in Kingston, Ontario, came B.W. Sargent, who had worked with Laurence on chain reactions and who was now granted a speedy release from teaching to go to Montreal. At McMaster University in Hamilton there was Harry Thode, a physical chemist, whom Mackenzie had favoured with the price of a spectrometer. Negotiations with Thode were long and difficult. Moving to Montreal would mean some financial sacrifice, given NRC's pay scales. An American attempt to prise Thode from Canada may have supplied the incentive necessary to a solution. Bill Bennett, Howe's executive assistant, and Mackenzie opposed losing any more promising young Canadians to the United States. In this case Mohammed was moved to the mountain, since it proved easier to establish Thode's laboratory as a division of the Montreal project, complete with RCMP guard, than to move his team to Montreal. Supplies and visitors rocketed back and forth along the railway between the two cities.[53]

Halban also hoped to recruit in the United States. This proved to be difficult since the Manhattan District had already vacuumed up most of the available supply of physicists and chemists. There were, however, a few left over, mostly of foreign origin and some already in British employ. The most notable was Bertrand Goldschmidt.

Goldschmidt was a refugee from France, forced to flee the anti-Semitic laws of the wartime Vichy regime. Trained in Madame Curie's laboratory, Goldschmidt was one of the world's few qualified radio-chemists. In the summer and autumn of 1942 he was working, as a British-paid scientist, in the chemistry division of the Chicago metallurgical project. Goldschmidt was present when, in August 1942, the American chemist Glenn Seaborg announced that he had actually seen a tiny fraction of plutonium – rose-coloured. In Montreal Goldschmidt joined Jules Guéron, another Parisian refugee, Colin Amphlett, a young Englishman recruited out of the university, Leo Yaffe from McGill, and, later, Les Cook, a Canadian trained in Berlin and in the Cavendish Laboratory in Cambridge.[54]

To complete the team Halban and Akers picked among physicists and mathematicians. Bruno Pontecorvo, who had fled from Italy to the United States via Paris in the 1930s, was fed up with teaching in Oklahoma; 'the British Security people,' Akers told Perrin in December, 'give [him] an unusually enthusiastic report.' And so Pontecorvo came, to add strength in physics and a gregarious personality to the team. Pontecorvo, however, came at British expense; Mackenzie resisted attempts to place him and other American-recruited staff on Canadian salary. It was probably just as well; the Canadian staff soon came to understand that their salaries were substandard as compared to those of the British, a grievance that Mackenzie steadfastly refused to confront.[55]

Despite difficulties, the transition by the end of November was moving relatively smoothly. Jackson reported that Mackenzie had provided an extra liaison officer, Lesslie Thomson from the minister's own department. A native of Montreal with a reputation as a jack of all trades, Thomson was intended to be a helpful fixer; as time would show, his talents were badly needed. It was now anticipated that there would be about seventy-five staff in all, thirty-eight of them professional scientists, and the rest assistants, clerks, draftsmen, and the like. Staffing was more, much more, than filling functional slots. Mackenzie was alive to the necessity of adding as many Canadians as possible, Jackson emphasized, because 'when the project has been brought to some suitable stage it will be possible for the English team to return [home] with complete information and experience while leaving a similarly self-complete team of Canadians to carry on the work here.'[56]

While Jackson and Thomson were squaring away the details of administration and supply, as far as NRC rules allowed, Wallace Akers was touring the United States and Halban was visiting Chicago. Their task was to carve a space for the Montreal laboratory in what would be a grand co-operative Anglo-American enterprise. The move to Montreal was an earnest of British good faith and desire to be helpful; the Americans, it was assumed, would now reciprocate with information, co-operation, and supplies. Supply centred, as it must, around uranium and heavy water. The British had bought some 20 tons of uranium oxide from Eldorado to go on with. To refine it to the desired purity and to convert it into metal, considered most effective for piling in the reactor, the British had a choice between shipping the oxide back to England for experimentation in their own incomplete metal fabrication plant and relying on an American factory, already functioning, in St Louis. Of equal importance was a supply of heavy water, which could only come from Trail, whenever that facility became operational. The estimated requirement there was 6 tons, from a production which the Americans thought would reach half a ton a month, or a year's supply. And if that were not enough, there was a requirement for 60 tons of graphite, which was also a useful moderator (and perhaps a necessary substitute for the preferred heavy water) in a uranium reactor or pile.[57]

PILES AND REACTORS

A *pile* was, originally, just that, a heap of uranium oxide (U_3O_8) inside a graphite matrix; the term was first used by Enrico Fermi in 1940, and was later applied to any apparatus designed to produce a nuclear-fission chain reaction. The term started to go out of use in the late 1940s and was replaced by *reactor*, which conveys the sense of a chain-reacting machine more accurately. Reactors vary widely in composition: they may employ natural uranium or enriched uranium or plutonium; while tons of natural uranium are required, much smaller quantities of enriched uranium, or plutonium, are needed to sustain a chain reaction.

The demand for a year's supply of American-contracted (although Canadian-produced) heavy water did not seem unreasonable to the British. They assumed that the principal centre of heavy-water research would be Montreal, and they argued that common prudence if nothing else (the Germans were known to be investigating heavy-water reactions) should make it an important line of research. Halban was sure that his boiler design would produce useful quantities of plutonium; in answer to an American question he agreed that a graphite reactor would also do it, though with great difficulty.[58]

> **PLUTONIUM**
> As uranium is named after Uranus, plutonium (Pu) is named after the planet Pluto. First produced by Glenn Seaborg and others from irradiated uranium in 1940, its main isotope, 239, is itself fissile; if it captures more neutrons it forms isotopes 240 to 244. Plutonium may be chemically extracted and, as Bertrand Goldschmidt has remarked, it is easier and less expensive to use this process than to separate uranium-235.

After Halban, Akers. Akers also wanted the terms of co-operation spelled out between the British and the Americans. He wanted, in particular, access to the secrets of bomb design, once the preliminary research was complete. The Americans protested, claiming that secrecy impelled them to enforce the strictest controls. It occurred to Akers's American interlocutors that he symbolized both the military interests of Great Britain and its economic interest. Should American money be spent to bolster Great Britain's industrial power?

A meeting in Washington produced only an impasse. The question of whether and how to deal with the British, and by extension the Canadians, was now passed on to higher authority. Meanwhile the shaping of the American nuclear program continued at the executive level.

General Groves had to sift among competing conceptions of nuclear production. Would isotope separation prove best, or nuclear reactors? If a reactor, what kind, graphite or heavy water, homogeneous or heterogeneous? Halban, talking with the general in Chicago, sketched out sixteen different kinds of reactor. Groves was not impressed, and as it happened, he did not need to be. Just as the general and Halban were meeting, a team of scientists under the direction of Enrico Fermi at the University of Chicago were watching a pile of uranium and graphite achieve a spontaneous chain reaction. Nuclear fission was an accomplished fact.

The Americans could feel pride in accomplishment. That was more, they reflected, than the British could do. Spurred on by what had already been done on American soil, and increasingly confident of what they could do next, Conant and Bush made some decisions. Groves already had. On 1 December 1942, the day before he met with Halban, and also the day before Fermi's pile went 'critical,' he issued a letter of intent (the 'intent' being to sign a contract) to the E.I. du Pont de Nemours company for the construction of a plutonium production pile. Du Pont engineers sat in on his conversation with Halban, who was annoyed by their sharp, specific questions.[59]

The American future included heavy water. The du Pont company was impressed by its properties, stimulating Groves to sign contracts for the construction of heavy-water plants inside the United States. This in turn offered the possibility, if all went well, of heavy water for a variety of experiments by the fall of 1943. With this prospect in the offing, Arthur Compton of the Chicago Metallurgical Laboratory (known colloquially as the Metlab), which had been sidelined by the industrial interest in plutonium production piles, began to veer towards his own heavy-water project.[60]

The Americans did not need Halban any more. His importance and that of his project were at best peripheral; a matter of convenience, no more. As for the Canadians, it was true that the production of heavy water at Trail might be jeopardized by a Canadian decision to embargo the stuff, but it was only a matter of time until the purely American heavy water came on stream. On balance there seemed little to gain and much to lose by working with the British. In Washington it was decided to revise the existing co-operation between the British and the Americans. With the support of the secretary of war and the acquiescence of President Roosevelt, Bush, Conant, and Groves determined to restrict collaboration with the British to what might prove useful to the American program for the production of nuclear weapons. Conant sat down to write a letter spelling out the terms.

Dated 2 January 1943, it was addressed to C.J. Mackenzie. It ostensibly responded to an earlier request by Mackenzie for the first 6 tons of heavy water to be produced from Trail, a request that Conant was pleased to grant, under certain conditions. First, while the Americans were willing to encourage the Montreal laboratory to study the design and functioning of a heavy-water–uranium pile, they did not wish the Anglo-Canadian group to concern itself with the actual production of plutonium. That, Conant said, was a matter between the du Pont company and the United States government. Indeed, he added, word had come 'from the top' that all interchanges of information between the United States and its allies on the subject were to be severely restricted, the condition of access being the ability of the recipient 'to take advantage of it in this war.' Because it seemed that the Montreal team might be able to contribute to du Pont's research, a limited exchange of information would be permitted. Conant implied that in exchange for consent the heavy water produced on Canadian soil would be available to the Anglo-Canadian atomic project.[61]

It was an audacious move and it was bound to produce a response. It only remained to learn what the response would be.

Wind from the South

Conant gave Mackenzie advance warning that something was up. In a telephone call on 2 January he revealed that a letter was on its way. It might seem rather strict, he added, but Mackenzie was to understand that it read more harshly than it was intended to be. This was ominous; how ominous Mackenzie and Akers (who was in Ottawa) did not learn until the letter finally arrived on the following Monday, the 7th.

Akers read the letter in the morning and, after conferring at the British high commission, had a response ready for Mackenzie to send off that afternoon. Mackenzie 'absolutely refused,' as he later told his diary. He had better things to do than repeat Akers's stale arguments over his signature and thereby lose whatever standing he had with Conant and Bush. He would instead go personally to Washington and talk it over.[62]

It was not until the 18th that Mackenzie walked into Conant's office. He had by then worked out his own position and cleared it with Howe; it was not a strong one. The Montreal laboratory desperately needed American information and American supplies, not least heavy water. The Americans, if pressed, did not need anything at all from either the British or the Canadians. The British, Mackenzie concluded, had little bargaining strength, as was evident from the trend of his conversations with Bush and

Conant. How could they justify spending American money to benefit England and get nothing in return? Congress would never stand for it. This argument struck a responsive chord in Mackenzie, as did what appears to have been their other argument, for which Mackenzie felt 'there is a great deal to be said.' There were too many foreigners in Montreal: French (Goldschmidt, Guéron, and Auger); Russian (presumably Volkoff); Austrian (Halban); Czech (Paneth and Placzek); Italian (Pontecorvo): in short the heart of the Montreal team. They did not stress that much of the high command of the American project was similarly derived (Fermi, Wigner, and Szilard were the most obvious examples). Bush explained that if there was 'anything that was really unfair' he and Conant would try to rectify it, and on that note the discussion ended. Dean Mackenzie and President Conant then went on to dinner – 'a pleasant evening,' Mackenzie recorded, following a 'very pleasant and profitable discussion.'[63]

Mackenzie's visit confirmed that the existing British and American positions were irreconcilable. Of course, the Americans said, they could be overruled by an appeal to higher authority, but Mackenzie did not put much faith in such a tactic. The British tried it anyway, after failing to move the Americans by the threat of withholding information. The Americans merely responded by telling the British that no information or supplies would be forthcoming for Montreal until the British accepted their conditions, and unless the du Pont company agreed that Montreal was doing useful pilot plant work for the design of its own heavy-water reactor.

There was nothing new in any of this. To Mackenzie the American terms were acceptable enough. He believed that the Montreal project had little military significance for *this* war, as the Americans argued. It benefited Canada technologically and scientifically, but if that had to be sacrificed so be it. His conviction that the Montreal laboratory contributed little of military importance would be crucial when he reported to Howe on 20 January.

The minister, he recorded, 'agrees with me that we should not become involved in any unpleasant controversy as this project is a relatively small one cast against the entire U.S.-Canadian contacts.' Howe meant the billions of dollars of Canadian-American trade, crucial to the Canadian war effort, and his carefully constructed contacts with US war-production leaders. There was no point in allowing an 'unpleasant controversy,' especially this one, to jeopardize Canada's larger interests.[64]

If Howe was unperturbed, Anderson was outraged. He communicated his indignation to his prime minister. Churchill quickly raised the subject with Roosevelt, but got no more than a cheery assurance in response. Anderson's indignation mounted. Messages and messengers moved back and forth across the Atlantic, and still the Americans stood fast.

At the beginning of March the Montreal laboratory was finally ready. The equipment had arrived. The staff was increasing, and by May would total a hundred: twenty-seven professionals from Britain, with six subprofessionals; twenty professionals from Canada, with twenty-two subprofessionals; and twenty-five administrative, secretarial, trades, and labouring staff. To assemble them the British and Canadian governments had scoured their scientific cupboards; and having brought them together, they now had to contemplate sending them home again.

Halban and his colleagues were not as worried as Anderson and Akers. Surely it was better to settle on the Americans' terms than to get nothing at all. Akers wrote on 30 January that he had reluctantly come to agree.[65] Unluckily for Akers and the scientists, the quarrel was now beyond their control; because Anderson believed that assault from above would prove more efficacious than gnawing from below, the British refused to consider surrender. Conant, Bush, and Groves continued to sit tight.

Goldschmidt took the impasse as a signal to try some diplomacy of his own. He and Auger set off for the Metallurgical Laboratory in Chicago where, flashing his pass from the previous summer, he was duly admitted. The result was everything he had hoped for. When the Frenchmen left, they carried with them not only memories of a happy return but data on Fermi's graphite pile and information about the precipitation of plutonium from uranium through the medium of bismuth phosphate.

'Goldschmidt,' said one of his friends, 'I gather that your visit may be the last for a long time, and I hope that you won't leave here with empty pockets.' He frisked the dubious alien who had wandered into his laboratory, but did not confiscate the two tubes, one with four micrograms of plutonium, which he found.[66]

It was enough to go on with, in addition to the carefully hoarded heavy water which now made a third hazardous journey, across the Atlantic to the Montreal laboratory. But without co-operation from the Americans, it was all they could expect to get. The future depended on Goldschmidt's ability to develop his own method for separating plutonium. There was, fortunately, enough uranium for the present, from British stocks lent to the Americans the previous year. Prodded by Mackenzie, the Americans promised to give it back, and even to throw in some low-grade graphite.[67] Graphite good enough for atomic piles could be had in Canada, at the Electro Metallurgical Company in Welland, Ontario. By scrounging it might be possible to assemble all the necessary materials for a pilot plant in Britain and Canada; but admittedly this was a large assumption and one that Newell, the engineering division chief, was rash to make.[68]

His rashness took some months to become manifest. At the beginning of April, Halban and his colleagues decided that they needed twenty additional tons of uranium oxide, over and above the oxide they had lent the Americans. Halban counted on Eldorado, whose product it was; since Eldorado was controlled by the Canadian government, it would be a simple matter to get some. Goldschmidt made inquiries, but this time his news was devastating. The Americans had bought all of Eldorado's production until some time in 1945. Unless Howe could be persuaded to break their contracts, no more uranium would be available for the Montreal laboratory.[69]

The news was no more welcome to Howe than it was to Halban. His success as chief of Canadian war production depended on his ability to be a reliable supplier to the Americans. The Montreal laboratory was apparently a marginal enterprise at best, and there was no point in ruffling American feelings on behalf of something that would probably not work before the war was over. Mackenzie was sent to England to tell the British that they must either put up on American terms or be shut out. If that occurred, Howe warned, 'he would not feel like supporting the project any further.' Canada 'had no interest in competition on such matters with the United States.'[70]

Mackenzie met Anderson in person on 11 May. It was not a happy meeting on either side and Anderson's attempt to convert the messenger to his own point of view, through what appears to have been a combination of geopolitical reasoning and personal abuse, did not succeed. According to Mackenzie's diary, Anderson 'fears the Russians and I fear is a little Englander. He is the type who sees only the good of England – people who don't act as good colonists should must be actuated by sinister motives.' Anderson had a gift for mixing contempt with hauteur, which raised Mackenzie's not inconsiderable hackles. 'He is the type of Englishman who makes me a good North American. The common people of England are simply marvelous [sic] – some of the rulers are *asses*.' To add insult to injury, Mackenzie learned – from what source we do not know – that Howe was thought to have 'sold the British Empire down the river.' The thought was attributed to no less a personage than Winston Churchill.[71]

Politics were not the only explanation for the predicament. The laboratory itself was floundering; as it did so it managed to create the impression that matters were worse than they actually were. The cause was political and beyond the control of anybody in Montreal; but the process was principally Halban's, and it derived from the laboratory director's inability to organize or inspire a large (but not very large) scientific workforce.

The outward sign was a stream of complaints issuing from Halban's office. He had insufficient administrative staff: the presence of Jackson, an experienced administrator from England, and the regular help of Thomson, Howe's aide, were not enough. Halban schemed to entice one of Mackenzie's senior staff down from Ottawa, and failed only because his quarry stipulated fantastically disproportionate terms. The senior clerks Mackenzie dispatched to Montreal did not impress; stories of bootlegging and Saturday morning blackouts were rife. When Halban tried to hire local staff, he ran up against low NRC salaries and manpower shortages.[72]

The director soon moved from expectation to disappointment. Mackenzie had no sympathy for him. What infuriated Halban seemed to Mackenzie to be 'petty annoyances which are a permanent part of Government.' After meeting Halban in February to discuss these annoyances, Mackenzie was moved to comment that the laboratory director 'as usual has many suggestions,' none of which he would act upon. But Mackenzie was nevertheless prepared to put his faith in Halban's scientific ability, until that faith was shaken by a further twist in the Anglo-American atomic war.[73]

The Americans remained intrigued by the possibilities of heavy-water piles. The best evidence for the feasibility of such piles lay in Halban and Kowarski's November 1940 experiments. With du Pont showing a lively interest in the subject, it became important to understand how accurate Halban and Kowarski might have been. The relationship between theoretical physics and design may not be immediately apparent to lay readers; but around such subjects as neutron capture, the behaviour of heavy water under neutron bombardment, revolved decisions on the size of a practicable reactor, the amount of heavy water that would be necessary, and the tonnage of uranium that should go into it.

Groves assigned Fermi, the man responsible for the original Chicago pile, and Harold Urey, the discoverer of heavy water, to check Halban's experiments, copies of which the Americans had had for some time under the old co-operative rules. When Fermi and Urey's calculations indicated some discrepancies, Halban was invited to visit New York and discuss the matter. Mackenzie was enthusiastic. It was an opening chink in the security fortress. Since it was Conant who made the approach this is unlikely, but it might have set a useful precedent. Halban wanted to go, and Malcolm MacDonald concurred. Anderson decided otherwise.[74]

The 'conference' went ahead anyway, while Halban sulked and Mackenzie fulminated. The results were sent to the British, along with the Americans' conclusion that the theoretical bases of Halban's results could not yet be taken as proven. More and more careful experimentation was necessary; could the British, and especially Halban, oblige? Kowarski, from his Cambridge exile, urged co-operation. 'I need not point out how much is at stake,' he wrote in June. If the Americans' doubts went unanswered, 'the possible usefulness of heavy water' would be judged to be far less than previously hoped, with the result that even the possibility of co-operation in that area would be discarded. There was another danger, Kowarski warned, and it lay in Halban's indignant response to the Americans' criticism. 'If Fermi and Urey were prosecutors in a trial for our lives I would hardly care to choose Halban as Counsel for the Defence.' A recalculation was made, showing that Halban and Kowarski's work was essentially accurate, if sloppy in some details.[75]

This gave the junior physicists something useful to do.[76] The chemists meanwhile ground six months' worth of work out of the radioactive samples that Goldschmidt had brought back from Chicago. Their time was hardly uneventful, though marked in Goldschmidt's (and other scientists') case by increasing frustration with Halban. Life had its moments nevertheless. Goldschmidt remembered that 1943 was the year when 'heavy water lost a great deal of the mystery that had surrounded it.' A glass carboy broke, spilling twenty litres of the precious stuff, about 10 per cent of the Norwegian supply. 'Attracted by the noise, I found myself face to face with a tall, thin British chemist, who had caused the accident, in his underwear, paddling in his bare feet in the precious liquid in the process of soaking it up in any kind of cloth, including cotton wool pads, that came to hand.'[77]

For the physicists a 2-million-volt x-ray machine was acquired from the United States; when Sargent's graphite arrived they were able to bombard it with neutrons from the machine to discover whether the material was altered, and in what way, by a process that would naturally occur under reactor conditions.[78]

The engineers had the difficult task of estimating the possibilities of constructing a reactor under widely varying conditions. There was a chance that co-operation with the United States might resume; under that circumstance, resources, not to mention information, would abound. Even on Conant's terms, limited supplies of heavy water and restricted information would appear. In the worst scenario there would be no co-operation; in that case supplies would have to come from Britain and Canada alone. This latter possibility promised to last for several years beyond the probable conclusion of the war.

The engineering staff had other matters to keep them busy. If there were to be a reactor, how should it be cooled? What kind of heat exchanger should be used? How would heat be dissipated? Which would be easier to cool, a homogeneous pile or a heterogeneous pile? This was a question that Urey and Fermi wished to discuss with Halban; it had to be contemplated in the abstract, given the shortage of heavy water. It became obvious that a heterogeneous arrangement consumed less heavy water than a homogeneous one. Goldschmidt contributed a uranium 'mayonnaise' – uranium oxide in solution in heavy water. The physicists and engineers finally decided to opt for a heterogeneous pile – in which the fuel and moderator were separate, rather than homogeneously mixed together in a mayonnaise or slurry. The shortage of heavy water and the success of the first Chicago pile encouraged attention to graphite as a moderator; perhaps when heavy water finally became available they could switch back.[79]

Newell, the engineering chief, was an able man with considerable experience. He enjoyed the closest co-operation with Halban, but in the minds of some of the more junior scientists Newell was tainted by his association with ICI and an unduly pompous manner. By the summer of 1943, he was also rendered suspect by too close an association with the director.[80]

Halban was handicapped. He was not seen to enjoy close relations with Mackenzie, and he was obviously unable to persuade the Canadians to deliver the goods. Mackenzie stayed away; his diary records only one visit in the first six months of 1943, in February.[81] Halban's technical committee, composed of his division heads, plus Laurence, seldom met; there was little feedback to staff that way. To reassure the staff, Halban summoned his British team for a talk. The prospects were poor, he admitted. But adversity should only spur them to further efforts, for it was obvious that the United Kingdom, and not the United States, was 'better qualified to take an interest in the Continent [of Europe] after the war.' Work on a British weapon for *after the war* was demonstrably in the common European interest. 'What do the Canadians think?' one of the scientists asked. Halban responded that they were 'not so dissatisfied about the speed, but [that] they feel they must not come into any strong opposition to the Americans. In this case it might be important to do the work, and we might have to finish the work in England' – without, we may presume, the Canadians.[82]

Halban was premature. Tube Alloys was simultaneously considering a variety of expedients for its program, including the construction of a full-scale plutonium-producing pile in Canada, to cost £5 million ($22 million). Construction was expected to take five years. It was not giving any special attention to a European partnership in the area. It did not, any more than Halban, tell the Canadians.[83]

What the British did not tell, the Canadians did not really want to know; but because they did not know their attitudes and policies were bound to flounder in ignorance and confusion.

Perhaps the British themselves did not know how little the Canadians at the highest level understood. Malcolm MacDonald, who as high commissioner was in a better position to grasp Canadian realities than a distant Anderson or a peripatetic Akers, commented at the end of June that it was high time to 'ascertain if [the Canadians] are prepared to co-operate with us in a 50/50 development of T.A. 2 [Halban's project] in spite of lack of Anglo-American co-operation. In view of Howe's general attitude I feel that I should put this question to Mackenzie King and Howe together.' MacDonald probably knew the answer as far as Howe was concerned, but he might have wondered whether Howe really understood the question.[84]

Canadian passivity in the face of Anglo-American bickering is not especially surprising, given the government's ignorance of what the Montreal project was about. Mackenzie King, who would doubtless have said that Canada could not hope to affect the outcome and might well be burnt if there were a conflagration, was occasionally brought up to date. Better, therefore, to hunker down and await events. In May 1943 Lord Cherwell, Churchill's scientific adviser, was detailed to brief King at a meeting in Washington. He informed the Canadian prime minister of the British point of view in the dispute; but there was no sequel to their conversation. The subject did not arise again until August, just prior to the first Quebec Conference.

The Americans in the meantime had not been inactive. Though Mackenzie had met with Bush and Conant, there had not been any contact between the military side of the American bomb project and the Canadian government. The contact now occurred. There were some difficulties over heavy water production, which was centred at Trail, BC. There were concerns over uranium deliveries, then being refined at Eldorado's Port Hope plant. Colonel Kenneth Nichols, who acted as General Groves's 'fixer' for problems, learned that the Canadian government might have a role in the difficulties. Whom should he see, he asked an Eldorado official. He 'advised that we should contact a C.D. Howe in Ottawa.' Nichols had his secretary put in a call, identified himself, and told Howe that he had a problem.

'In a pleasant voice,' Nichols later wrote, 'he responded that he was happy I had called. He did have many questions and had been patiently waiting for someone in the US government to contact him. He asked how soon I could come up to Ottawa to discuss the matters.' Nichols said he would arrive the next day. In Ottawa, he 'went to the address Howe had provided and discovered to my surprise that he was the minister of munitions and supply.' The two men chatted amiably, as Howe gently but firmly told his visitor that he did not wish the US army to act independently of the Canadian government on Canadian territory. With that difficulty out of the way, Lesslie Thomson was assigned as liaison between the Manhattan Engineering District and the Canadian government.

Nichols's visit, on 15 June, was only the first meeting between Howe and the Americans. After a visit from C.J. Mackenzie in mid-July, General Groves told Nichols that it had been suggested he get in touch with C.D. Howe. Nothing easier, Nichols replied, and told the story of his recent visit. 'It is wonderful,' Groves commented, 'to be young and not bothered by protocol.' Since lack of a sense of protocol was one of Howe's characteristics, Nichols's visit had done no harm: rather the reverse. On 26 July Groves and Nichols visited Howe in his office. Uranium was the main topic of discussion, but as important as the subject-matter was the atmosphere. Co-operation was, as far as the three engineers in Howe's office that day were concerned, assured. Howe, the Americans decided, could be trusted.[85]

Not long afterwards Mackenzie King had Sir John Anderson to tea at his country estate. The first Quebec Conference was impending, and Anderson had accompanied Winston Churchill to North America. In response to MacDonald's earlier plea, Anderson may have decided to take the opportunity to explain as best he could the military potential of nuclear energy, and he arranged to see King without either Howe or Mackenzie being present. He would probably have been disconcerted had he known what the prime minister would make of his explanation. 'The root secret of it all is some atomic power,' King told his diary. 'As he developed the matter scientifically, my own mind made a parallel between power of the mind that comes from tapping deeper sources of energy as explained by [the philosopher] William James.' But King grasped the political point that whatever country developed atomic weaponry first would be in a position to 'control' the world.

There had been some difficulties between the British and the Americans, Anderson continued, because of the Americans' belief that the British invariably took advantage of their bucolic transatlantic cousins. The British had now made plain that they were not seeking any special advantage, and the matter was on its way to solution. Mackenzie King, who had not known that Canada was unwittingly part of the problem, could not have guessed that Canada would now be part of the solution.[86]

The British, responding to American fears, were about to promise that they would not appropriate the atomic secret for industrial uses, so as to shore up Britain's postwar competitive position. Their intention was purely military. Whether the weapon was developed during the war or not, the British wanted it and would have it. That broke the log-jam. If the British wanted to participate in the atomic program for military reasons, the Americans were prepared to concede them a place, not merely at the bargaining table but in the laboratories. An agreement was struck at a meeting at Roosevelt's country estate at Hyde Park, New York, and ratified in a conference between the British prime minister and the American president at Quebec City in August 1943.[87]

The resulting Quebec agreement was signed on 19 August. It stated that victory would be the speedier and more certain if all available allied brainpower were pooled in the great atomic project. Thus there should be 'full and effective interchange of information and ideas between those in the two countries engaged in the same sections of the field.' The agreement recognized that good wishes were not enough, and it set up an implementing and supervisory body, to be known as the Combined Policy Committee (CPC). On it would serve three Americans, two Britons, and one Canadian, in recognition of Canada's part in the British project. The Canadian would be C.D. Howe.[88]

The CPC was news to Howe, as was the subsequent news that nobody had thought to invite him to its first, informal, meeting. Informed by telephone from Washington of what had occurred, the Canadian minister replied easily that it did not matter very much, for even if 'he were unable to come to Policy Committee meetings, [he] would be quite content to leave the Canadian side of any question in the hands of the British delegates.'[89] Instead of Howe, Mackenzie attended a meeting of a subcommittee struck for the purpose of examining existing programs and recommending new means of co-operation. To that end it discussed the progress the Americans had made with heavy water and graphite.

The Americans by now had a second graphite pile in Chicago and a third, a huge cube of graphite with tunnels containing natural uranium fuel sheathed in aluminum, at Clinton, Tennessee. A reactor complex was also being built at Hanford, on the Columbia River in Washington State. The latter was cooled by water straight from the river, which, after performing its function, was subsequently pumped back. Construction had started in June 1943. There was thus a certain amount of experience to share.[90] When Mackenzie's subcommittee, meeting in Washington on 9 or 10 September, agreed that the Americans would 'discuss freely all the scientific work on the Heavy Water Pile' and at least some of the work on the graphite pile, the benefits to be anticipated for Montreal were very considerable. There was no need to exchange information on the production of heavy water, since the Americans 'had already arranged a production programme of 3 tons a month, and some of the plants were, in fact, just starting.'[91] The heavy-water drought would soon be over. It was the first time, but not the last, that optimism prevailed over caution in the field of heavy water.

The next months promised well. Anderson himself visited the Montreal laboratory on 9 August to tell the scientists that they could proceed with their work.[92] Anderson's impression of the laboratory is not known, but succeeding visitors were uneasy.

James Chadwick, who had just been appointed to manage scientific liaison with the United States, in one of the fruits of the Quebec agreement, visited on the 27th. Akers showed Chadwick and Mackenzie around. Delicate, inclined to hypochondria, and pessimistic, Chadwick may have understood Halban (whom he already knew) all too well. In any case, he did not like what he found. To Mackenzie, he commented that the administrative staff was excessive for the size of the laboratory. The NRC director for his part was alert to signs that Canadians in the laboratory were being excluded from any real understanding of what was afoot; for that he blamed Halban's 'aggressiveness and acquisitiveness.' Later, in Ottawa, Chadwick revealed that he did not 'hold a high regard for Halban.' Mackenzie promptly decided that he liked his new British colleague very much.[93]

The re-establishment of cordial relations with Mackenzie was worth something, and it spoke highly for Chadwick's objectivity, as well as for the importance the British attached to his mission. Equally important were relations with General Groves, who now assumed a critical role in the future of the Anglo-Canadian project. Mackenzie had met Groves in July and found him 'a man we know how to talk to'; Howe's visit from Groves we have already discussed.[94]

Chadwick and Mackenzie understood that Groves would be the man to decide where the Montreal lab fitted in. Unfortunately, Mackenzie could not ensure that Halban made the best possible impression. When Groves and his staff visited the Montreal laboratory to discover what the British contemplated its role to be, Halban disappointed them. Instead of presenting a clear proposal, he ranged over the possibilities, from homogeneous slurries to a uranium hexafluoride 'boiler.' Groves was not impressed. If Halban did not know his own mind, Groves did not propose to waste time and money on him. Meeting separately afterwards, the Americans resolved to offer little or no help to the Montreal project.

Second thoughts were less drastic. Information was exchanged, though not on plutonium separation, and some materials were lent to the Canadians. But the Americans had not been convinced that the Anglo-Canadian project should proceed beyond the laboratory stage. This was a disaster for Halban, whose standing with his Canadian partners sank further. 'In relation to Chadwick he is a mere child,' Mackenzie wrote, 'and a temperamental one at that.'[95]

So thought Chadwick, and so thought Halban's staff. Because of Chadwick's custom of dining with junior staff on visits to Montreal, he soon had a large repertoire of Halban stories, all tending to reinforce the impression that the laboratory director was arrogant, uncomprehending, and unfeeling. However, there was no point in dismissing him unless and until it was known what to do with the laboratory.

It must progress or stagnate. The next stage was a large reactor, similar to the experimental ones the Americans were building. As far as the Americans were concerned, it was up to the British and Canadians to pay, somewhere in the range of $25–50 million, depending on which option was selected. 'It seems that either we undertake the fifty million dollar project or we close up Montreal,' Mackenzie wrote in December, after a conversation with Chadwick. If only the Americans would give the Canadians the responsibility for the entire heavy-water scheme, it could all be justified, he lamented. But as Halban learned in January 1944, on a visit to Chicago, the Americans had designed and were building their own heavy-water reactor (known as Chicago Pile 3, or CP-3). They expected to have it operational in six months.[96]

If the Montreal laboratory was not to be overtaken by this event, it had to do something, fast. But doing something cost money, which only the Canadians would or could supply. They would not give money unless they were certain the Americans were co-operating. The Americans would not assist unless they thought the laboratory would be useful in the war effort. Every passing month brought the end of the war closer, and made the military argument for contributions less plausible.

To Chadwick the first step was obvious. Halban must go. Difficulties in the laboratory, between the laboratory and the Canadians, and between the British and the Americans were all bound up with Halban's personality. After talking to Chadwick in December, Mackenzie expressed the thought to his diary: 'if Cockcroft will agree to come to Montreal Groves will agree to the proposal.' John Cockcroft was a high card. His reputation as a physicist was excellent. He knew the Americans, and they liked him. Best of all, Cockcroft was willing to come – provided he could get his hands on American research and supplies.[97]

In Montreal, morale was plummeting. Halban's reaction was to isolate himself. His enemies claimed that, if cornered and questioned, he would flop back and feel his pulse to see whether disagreement was bringing on heart palpitations. At the beginning of February George Laurence, irritated by a fresh disagreement with Halban, sent Mackenzie a letter detailing Halban's lapses as a laboratory director and demanding that something be done to resolve the situation before the laboratory collapsed under the weight of accumulated tensions. The French scientists were restless; one was leaving to join General de Gaulle's headquarters, while the others were showing signs of boredom and frustration. The knowledge that the Americans were building their own heavy-water reactor, the world's first, seemed to confirm that Montreal's mission was utterly meaningless. Without the Americans they could see no future.[98]

The future meant Groves. To Chadwick, Groves seemed a better bet than Conant, who true to form plumped for the strictest possible construction of the Quebec agreement and demanded that CPC sponsorship be limited to projects relevant to the current war. And if Halban had not made up his mind as to a proper design for a large reactor (the director was at that moment considering only a pilot plant), Chadwick had: it should be heavy water and natural uranium, and its cost, as he and Mackenzie had discussed it, would be about $50 million. This proposal was put on the table at a CPC meeting on 17 February 1944, with Mackenzie rather than Howe representing Canada (a snowstorm had prevented Howe from landing in Washington). Chadwick did most of the running. He considered the proposal for a heavy-water, natural-uranium reactor to be the best bet available: 'it was the best and cheapest method and,' in Mackenzie's words, 'we could do it in quicker time.'[99]

The meeting could agree only in disagreeing. A subcommittee was struck, to consist of Chadwick, Groves, and Mackenzie. Mackenzie promptly bowed out. Anything that the British and Groves could agree on, the Canadians would accept, and he would therefore 'not take [an] active part in formulating [a] report.'[100]

Groves deputed one of his officers to investigate the Montreal project and to make recommendations. The report was negative. And so Chadwick, in a tour de force, rewrote the report, altering its conclusions to support his case, and took it to Groves. Though the American reactors at Hanford, Washington, would produce enough plutonium for 'essential military needs' the heavy-water reactor was not to be 'wholly neglected.' It held possibilities, possibly brilliant ones, and in the interest of the future it should be investigated. The general was not personally disinclined to help the Anglo-Canadian project, and he respected Chadwick's judgement. He swayed, then veered round to accept the British argument.[101]

It was just in time. Chadwick knew that the British government had decided to abandon the Montreal project, and preliminary steps were already in train to brief the staff and offer them the alternative of returning to England to continue their work. The Canadian staff were to be invited to go with them, if they desired.[102]

There would, instead, be a new director and a new program. Rather than being terminated, the Canadian venture in atomic energy would expand. In expanding, it would leave a permanent mark on the landscape: Canada's first nuclear reactor.

An Idea in Exile

Halban's frustration was complete. He got the bad news from Mackenzie and Howe as they passed through Montreal on 15 April 1944, on their return from the CPC meetings in Washington. Around the laboratory the news of Cockcroft's arrival was greeted with ill-concealed joy.

It was preceded by a visit from Howe himself. With Mackenzie in train, the minister visited the laboratory on 15 April to give the scientists a 'pep talk' and to tell them that all the obstacles had been cleared. Then Howe went down the mountain to talk to the executives of CIL. The Montreal project would soon need more engineers. Defence Industries Limited, CIL's wartime subsidiary, would provide them, in a team headed by B.K. Boulton, an experienced manager from Montreal, Light, Heat and Power. (It had just been nationalized, and Boulton was out of a job.) With

Boulton's appointment, Howe and Mackenzie were ready for Cockcroft's arrival.[103]

The great day was set for 25 April. Cockcroft arrived at Dorval airport in pouring rain, and was warmly greeted by George Placzek. (Placzek, who had married Halban's divorced wife, was known around the laboratory as the 'husband-in-law'; nevertheless, he did his best to support the unfortunate ex-director.) The next day Cockcroft arrived at the laboratory. Halban was not present, having gone off to inspect a possible reactor site. So it was not until the 27th that Halban introduced his successor to the laboratory staff. Cockcroft, he said, had come to supplement his own talents, which would continue to be employed at a high, indeed an equal, level. It was an embarrassing situation. Those present knew that Halban had failed, and was on the way down if not out. Would Cockcroft by his silence ratify Halban's version of events? Cockcroft would not. Clearing his throat, he told the meeting that he had come as sole director and not as a supplement to his predecessor. Auger had left, creating a vacancy as head of physics, and that was where Halban would go.[104]

Relations with Cockcroft were cordial, and there was now plenty to do. Kowarski had finally been imported without apparent objections from Halban. He got an autonomous project that kept him from daily contact with his former patron. Halban's other bête noire, Laurence, remained a member of his division; neither man liked the other any better than before.[105]

The Normandy invasion and the subsequent liberation of Paris on 23 August gave the ex-director food for thought. He was insufficiently employed in Montreal; as he told Cockcroft on 19 September, he could now 'easily be replaced.' It was time to go back to England and perhaps on to Paris, 'to put myself at the disposal of the French Scientific effort for any even minor job I could do in France.'[106] Halban did not envisage himself in a minor job. His British contacts were still good. He felt secure in Anderson's esteem, Akers was friendly, and Lord Cherwell, who visited Montreal at the end of September, was well disposed. They might help Halban to re-establish contact with Joliot, who had survived the war and was still living and working in Paris. The Joliot-Halban team had its 1939 patents, and the British were mindful of their obligations to France. In the short run, Halban secured permission to visit London and, despite Chadwick's misgivings, telephoned from Washington, he went.[107]

Chadwick had been listening to what the Americans had to say. They were not bound by any debt of gratitude or previous alliance with the French. Joliot was a Communist, whatever his standing in nuclear physics or with the French government. There might already have been some apparent leakage from Auger and Guéron, who had visited Paris earlier in the fall; there was absolutely no guarantee, in Groves's mind, that Joliot would not pass on what he learned straight to Moscow.[108]

Halban paid no attention. He wanted to go to London, and then on to Paris, and he counted on his friends in the British government to arrange it. They did. Halban duly reported to Joliot, only to discover that his mentor was highly displeased with him. If the Americans had no use for Joliot, Joliot had no use for them either. He wanted an Anglo-French collaborative atomic program, and would settle for nothing less. He would say the same thing to Anderson, when they met.[109]

So much for Halban's liaison role. It must be admitted that he had more immediate problems. General Groves, when he learned what Halban had done, had an easy answer: Halban, and all the other Frenchmen, should be put in jail. If not the others, he conceded, then at least Halban, who was sailing back to Montreal via New York and who would shortly be in Groves's power.

Chadwick had the ungrateful task of negotiating terms on Halban's behalf. Groves finally conceded that Halban need not be imprisoned, but that never again should he be allowed near a secret project. Furiously the general blamed the Canadians for letting Halban out of their country in the first place. Mackenzie passed the heat on to Howe, and 'he went off the deep end immediately,' as Mackenzie put it. Malcolm MacDonald was summoned to receive the minister's wrath. Told that Halban was en route to Canada, Howe insisted that he should not be let back into the laboratory. Chadwick at first demurred. Banishing Halban from the laboratory would merely confirm that he should go back to Britain, as he wanted. Goldschmidt was summoned to Washington to help corral Halban as soon as he landed. According to him, Chadwick would have preferred to see Halban's transport sunk, with the scientist on board. Failing that, Goldschmidt was to spirit Halban off his boat and back to Canada before any more damage was done.[110]

Now it was Howe's turn. Howe had had an earful from Groves; he had not known what had been going on but had still to take the blame. To Howe the British were responsible for a breach in Canadian-American relations; it was up to them to make reparation. At first the British denied that anything much had happened; later, as Groves's reaction developed, they concentrated on the immediate problem of keeping Halban out of jail. After a tense meeting of the CPC on 25 January, Howe undertook to produce an acceptable solution. Instead of jail, Halban could merely be sent to Toronto for the duration of the war. Eventually, he agreed that Halban should only be banned from the laboratory; he would also be forbidden to leave Canada.

Later, speaking to MacDonald, the minister drew a larger conclusion. He had intervened because, he stated, 'some of the discussion [in the CPC] seemed to be based on the assumption that what happened in the Montreal laboratory was a matter for decision for the authorities in Washington. He objected to this, and desired to assert his position as the Canadian Minister responsible for ultimate control of a project being pursued in Canada.'[111]

Halban was through. Cockcroft informed his unfortunate predecessor that he would henceforth be working at home, nominally as a consultant to the laboratory. His colleagues were officially told that Halban had resigned. Few mourned his departure.[112]

Halban was the principal author of his own misfortunes. Talented, if arrogant and self-willed, he had until 1944 led a charmed life. Scenting personal failure after his replacement by Cockcroft, he attempted to reinvent a role for himself through his mission to Joliot. He only succeeded in undermining what little prestige was left to him in Canada, while ironically failing to impress Joliot. The British gave him a professorship at Oxford after the war. He enjoyed an Indian summer in England, plentiful in experiments and fruitful in publication. Eventually he returned to France where he resumed his interrupted career; he died in Paris in 1964, aged fifty-six.[113]

Halban's time in Canada had not been entirely futile, nor

completely misplaced. On the contrary, much valuable work had been done. The downgrading of homogeneous reactors (they demanded too much heavy water and absorbed too many neutrons) left the laboratory free to concentrate on the design of a single reactor type. Thanks to experiments in 1943–4 much more was known about the performance of various materials – including graphite, light water, and heavy water, all potential moderators – under radioactive conditions. Halban's departure, of course, ensured that American-supplied heavy water and not graphite would be the moderator.[114]

The conception of a heavy-water reactor was no small accomplishment and as far as Canada was concerned, it was a lasting one. Every subsequent Canadian or Canadian-designed reactor is, properly, a descendant of Halban's original conception and experiments. For the Montreal laboratory was nothing less than Halban's own creation. He was determined, above all, to be his own man, to head a major research team, and to prove the validity of the line of research that he and his colleagues had undertaken at the Collège de France. A more modest or more realistic man might not have persevered, but Halban's persistence, charm, and force of intellect had made Sir John Anderson, according to a famous story, decide 'That is the horse that I will back.' Ever afterwards Hans von Halban was known as 'Harry the Horse,' at least to his many detractors. But Halban was the prime mover in establishing the Montreal laboratory; and it is to him that Canada owes the origin of its first nuclear project.[115]

2

Errand in the Wilderness

The fact that the Montreal laboratory survived Hans von Halban was, in the eyes of the survivors, the principal accomplishment of its first year. That it had also survived President Roosevelt, James B. Conant, General Leslie Groves, C.J. Mackenzie, and C.D. Howe was more accurate, but the long odds against its achieving its second year were not stressed after the fact. It was not, and could not be, the subject of prolonged self-congratulation. Compared to their American counterparts, the Anglo-Canadian team had little or nothing to show for their work, while the Americans were building – and completing – reactors at Chicago, Oak Ridge, Tennessee, and Hanford, Washington. The task of reversing fate lay in the ability of one man, the new laboratory director J.D. Cockcroft.

J.D. Cockcroft

John D. Cockcroft was a product of the Cavendish Laboratory at Cambridge, a student of Lord Rutherford who, as a young professor, had conducted some of his fundamental work in radioactivity at McGill. What Ernest Rutherford had transported across the Atlantic in 1907 now returned, amplified, in 1944.[1]

Cockcroft was not quite forty-seven years old when he landed at Dorval airport outside Montreal. Like others of his generation, his education had been interrupted by the First World War; afterwards he passed through the Metro-Vickers electrical engineering company en route to Cambridge and the Cavendish, where he took his PhD in physics in 1928. It was not long before Cockcroft acquired a national and then an international reputation for his work on the disintegration of atoms; and he early on concluded, unlike Rutherford, that atoms were a tremendous reservoir of energy.[2]

Cockcroft became a figure of note at conferences; rotund, cheery, bespectacled, he was an attractive personality – and ambitious. After his first American tour, in 1933, he visited Montreal where he 'gazed with appropriate reverence on the rooms where Rutherford made his reputation 25 years ago or so.' Cockcroft would not need to make his reputation in Montreal; it was already secure by the outbreak of war. That year, 1939, he became Jacksonian Professor of Natural Philosophy at Cambridge, a post he would hold for the next seven years. During those seven years, as he later wrote, he 'delivered not a single lecture, a record probably unequalled even by the 18th century professors.'[3]

Nuclear physicists were deemed good material for radar work, and a variety of Cambridge physicists, nuclear and non-nuclear, were swept up in war work even before Hitler's war had officially begun. From the Cavendish, Sir Edward Appleton, W.L. Bragg, John Cockcroft, and W.B. Lewis took a variety of posts, according to seniority, with the government. Cockcroft, whose American connections made him an ideal ambassador for war science, was sent to North America in 1940 to establish contacts. On the boat trip over he entertained the crew by lecturing on the uncontroversial subject of atomic science; his secret work, which he was forbidden to mention, involved radar.

During his trip Cockcroft visited Canada, where he met C.J. Mackenzie. At first the voluble Mackenzie was disconcerted by Cockcroft's failure to respond immediately – and favourably – to his offers of Canadian scientific help. The next day, however, Cockcroft repaired the omission, and Mackenzie later reported that the basis of harmony had been laid. Cockcroft recorded that he found the Canadian scientists 'extremely able and willing to help.'[4]

Mackenzie remembered, as, presumably, did Cockcroft. It was

not Cockcroft's only Canadian experience; he was over again in 1943, discussing proximity fuses and radar with the Americans, and he took the occasion to visit the Montreal laboratory. The visit was instructive, but it was not innocent; Cockcroft had already been told that he was regarded as a possible successor to the floundering Halban. It was not a task he relished or even desired, but if necessary he would go. And so, in April 1944, he did.[5]

By this time most of a year of reactor development had been effectively lost. The British project was so far behind the American that there was no possibility of putting it to work during the war for any conceivable immediate military purpose. The Americans were able to rely on what they had already built – a uranium separation plant, for uranium-235, and the Hanford complex, nearing completion, for the production of plutonium from graphite reactors. The British were contemplating the postwar possibilities of reactors, and the Americans, to accommodate their ally, were willing to go along with their desire for at least a quasi-independent project. The Americans were not being especially altruistic, for their interest also dictated that they pay attention to the possibilities of heavy-water reactors; from Montreal they would receive details of Anglo-Canadian technical progress in exchange for briefing the Montrealers on American knowledge and experience.

Interchange of information was relatively free, except in one area: the chemical separation of plutonium from used fuel from the future reactor. The Americans suggested that spent fuel could be sent south for processing in the United States; there the Americans could remove whatever plutonium it contained and return the rest.

Cockcroft's first task was to negotiate whatever details Chadwick's agreement with Groves left unclear, and these were many. His second was to do something about the laboratory's administration. One of his initial actions helped accomplish both ends. By delegating some of the laboratory administration to E.W.R. Steacie of the National Research Council, a senior though non-nuclear chemist of international standing, he cleared enough time to move about, whether to Ottawa to see Mackenzie, or to New York to contact General Groves and his staff. Out of these negotiations he secured bars of irradiated fuel from Oak Ridge, delivered under armed guard to Montreal's loading dock.

Cockcroft brought a strikingly novel personal style to the direction of the laboratory. It made such a profound impression on his staff that it may be said in itself to have helped restore an authority that under Halban was held increasingly in disrepute. Unassuming and taciturn, Cockcroft would listen patiently to whatever proposition was being put to him; then he would reply 'yes' or 'no,' sometimes recording the fact in a little black notebook in his minuscule handwriting.[6]

A Technical Physics Division was established under George Laurence to consider what equipment would be needed, how it should be designed, and what standards it should embody. Health and safety, which had not hitherto been a major concern, now came under scrutiny; to direct studies in this area Cockcroft recruited Commander C.B. Peirce, a Royal Canadian Navy doctor, and Dr J.S. Mitchell, of Cambridge. Safety implied current precautions and standards of radiation exposure; but health also involved the investigation of new isotopes, to be created inside the reactor, for the treatment of disease.

This new concentration on a single experimental pile did not completely exclude alternative lines of investigation. A Future Systems Group was also established, to contemplate the possibilities, in a more nearly ideal state of supply, of heat exchange, novel coolants (gas), and breeder reactors, which, as their name implies, 'breed' more fuel out of an existing fuel supply. At a time when the world supply of uranium was believed to be strictly limited, this was no small consideration. Because uranium was scarce, the scientists made provision for experiments converting thorium into uranium-233; the future reactor would be used, in part, for this purpose.[7]

Attention could now turn to the actual construction of the reactor. C.D. Howe said in April that Canada would 'stand the racket,' that is, gladly assume the full costs of the operation; but Howe contemplated less rather than more cost. He had, in fact, a location already picked out – a munitions factory run by Defence Industries Limited (DIL), a subsidiary of Canadian Industries Limited (which in turn was a subsidiary of ICI and du Pont), at Nobel, Ontario, outside Parry Sound and near the clear, cold waters of Georgian Bay. Howe did not like to see the buildings and houses constructed at government cost in this isolated facility going to waste, and he urged Groves to come and inspect the site.[8]

The problem was that its advantages were more apparent than real. It was true that there were houses, shops, sewers, and buildings. But there was not enough power, the main factory building was unsuitable, and there was not enough housing, in the proper proportions, for the staff. Investigating teams moved on, to Lake Nipissing, to Lake Superior, and to the North Channel of Lake Huron. This took up most of May. In June, searches were made up the valley of the St Maurice River in Quebec; only at the beginning of July did attention turn to the possibilities of the upper Ottawa Valley.[9]

Attention focused on the Ontario side of the Ottawa River, a few miles north of the railway village of Chalk River. The place was remote; Chalk River was not large; there was transportation by highway and railway; Ottawa was not especially distant; and Montreal was only a few hours further on, by train. An adequate power supply was possible and, most important, the river ran strong and clear at that point. Only sampling would reveal just what was in the river as it flowed past Indian Point, but there was no reason to expect that it would be prohibitive in terms of possible corrosion. There were several sites for a village at a suitable distance, in case of accidents. The choice was obvious; obvious, that is, to everyone but Howe.

Howe stuck to his guns. Modest estimates for Chalk River showed a cost of $6 or $7 million. Much of that was housing and services, not to mention land; surely Nobel deserved another look. Cockcroft resisted, stating 'that, if necessary, he would not mind living at a distance of 2 miles from the NRX Plant, but that he would not feel safe in having his family in the Village at this distance, and further stated that in his opinion the plant should be located at least 4 miles from the Village.' This would exclude Nobel, but Howe was unmoved. Another look and another round of argument were necessary before the minister capitulated. But it was now the middle of August.[10]

If a reactor was to be built, someone must build it. Howe had turned to DIL. DIL had a strong engineering staff; it also had experience in building and running large munitions plants in isolated or at least rural environments. Only the design and construction of the reactor building and its attendant safeguards would be novel. With Mackenzie's agreement and British support, DIL received a 'go ahead' letter on 7 July. The company accepted on 17 July.[11]

The contract called for the company to handle all construction and operation of the new 'NRX Project' – for National Research X-perimental. The government reserved the right to approve any arrangements the company made, and directed the company to maintain all necessary secrecy and security, to use labour and materials only from the British Empire (unless no such items were available), and to employ veterans first. There would be no fee for the work, though all expenses would be reimbursed. DIL, which like its parent firms was sensitive to the charge that munitions firms profited from war, stuck to a no-profit–no-loss principle.[12] Next came the construction contract, which fell to the Fraser Brace Company. Approached for his approval, General Groves gave it on 25 July, subject to strict security precautions.[13]

Senior members of the DIL engineering staff were sent down to Oak Ridge to see what the Americans had done. It was not, David Kirkbride recalled, very surprising. He knew that the possibility of nuclear weaponry existed, and here it was. A survey of the Canadian plant site was already underway, even though neither Howe nor Groves had as yet given final approval. Expropriation of land was done through the Canadian National Railways, which did not run anywhere near the site. On 23 August 1944 approval was given for a townsite, up the Ottawa from Chalk River. On the same day Fraser Brace's construction supervisor moved to Chalk River. The layout of the plant was definitively approved on 1 September. Relieved, Cockcroft took a trip back to England, to report to his superiors and attend to his family.[14]

Seen in retrospect the speed of the work was remarkable, but that was not everyone's impression at the time. Having approved the project on what he knew to be a highly speculative basis, Groves was sensitive to delays that would push Chalk River's completion not only beyond the end of the war but beyond any plausible projection that it might be ready in time to contribute to the war effort. Groves was provoked to tell Chadwick that Cockcroft should spend more time with the project, since 'his US people were [already] disappointed with the progress in Canada.'[15]

Cockcroft was not, nor were his staff. Time was an enemy, to be sure. The length of time that it would take to get NRX built and running might well leave some of the scientists with energy to spare; an intermediate goal might well, therefore, be suitable. As NRX began to take shape on the drafting board, Cockcroft turned his attention to another, quicker goal.[16]

ZEEP

Groves might have been more 'disappointed' had he remembered a conversation with Cockcroft the previous spring. Then the new director had mentioned the desirability of a facility to study physics problems, a zero-energy reactor. Perhaps Groves could supply the necessary extra heavy water? Groves apparently agreed.

Cockcroft does not seem to have mentioned the subject again for several months. On 25 July he raised among his colleagues the possibility of giving the pilot plant NRX its own pilot plant – a Zero Energy Pile, or ZEEP, which, because it would produce no more than a few watts of power, would be far less radioactive than NRX. It could use heavy water destined for NRX, speeding the latter's design by testing materials and solving problems on a practical rather than theoretical basis. ZEEP soon became a certainty, with a team of its own, and Kowarski as project chief.[17]

Kowarski had arrived in Canada in the wake of Halban's failure, to find work already underway on the design of ZEEP.[18] Kowarski's assignment was organizationally desirable, for it gave him an independent mandate outside the usual line of command. The idea of a project chief ('chef de projet') was an 'innovation,' according to Jules Guéron; at Montreal it allowed Kowarski easy access to Cockcroft and at the same time permitted him to peer into others' projects while concentrating his energies on his own team.[19]

ZEEP, Kowarski reported on 23 August, had many specific advantages. It could be used to manufacture fission products for further experiments; it could be used to test materials for NRX; and because of its low output it could be used to experiment with alternative ideas for design and construction. A prototype already existed (CP-3) in Chicago, and it could now be improved on.[20]

General Groves was not very happy. Montreal, he told Chadwick in November, was 'splitting its effort' to no very good purpose. But grumbling or not, Groves would supply the heavy water: twenty tons of it, as soon as the pilot plant was ready.[21]

ZEEP
The Zero Energy Experimental Pile was designed in Montreal and built at Chalk River in 1944–5. According to Lew Kowarski and C.M. Watson-Munro, it was to have 'an aluminum cylinder of diameter 6′ 9″ to contain a maximum of 9 long tons of polymer [heavy water] liquid and have suspended in this liquid an absolute maximum of ten long tons of [uranium] metal in the form of a lattice of vertical cylindrical rods of diameter 1.1″.' ZEEP, which began operating in September 1945, was the first functional atomic reactor outside the United States; early experiments in ZEEP were crucial to the determination of fuel geometry and the disposition of fuel and control rods for later reactors, NRX and NRU. Because of its low power rating it required little in the way of shielding; but because of its small size (only 8′6″ high) it was of limited value in studying later and larger fuel rods. It was later removed from service, and is now a museum-piece.

ZEEP needed, and received, the highest priority of any NRC project, but its completion was dogged by shortages of labour and supplies. Costs soared, as site managers at Chalk River enticed scarce craftsmen with higher pay. If someone had a heartbeat, a joke ran at the time, you pressed a travel warrant into his hand and shipped him up the Ottawa. Even after a worker arrived, he frequently had to wait outside the gates while a security check was run; and that could take days. At any given point, according to one story, the workforce could be divided into three equal parts: some coming, some going, and the rest making up their minds.

Much of the work done was good; but much was not. When the installation team arrived to prepare for ZEEP, they found the concrete floor of the reactor building 3 inches too low. Stainless steel pipes and tanks arrived dirty; the site engineers, and some of the scientists, had to take time off to scrub them down. What went into a reactor had to be precisely as planned; the formula for the chain reaction did not allow for the presence of dirt or rust.[22]

As the scientists scrubbed and the summer heat blazed above them, the war was coming to an end. The European war ended in May 1945: there would be no German bomb. The Americans tested an atomic bomb in the deepest secrecy in the New Mexico desert in July; soon afterwards, rumours reached Montreal that the bomb was, after all, practical. The Japanese war ended in August when the Americans dropped atomic bombs on Hiroshima and Nagasaki. A week later the Japanese announced their surrender; it was formally signed on board the US battleship *Missouri* on 2 September.

Harry Truman, the American president, announced that the weapons used over Japan were atomic and of unprecedented power. In Ottawa, Howe issued a statement, drafted in advance, explaining that Canada had had a role in developing the new weapons. Then, on 15 August, he and Mackenzie, with Cockcroft, Laurence, and Nicholas Kemmer, a physicist, representing the scientists, held a news conference to describe, as best they could, what had happened. It was, Cockcroft later wrote, a useful occasion, which could have introduced the Canadian public to ideas about nuclear power, but the dominant news of the day was the Japanese surrender.[23]

ZEEP was still not ready. The uranium was in place, the heavy water had arrived, the plumbing had had one final scrub. The reactor crew, scientists and technicians, assembled on 4 September and began to pour in the heavy water that would, it was hoped, start the reaction. Drum after drum was poured in, until there was only one in reserve. A gauge outside measured what was happening, and the measurement was slow – so slow that it took another day. On 5 September the last drum was poured in and, at 3:50 in the afternoon, 'a divergent chain reaction was finally started.'[24]

As Groves, Conant, and Mackenzie in their different ways had predicted, Canada's first reactor had not come in time to make any difference to the war. Yet the reactor had been built; small as it was, it was a success. It was the first successful reactor outside the United States. It pointed, as reporters had sensed the previous month, towards nuclear power; but it pointed as well towards a military application.

The dropping of the bomb was a nine-days' wonder. Newspapers were filled with speculation of what it might mean. Could Hiroshima herald the end of civilization? It was, a historian has concluded, 'a psychic event of almost unprecedented proportions.'

But although the scientists in Montreal and Chalk River were concerned at what the actual use of the bomb might mean, their reactions were muted. There was some agitation among the British staff, even a petition, but work went on. The attempts by some American scientists to mitigate American military strategy were not duplicated in Canada; the scientists in Canada, unlike the United States, were not inclined to interfere in either politics or administration.[25]

"Many people,' Cockcroft wrote on 10 August, 'are very troubled about this new power and most of us working on it have been equally troubled.' The choice offered the world was between annihilation and living peacefully together. It was, he thought, a responsibility that bore heavily on the three great surviving empires – the United States, the USSR, and the United Kingdom. 'No other nation,' he added, 'can come into the picture as potential dangers to the world.'[26]

C.J. Mackenzie

Chalmers Jack Mackenzie was Jack to his friends, C.J. to his colleagues, and 'Dean' or Doctor Mackenzie to those more distantly situated, either geographically or in status. He had moved up from acting president of the National Research Council to full president in 1944. He would remain president for the next eight years, the federal government's senior scientist and a power in his own right among the bureaucracy. Nor did he suffer from being the friend and confidant of C.D. Howe; where Mackenzie spoke, he spoke with the authority of a senior minister behind him. To his listeners, Mackenzie conveyed charm, simplicity, and decisiveness. He left a disarming impression of modesty: an engineer before scientists and a scientist before engineers, few bothered to wonder what, in fact, Mackenzie really was.[27]

Perhaps they did not need to. Mackenzie's accomplishments were manifold and manifest. To his admiring NRC staff, he bequeathed a scientific establishment vastly expanded from its modest prewar beginnings and a complex of glistening white laboratories in Ottawa's eastern suburbs. All this, Mackenzie liked to say, with a minimum of memos and paperwork. He relied instead on delegation of authority and an instinct for identifying the 'right' personality; his pronouncements on the subject bear a striking resemblance to those of his friend and minister, C.D. Howe. Howe liked to fly his departments by the seat of his pants, trusting in people while mindful of their frailties. And so Mackenzie ran NRC.

Anyone, Howe believed, could make a mistake; but those who made a second were soon gone, forgiven but also forgotten. Given the resonance of Howe's personality in Ottawa at the time, the resemblance to Mackenzie was probably more than coincidental. A friend to science and a foe to red tape and 'administration,' the NRC president collected little but kudos. Mackenzie's advocacy counted for much among 'the seniors' (his term) in the bureaucracy as well as with the minister, and there is evidence that they relied on his judgement independent of the influence of his minister. He would leave a lasting impression in Ottawa, and on Canada. Among his legacies is counted atomic energy.[28]

Later in life Mackenzie liked to reminisce about the dash and decision of Canada's acquisition of atomic energy. Canada's nuclear industry had begun with a characteristic expression of Howe's 'let's go' enthusiasm, and the aim was clear: the advancement of Canadian science and an entrée into the most prestigious region of modern science – and all for an expenditure that was by American and even British standards exceedingly modest. It was not in Mackenzie's character to brood over the dark places of his experience; he was a man of peaks, not valleys. By the same token, his ability to deal with the valleys was extremely limited; not himself an administrative specialist, Mackenzie relied on others to untangle knots and open vistas. So it had been with Halban; so it was, in 1944, with John Cockcroft.

It was true that Mackenzie had not left everything to Cockcroft. Through E.W.R. Steacie, who combined duties in Ottawa with liaison in Montreal, he had a window into the atomic laboratory. Steacie and Cockcroft managed to get on well enough, as did Cockcroft and Mackenzie. It was natural for Mackenzie to solicit their advice in the summer of 1944. The Canadian government was assembling plans for the postwar future, a practical future in which, it was assumed, science would have a place in stimulating the economy. Knowledge of science, pure and applied, might allow

Ottawa to plot an industrial future for the country, and Mackenzie was hard at work defining what science could do for Canada – and for Howe.[29]

The two men had already discussed what should be done. Science should have a larger role in postwar Canada, Mackenzie believed. To the board members of NRC Mackenzie explained that Canada, 'through the National Research Council, should maintain a post-war research program equal to or greater in scope than its present [December 1943] war research program.' To the Canadian Manufacturers' Association the NRC president argued that if Canada were ever to be 'a full-fledged, well-integrated and rounded-out industrial nation' there would have to be more research. 'I would ... urge the point,' he added, 'that while we got along fairly well on imported research in the between-wars era, it will not be possible to do so in the future.' Mackenzie had already said the same thing to Howe, and Howe agreed. Canada's reconstruction policy would include a commitment to science.[30]

Cockcroft and Steacie called on Mackenzie in his office on Ottawa's Sussex Street on 1 September. Mackenzie, of the three, had the greatest responsibility, and it was he who set the agenda. As Cockcroft reported back to Britain, Mackenzie was 'getting restive with no future plan – feeling that he might be left to "carry the baby" on a sudden withdrawal of the UK staff.'[31]

To dread of the future Mackenzie added a sense of ambiguity about the present. He had never been fully convinced that the Anglo-Canadian atomic project was really military. True, it was military in form; that was where it got its budget, not to mention the rationale for Groves's co-operation. The Canadian government had already agreed to spend $4 million at Chalk River building a heavy-water atomic pile, essentially on Mackenzie's recommendation. But when Howe justified the expenditure to the small number of ministers who needed to know where the money was going, he also mentioned that atomic energy had electrical possibilities.[32] In terms of the overall war effort, $4 million was a small amount, but it was the largest single sum ever approved for a scientific project in Canada. The project, if it succeeded, would outlast the war. If it were to outlast the war there must be a new understanding. Without one, 'nearly all the senior men will go back to England.'[33]

There was an alternative, Cockcroft and Steacie reported on 27 September. First, it was well to remember that NRX would be a most important research pile. It would assist in the design of future atomic plants and permit fundamental experiments in nuclear physics. It would also prove useful in radiation chemistry and for the production of isotopes. NRX need not stand alone, however. It might be possible to respond to what was believed to be a world shortage of uranium by building a chemical plant for the separation of uranium-233 from irradiated thorium, a much more common substance (it was called, in code, 'the 23 plant'). Such a plant might tempt the British, and could form the basis for a co-operative atomic program; in any case the Chalk River program 'should be a joint one between Canada and the UK with some British based staff.' It was only to be expected that the British would set up their own atomic establishment after the war, but that would permit 'the British staff [to be] regularly interchanged.'[34]

From London, Sir John Anderson added his own blandishments. He assured Howe that at the end of the war something would be done to keep the Chalk River project afloat and to maintain the necessary expertise in Canada through exchanges of British personnel. Sir John was doubtless keeping his options open; while the trend of British policy was clear, the details had yet to be settled.[35]

Mackenzie seems to have been comforted. No new agreement was concluded, but all might still be well. Deeds spoke louder than words, and the deeds were most evident during the summer of 1945. Chalk River took shape, ZEEP was assembled, and a start was made on NRX. And finally, in August, the power of the atom was made shatteringly apparent. Having 'won the battle of the laboratories,' as the *Toronto Star* claimed, it was time to decide what to do with them.[36]

Mackenzie decided to anticipate events. He offered Cockcroft $10,000 a year and the vice-presidency of the National Research Council, if he would agree to stay and run Canada's reactor project.[37] The offer, Cockcroft reflected, would allow him to 'live comfortably enough and the work would be interesting. I wouldn't mind taking on such a job since I like new places and changes of work. But I don't suppose I *will* take it on, for I do feel that there are important jobs to be done at home still.'[38]

The important jobs to be done included creating, and then running, Chalk River's British equivalent. The site, at Harwell outside of Oxford, was quickly chosen. It was unclear for some time that the principal role would be Cockcroft's; what was clear was that there would be a large British atomic establishment, just as predicted in September 1944. The Montreal laboratory and the Chalk River enterprise were organized for just such a contingency, acquainting those who would have to run the future British laboratories with the problems they would encounter, and filling them up with Canadian and American research data.[39]

The actual authorization of the British atomic program was interrupted by the British general election of July 1945. Winston Churchill and the Conservatives were swept out of office; Clement Attlee became prime minster at the head of a Labour government. Sir John Anderson, though a member of the outgoing government, was asked to stay on to advise on atomic policy, and it was therefore under his influence that the future of Britain's atomic program was finally shaped.[40]

The crucial decisions were taken in London in September and October 1945. There was no doubt as to their importance. Britain's position as a great power depended on independence from American weaponry. What the Americans had, the Russians would want and would get. It seemed unwise to depend on American altruism for Britain's defence, and it seemed inevitable that any future strategic balance must involve atomic explosives. A British atomic program there must be, even if that made American co-operation unlikely. And if the Americans' co-operation was doubtful, so was Canada's.

Anderson was the crucial actor in British atomic policy in the fall of 1945. He was employed by the Labour government precisely because he had a long and detailed memory of British policy in the area; among his recollections were his disappointment at Canada's failure to support the British atomic position in 1943 and his specific mistrust of Mackenzie, whom he viewed as slavishly pro-American. Mackenzie had to be swallowed in 1944 and 1945; his was a bitter, but necessary, tonic to flagging British resources. With peace, those resources could be redirected and reconcentrated. Under the circumstances, helpful though Canada might be, it was no longer essential.

Interestingly, as we have seen, Mackenzie was at the end of the war more disposed to co-operation with Britain than he had been two years before. It was his minister who had his doubts. In August and September, Howe had participated in a general exchange of self-congratulation among allied atomic personalities. General Groves observed to Howe's Washington representative, George Bateman, that as far as he was concerned, 'the U.S. and Canadian partnership is much more important to the U.S. than the U.S.-U.K. partnership.' Absolutely, Howe replied, Groves was 'right about the relative merits of the two atomic partnerships.' To Bateman, the minister added that 'we intend to strengthen our own staff and be independent of outside help,' a rash promise that would have been news to Mackenzie.[41]

Without knowing precisely why, the British had reason to suspect that an atomic partnership with Canada might suffer from the same constraints and concerns that had plagued the first two years of the Montreal laboratory. Besides politics, there was the chronic British difficulty of scraping together enough dollars to spend on North American goods and services of any kind. Against that must be set the fifteen months of happy collaboration under Cockcroft, and the Canadians' obvious desire to see it continued in some form. There were advantages in adding Canada's strengths to those of the United Kingdom, and the overall political benefits might not be negligible. The terms of collaboration would not be easy, but it did not necessarily follow that they would be impossible to negotiate.

Impossible or not, Anderson did not wish it to happen. Although Cockcroft made an effort to persuade Anderson's official committee (the Advisory Committee on Atomic Energy) that it was desirable to locate a plutonium plant in Canada, it was not long before suitable sites inside the British Isles were being investigated.[42] Attlee was informed of his committee's recommendations, and agreed. His explanation, conveyed in a telegram to the British ambassador in Washington, deserves to be quoted at some length:

> We feel that it would be quite wrong for the United Kingdom to be dependent on an outside (even though a Dominion) source for the supplies of fissile material which will be required for the research and development work to be conducted at Harwell and for the application of

atomic energy to other needs.... If the pile is to be built and paid for by the Canadian Government it becomes a Canadian enterprise, and even though we might be ready to help by lending staff, the plant would necessarily be owned and to a large extent controlled by the Canadian Government.

This last was a point that Howe had felt compelled to make early in 1945, and the British cannot be faulted for reflecting upon it; it does, however, seem that neither side made any effort to discover how British independence and Canadian sovereignty could be made to work together instead of at cross purposes.[43]

The British government did not tell the Canadians what it was doing. Though a prudent observer might have troubled to inquire, there is no evidence that Mackenzie did so. No Canadian pressure was therefore brought to bear, for while the Canadian government was beginning to take a serious interest in atomic affairs, its attention was concentrated on a forthcoming conference between Mackenzie King, Prime Minister Attlee, and President Truman, to take place in Washington in November, and, beyond that, on the possibility of an international control over the subject.

The British government therefore decided what it would do on its own. The decisions that directly affected Canada were three in number: there would be an atomic research establishment at Harwell; Cockcroft would be its director; and there would be a plutonium plant in a remote corner of the United Kingdom. The decisions were made in the strategic interests of the United Kingdom, interests which could not be compromised by mortgaging crucial facilities to a whimsical dominion whose leadership was all too susceptible to American pressure.

Chalk River Becomes Canadian

A snapshot of the Canadian atomic program in the fall of 1945 would have disclosed a curious mixture of confidence and pessimism. The confidence was largely Mackenzie's; the pessimism was located in the Department of External Affairs and in the Privy Council Office which now, for the first time, began to play a role in determining Canada's atomic policies.

It is worth pausing briefly to describe these two organizations.

Both were small; including all its embassies, legations, and high commissions, External Affairs had fewer than five hundred employees in 1945, while the Privy Council had just over fifty. Its permanent head, the clerk of the Privy Council and secretary to the cabinet, was Arnold Heeney, a forty-three-year-old Montrealer. The cabinet's business flowed through and past Heeney; as the prime minister aged, Heeney's influence, always discreet, grew in proportion to King's inattention. His counterpart in External Affairs, Norman Robertson, was just forty-one, but already high in the confidence of his minister, who happened to be Mackenzie King. Robertson's peers and associates in the diplomatic service were also generally in their forties. Two of them bear special mention: Hume Wrong and Lester B. Pearson.[44]

Wrong in 1945 was associate undersecretary for external affairs; a waspish, balding man, he was known to have tried his wit on the prime minister, who did not love him for it. Wrong had, during the war, experimented with the idea of functionalism, a political theory that held that a nation should have influence in proportion to its ability to contribute to international affairs. Canada, though raw and underpopulated, would by this token exercise influence in international economic discussions because of its wealth and productivity, and in security matters by its recently demonstrated ability to mobilize a high proportion of its population to warlike activity. As for Pearson, he was Wrong's senior in age and rival in status; in 1945 he was serving as Canadian ambassador to Washington. What Wrong lacked in charm overflowed in Pearson, whose affability was a byword; but like his Ottawa counterpart, Pearson was a shrewd and far-sighted political animal. Both men knew, in the fall of 1945, that the political foundations of the world had suddenly shifted, and that science had become a primary concern of international politics. This shift had occurred without warning, and without their knowledge.

Even had it been with their knowledge, it is doubtful whether they would have, or could have, shown an immediate sympathetic understanding. Few of the Canadian higher civil service had much training in science; the exception, R.B. Bryce, an assistant deputy minister of finance, had started out as a mining engineer before switching at Cambridge to the discipline of economics. Well-educated men, like Heeney, Wrong, Pearson, or Robertson, were at an immediate disadvantage for which their legal training or graduate studies at Oxford or Harvard had not prepared them. But, like everyone else, they well understood that after Hiroshima nothing could be quite the same. The first thing they had to learn was where Canada stood in the matter, and what had changed inside their own country. With Heeney's assistance, they set out to discover what was different. Naturally, the first person they asked was C.J. Mackenzie.

Mackenzie's replies to inquiries from Wrong betray a persistent though not unqualified optimism. Canada had the only atomic plant in the British Empire. NRX, whenever it became operational, would be a uniquely valuable pile because of its high neutron flux, just as Cockcroft and Steacie had reported the year before. The Americans did not especially need it, but the British did. They had no pile of their own, and could have none for at least a year. Over the longer term the British could establish their own program, with facilities inside the British Isles; and Mackenzie admitted that some suggestion was under consideration. Fortunately, a 'better informed' and 'wiser' school of thought favoured collaboration with Canada, meaning 'a large Commonwealth plant in Canada, which could supply the material for the other parts of the Empire, and to the research laboratories of [sic] which teams from the various Dominions and Britain could come for research work.'[45]

Mackenzie had good reason to be optimistic. Cockcroft had sketched a plan of co-operation a year before. Chadwick, stationed in Washington, was well disposed. So were the British members of the Combined Policy Committee. But by the time Mackenzie sent his letter to Wrong, on 26 October 1945, all these views had been weighed and found wanting.

C.D. Howe learned that matters were afoot early in the morning of 5 November, when his dog trotted into his bedroom with his morning paper. The *Ottawa Journal*, repeating a story in the London *Daily Express*, broke the news that Cockcroft was being appointed head of the new British atomic energy establishment. When the British high commission cabled for the details, London lamely explained that it had wanted to get Cockcroft on board before informing the Canadians, with the result that Howe was not officially notified even after the story had become public knowledge.[46]

Sir John Anderson never shirked the task of dispensing bad news in person. The Washington conference in mid-November, attended by Attlee, Anderson, Mackenzie King, and Howe among others, with President Harry Truman playing host, afforded an opportunity. Mackenzie recorded Howe's 'chat' with Anderson. 'We forced our view very firmly and diplomatically,' he wrote, by demanding an assurance that Cockcroft not be withdrawn from Canada until NRX became operational. The absence of a 'senior director' would force Canada to appeal to the Americans for help, 'and the Empire connections would be lost.' It was not an argument calculated to move Anderson, whose view of Canada's imperial potential remained rather limited. To Anderson, Howe's and Mackenzie's enthusiasm for the empire must have seemed opportunistic; in the light of the experience of 1943, Canada's empire patriotism appeared to be an infirm foundation for a fundamental British interest.[47]

The Washington conference established a minimal co-operation among the United States, Britain, and Canada, pledging the three countries to consultation over the use of atomic weapons and the exchange of 'basic' information. The British and the Americans agreed to perpetuate the CPC, on which Howe sat, and the Combined Development Trust, which sought out supplies of uranium outside the British Empire, Canada, and the United States. Howe, who suspected that the British were attempting to raid Canadian uranium supplies or the Canadian treasury, or both, balked at signing the 'Anderson-Groves memorandum,' which embodied these points.[48]

That did not matter very much. What did matter was that the Washington conference failed to clear the air between London and Ottawa on the subject of atomic energy.[49] Even on the level of higher policy – international control – the three powers remained vague. A special forum of the United Nations was established, the United Nations Atomic Energy Commission (UNAEC), which would meet in New York the following spring. Canada, as an atomic power, was to be a member: functionalism in practice.

With the conference over, the Commonwealth conferees, King, Howe, Attlee, Anderson, and company, adjourned to Ottawa for a round of visits and feasting. Anderson and Howe decided to fly on to Chalk River, landing among the log booms in a flying boat to the consternation of the project staff, who could easily envisage the untimely demise of their political directors in the river's freezing waters.

Goldschmidt was detailed to show Howe around the new chemistry laboratory. 'So as not to waste time,' Goldschmidt wrote, 'he announced that he was going to inaugurate it officially, and, after ascertaining that there was no bottle of champagne and that no women were present, he christened it himself.'[50] Mission accomplished, the minister emplaned for Ottawa to supervise the inauguration of UNAEC.

Functional representation on UNAEC required that Canada establish what its atomic policy actually was. Howe viewed this prospect with some discouragement. As early as October he confided to Bateman that it looked 'as though the atomic bomb is moving into a sphere of diplomacy which is all right with me. The only danger,' he added, 'is that if Mackenzie and I get too far away from policy matters, the operation in Canada may suffer.'[51] If distancing Howe from policy was a danger, it was not immediately evident. In October, with King away in Europe, the cabinet defence committee dutifully resolved that 'at the present stage, recommendations on policy should be formulated by the Prime Minister and Mr. Howe.'

In this case it was not the prime minister who made policy, but Howe. As nearly as can now be determined, his policy was both spontaneous and inadvertent. Responding on 5 December to a question from a Progressive Conservative MP, the minister of reconstruction announced Canada's basic policy on atomic weapons. 'We have not manufactured atomic bombs, we have no intention of manufacturing atomic bombs,' he told the House of Commons. That was true, as far as it went. Howe saw no reason to add that the original purpose of the Anglo-Canadian project had been to perfect ways of making atomic weapons and that the success of the atomic project would materially assist the British in their bomb-making enterprise. It was also true that in default of an agreement with the British no consideration was currently being given to any bomb-making facilities in Canada. These two facts were converted into a prescription for the future. 'We,' the Canadians, had not actually made a bomb, and now 'we' would not do it. That others might, Howe left unsaid. And that was all there was to it. In the event, no bomb-making divisions would ever be established or even contemplated at Chalk River, but not, as it happened, because of Howe's policy statement.[52]

That statement concealed an ambiguity. Much of the British interest in Chalk River lay in its suitability for plutonium production. Plans had been laid for a thorium conversion unit whose importance lay in increasing the stock of fissile materials available to the British and Americans. Their principal interest, as Howe knew from his membership on the Combined Policy Committee, was in weapons. The real answer, and the answer that would determine whether Howe's remarks were more than a temporary policy, depended on others. Howe must wait on events outside his control: at the United Nations, in London and Washington, and in the Privy Council Office and the Department of External Affairs.

The timetable was established by the approaching meeting of the United Nations Atomic Energy Commission. For that, if for no other reason, Canada must have a policy. Arnold Heeney, King's official factotum, suggested to the prime minister an advisory panel of senior officials to advise on atomic energy. It was obvious to Heeney that if the United Nations meeting concocted a policy, there must be Canadian institutions to respond to it. Atomic energy was a matter of strategic significance; to Heeney, and even more to Hume Wrong, Howe was not the man to cope with large strategic issues. To outflank this obstacle, Heeney and Wrong suggested that they themselves, with the addition of Mackenzie and Dr Omond Solandt from Canada's defence research establishment, be constituted as an advisory panel on atomic energy, rather like the British official committee over which Sir John Anderson was presiding. On their agenda was Howe's replacement as minister responsible for atomic energy by a more appropriate appointment, but before that could happen, Howe had to agree both to the panel and to its membership. The panel, he realized, was inevitable. He would, however, alter its membership by inserting George Bateman.[53]

The panel got under way in April. The Americans had set an example for its deliberations by proposing, and then noisily debating, the principle of civilian control over atomic research.

Politicians, soldiers, and scientists all joined the squabble. The resulting legislation created a civilian agency, the Atomic Energy Commission, to replace the army's Manhattan Engineering District. But if the American act served as a model, the manner in which they devised it did not. As far as possible, there would be no public debate in Canada.

And, to a remarkable degree, there was none. Though the British scientists at Montreal and Chalk River had their own views on what their future should be, they did not ask their Canadian colleagues to join them in protest or supplication. Their Canadian colleagues had remarkably little, if anything, to say. Perhaps the absence of an ordnance division at Chalk River or in Montreal and the real possibility of peaceful use for atomic energy persuaded them that there was no point in making representations. Perhaps, too, the prospect of United Nations action on atomic energy made them hold their peace. Finally, there was no individual scientist who was willing to act as leader on the Canadian side. They were, most of them, still young; of the older scientists, George Laurence was quickly pre-empted at Mackenzie's order to serve as special adviser to Canada's UN mission.[54]

Lesslie Thomson, still Howe's dogsbody for atomic energy, barely waited for the Japanese surrender before proposing a special act of parliament to cover the subject. With the end of the war the government's sweeping emergency powers would expire, and given the importance as well as the sensitivity of atomic energy it was advisable to re-create them in statutory form to safeguard both uranium supplies and atomic research. Mackenzie forwarded the suggestion to Howe on 20 September 1945, with his approval. Howe was reluctant. To his mind direct ministerial control was best, with as few intervening layers as possible; eventually, however, he agreed that a supervisory board might be a good idea. A rough draft emerged from the Department of Justice in October, and moved majestically around Howe's circle of advisers through the winter. It arrived at the advisory panel in April, just as it was considering what to do when the UN commission met in June.

The prime minister considered it essential to place a distinguished personality at the forefront of Canada's atomic diplomacy. Howe would not do, and not merely because he was notoriously impatient and undiplomatic; as a valued minister, King needed him in Ottawa, not New York. A senior civil servant would not be suitable either. The prime minister had firm views on *that* subject. Yet because atomic energy was, after all, a matter of science, it would help to appoint someone who could at least begin to grasp the terminology of the new age.

Fortunately there was a happy compromise. General A.G.L. McNaughton, the former president of NRC, the former commander of Canada's European army during the war, and latterly an unsuccessful Liberal politician, was available. He was the man to stand up to Howe, if circumstances required it; as it happened, they did not, for Howe was busy with other problems. It was McNaughton whom Laurence would serve, for a year of meetings in New York City. Laurence was the senior Canadian associated with Chalk River, but he would not be there, or even in the country, while Mackenzie and Howe wrestled with the problems that confronted them.[55]

Only one of these problems need concern us here. Eldorado Mining and Refining, which handled Canada's uranium production, had been nationalized in 1944. Howe had left its management in place, an action that turned out to have been a serious mistake. To reorganize the company and to preside over its resurrection (or its liquidation) the minister appointed his executive assistant, Bill Bennett, to be managing director and vice-president; the next year, 1947, Bennett assumed the title as well as the responsibilities of president. He would, in the years that followed, become increasingly prominent in Canadian atomic policy.

The advisory panel produced a version of an atomic energy bill in May 1946. It was revised at Howe's insistence to give it an all-purpose preamble, and presented to parliament in June. The Atomic Energy Control Act created an Atomic Energy Control Board (usually abbreviated as AECB) to consist of four part-time members and a full-time president. The board was to have 'control and supervision' over 'the development, application and use of atomic energy' in Canada. It could sponsor research into atomic energy, either on its own account or through an agency. It could acquire radioactive materials and the mines that produced them. It could, with the approval of cabinet, make regulations governing all aspects of atomic energy in Canada. Finally, it could establish companies in the field, or acquire existing ones. It could not, however, exercise its powers outside the country; diplomacy would be left to the diplomats, or to McNaughton. Since McNaughton was also appointed the board's first president, observers could be forgiven for missing the distinction.[56]

All this might seem straightforward, as far as a bureaucratic maze can ever seem straightforward, but it was in fact a considerable constitutional innovation. The federal government, through the Atomic Energy Control Act, was asserting its jurisdiction in a field, indeed several fields, previously reserved to the provinces. Mines and mining, outside federal territories, belonged to the provinces; so did such matters as health and safety and labour relations. All of these would be lively issues in the years ahead; at the time, except to the uranium-mining province of Saskatchewan, the intrusion of federal authority was minor if not entirely symbolic. The justification for the federal excursion was, of course, the atomic bomb.

The AECB was, to be sure, largely symbolic. McNaughton was usually absent at the United Nations, conducting fruitless negotiations. In 1948, when it was clear that the United Nations was not going to solve the world's atomic dilemma, he resigned and took up other duties. The board passed under Mackenzie's control, as president; Bennett joined it in March 1948. The board was small, and it stayed small; indeed it shared some of its staff with its two principal clients, Eldorado and Chalk River. Those two organizations reported elsewhere, and ultimately to Howe. The AECB became a ratifying authority for decisions taken elsewhere, and its functions remained latent rather than active. Although the board formally assumed responsibility for Chalk River on 1 December 1946 under the act, it delegated its authority almost immediately to the National Research Council of which Mackenzie was, of course, president. Just over a year later he was also president of the AECB. No one thereafter checked to see just which hat he was wearing when he reported to C.D. Howe.[57]

The effect of these adventures was to clarify the legal status of atomic energy in Canada; they also confirmed Howe's and Mackenzie's direction of the research side of the Canadian atomic project. They did nothing to advance the larger question of what the atomic project was supposed to be doing. That question would not be answered until or unless the Canadian project's relations with its British parent and American cousin were sorted out.

This was no easy matter. The impending withdrawal of Cockcroft left Howe and Mackenzie in a sour mood. Although in his lighter moments the minister continued to hope for a functional connection between Chalk River and the British, he eventually concluded that such an eventuality was extremely unlikely. If Cockcroft left, therefore, relations with the British would be severed. He left it to Mackenzie to discover whether the Americans were willing to take their place.[58]

The American connection had not ceased with the end of the war. In the fall of 1945 General Groves was still worried about the presence of foreigners in the Canadian project; sufficiently exercised, in fact, to demand the departure of all those who would not immediately take out appropriate citizenship, or at least sever their ties with the unreliable French government. Bertrand Goldschmidt would not, and departed, though with Groves's good wishes ringing in his ear; Bruno Pontecorvo would, and stayed. The general was appropriately 'pleased.'[59]

Security was already becoming a lively issue. In September, a cipher clerk at the Soviet embassy in Ottawa defected to Canada, bringing with him evidence of a lively and unofficial Russian interest in the activities of the Montreal laboratories. One of the British staff in Montreal, Alan Nunn May, had assisted the Russians, as had a number of Canadian civil servants. The existence of the spy ring was kept secret from all but Cockcroft. Nunn May, unsuspecting, was arrested in London at the beginning of March 1946, just as his Canadian confederates were being rounded up.

It is difficult to judge how much the Canadian information helped the Russians (compared to what Klaus Fuchs, a more important spy, was supplying from the American bomb project at Los Alamos, it was minuscule), but May did include a tiny plutonium sample that he had purloined from his colleague Jules Guéron's laboratory. Guéron had been exercised at the time; it was slight consolation finally to learn what had happened to a fraction of his precious plutonium stock. May was given ten years of penal servitude to contemplate what he had done.[60]

In Canada, twelve men and women were arrested at 7 AM on 15 February 1946, including three employees of the National Research Council. Further arrests followed. There was a sensation in the media and in parliament. Prime Minister King sadly reflected on the duplicity of the Russians in his diary and wondered where the world was tending.[61] In Chalk River the affair was a nine days' wonder, as scientists traded stories about May and his activities; for some of them, May's trial and sentencing coincided with the opening of a lively and enjoyable tennis season, the first in the NRC's company town at Deep River. Pontecorvo, who moved there that winter, was by common consent the star player.

For Cockcroft the season was definitely the last. Anderson had made the concession that Cockcroft should stay on for the present, at first to see in NRX and then, when delay followed delay and the reactor's operational date receded, until a proper successor could be found. Cockcroft in the interim divided his time between Harwell and Chalk River, to the annoyance of the British, who wanted him home, and Howe, who resented his lengthy absences.

By the spring, however, Howe and Mackenzie had a successor in mind. He was Walter Zinn, a serious-minded Canadian scientist who had taken his PhD at Columbia in 1930 and had subsequently stayed in the United States. Zinn had worked in the Chicago metallurgical laboratory during the war, and was familiar with the Canadian project; in 1946 he was working on research on breeder piles for General Groves.

Scientists at Chalk River were not surprised to see Zinn paraded by in April 1946. He was not, admittedly, Cockcroft. To Zinn, Mackenzie offered the direction of the Chalk River laboratory and a salary of $8000. As director, he would have a relatively free hand in selecting research projects, with the proviso that the work would be civilian in nature; no military projects were contemplated. There was, in addition, the 'pleasant' site at Chalk River. 'The intellectual atmosphere at the village,' Mackenzie suggested, 'you would find congenial.' Zinn weighed the offer against an alternative proffered by General Groves: direction of the Argonne National Laboratory outside Chicago. It would be bigger and richer, it would be close to a real metropolis, and that made all the difference. On 6 May Zinn turned Mackenzie down.[62] Another Canadian candidate, R.L. Thornton, also seems to have refused an offer.

Mackenzie did not quite give up. He seems to have sent another inquiry through Howe to the Americans – possibly to Groves in person. He then departed on a trip to England. And so it was in England that he received a telegram from Howe saying that there was no hope of getting Zinn. Mackenzie should now make arrangements with a suitable man in the United Kingdom.[63]

The British had been ready for some time with their own candidate. Cockcroft did not wish his successor to be an American, or an Americanized Canadian; as he wrote Sir Edward Appleton on 7 February 1946, 'it would be against our [British] interests if such an appointment were made, since we desire to retain the closest ties with the Chalk River laboratories.' He had therefore ventured the name of W.B. Lewis, with whom he had worked at the Cavendish Laboratory in Cambridge, but who had not, miraculously, been engaged in atomic work during the war.[64]

Chadwick, whom Mackenzie trusted, met with the NRC president, Howe, and Malcolm MacDonald at the end of March. His task was to 'smooth things over,' as he put it, and he seems to have done well enough at the job. He proposed a new working arrangement, by which any new director would be a Canadian appointee. Although the British would continue to pay for 'a substantial team' in Canada, they would work on Canadian terms and under Canadian overall control. Co-ordination of effort, however, was left to a future meeting.

With these general points, as he hoped, out of the way, Chadwick turned to the specific question of the new director. Lewis, he told Mackenzie, was currently the chief superintendent of the Telecommunications Research Establishment (TRE) in England, and had a background in nuclear physics in Rutherford's laboratory at Cambridge; he was, moreover, 'the most suitable man in England who was available for the task.' He had mentioned Lewis before, over the telephone, and had told Mackenzie that Lewis was 'very good and to my surprise [said] that he is better than Cockcroft.' That cut no ice with Mackenzie until after Zinn's refusal. 'In Lewis' favour,' Chadwick reported to Anderson, 'is his high standing and wide experience; against, his lack of knowledge about this project, doubts that he might not wish to remain long in Canada, and also the feeling that he ought really to be engaged in Radar work. I think,' he concluded, 'the decision may turn on the evaluation by Howe and Mackenzie of Lewis' ability to appreciate the Canadian point of view and to co-operate with the Canadians and the Americans.'[65]

The British were not willing to leave anything to as uncertain a commodity as Mackenzie's moods or Howe's affections. Sir Stafford Cripps, the president of the Board of Trade, should be mobilized to write to Howe recommending Lewis; if that failed, Prime Minister Attlee should be urged to send a personal message to Mackenzie King to the same effect. No action seems to have been taken, but Zinn's refusal opened another opportunity.[66]

That reckoned without Mackenzie. Faced with Zinn's refusal, Mackenzie mused about making W.V. Mayneord, a British physicist inclined to medicine, the project director, with K.F. Tupper, a senior engineer from DIL, and Otto Frisch, a British physicist of considerable distinction, as associate directors. This troika proposal caused considerable surprise in the United Kingdom; it would have surprised Mackenzie's nominees too, had they known. But the troika was nothing more than a will-o'-the-wisp. By the time Mackenzie arrived in Britain for a visit in June he was willing to listen to blandishments on the subject of Lewis, and on 7 July he announced to his greatly relieved hosts that Lewis it was.[67]

Cockcroft packed and made ready to leave. Lewis landed in Canada in September 1946 and proceeded to Chalk River, where Cockcroft showed him round. It was, sadly, not yet an operational site, for NRX was taking far longer than anyone had imagined it would. But Chalk River was otherwise complete, with a reasonable complement of scientists and engineers. Though it was still a construction site, its work had lasted through one winter and was ready for its second.

Cockcroft's own achievement was obvious in the fact that Chalk River existed at all. That it worked was even more to his credit. It had been no easy feat to transform a dispirited team of also-rans into a functioning, large-scale scientific enterprise. Cockcroft expected great things from Chalk River; though he did not say so, in conception and design the NRX reactor was a farewell gift, a vote of thanks, from British atomic technology to Canada.

Cockcroft was the second and last British appointee to direct the Canadian atomic energy project. But his departure was not quite the break that it seemed. While Lewis was a Canadian appointment, he might still prove to be a bird of passage; he might also prove to be a strong link with the British atomic project and with Cockcroft, his old colleague from the Cavendish. That may well have been what Howe and Mackenzie expected; it was what Cockcroft hoped for; only Lewis's thoughts on the subject are unknown, and incalculable.

Atomic Dreams

The end of the war meant the end of secrecy. One of the first, triumphant acts of the Montreal team was to schedule a series of lectures at McGill to tell the world what they now knew, and what they had done. But there were limits and the limits were set by inter-allied agreement; effectively, what could be disclosed was what the Americans revealed. The news was exciting enough. 'It is a particular pleasure for me to announce that Canadian scientists have played an intimate part, and have been associated in an effective way with this great scientific development,' Howe announced. While the atomic bomb itself occupied most of the public's attention, energy was also mentioned. Professor E.F. Burton of the University of Toronto's physics department told the press that 'within 25 years, this will probably be our source of all energy.... We'll probably be working only three days a week, six hours a day.'[68]

The openness was deceptive. While it was permissible for government scientists to tell their wives what they were doing, and what they had been doing for the past three years, they were still forbidden to reveal details of their work to the press. Photographs were carefully controlled, and reporters were kept at a distance. In some respects the government's atomic project was becoming more secret, not less. For while in Montreal the scientists lived in the city, among the general population, they were soon to move to a much more strictly controlled environment. Their workplace would be large and well-equipped, but it would also boast a security fence and guards. The same would be true of their homes: like a military base, the townsite enjoyed controlled access: gates, barriers, and guards halted casual visitors until the purpose of their visit could be scrutinized and approved. Deep River was, from the beginning, a place apart.

Most of the documents that were crated in the Université de Montréal building for shipment to Chalk River in the winter of 1945–6 were marked with varying levels of security, from confidential to top secret. So were the people who transferred up the valley of the Ottawa River.[69]

Some scientists moved in other directions. The end of the war marked an end to war priorities. Opportunity the war might have been; but to some it had been an interruption. George Volkoff was among those who preferred not to allow their war service to become a permanent detour. Mackenzie tried to entice him to stay. Surely, the NRC president asked, he would prefer to stay and work rather than become a lotus-eater on the west coast? But Volkoff's agenda was different from Mackenzie's, and he preferred Vancouver to the Ontario bush. As soon as he could, he left.[70]

Those who chose to stay could take the train to latitude 46 north, longitude 77 west, and alight at Chalk River station. Chalk River was a divisional point on the CPR; crews changed there, and to service them a small settlement had grown up around the station, taking its name from a local stream. Beside the railway tracks there was a two-laned paved road, Ontario Highway 17, which snaked through the towns of the Ottawa Valley to the shores of Lake Superior, three hundred miles beyond Chalk River. Train schedules were designed to accommodate the intercity traveller, the cities in this case being Montreal, Ottawa, and Winnipeg. Chalk River did not rate as far as the schedulers were concerned, and so travellers to the atomic project had to accommodate themselves to getting off the train after midnight and arranging transport as best they could along the highway to the townsite at Deep River, ten miles away.

Chalk River was at the northern end of Renfrew County, which in turn is Ontario's northernmost county. Beyond Renfrew, settlement petered out into mining camps and pockets of arable land, none of them sufficient, in the eyes of the province's administrators, to support the kind of municipal government –counties and townships – enjoyed in the south. Renfrew's county seat, with courthouse, jail, hotels, stores, and other signs of urban life, was the lumbering town of Pembroke. It had a government liquor store, the haunt and resort of bachelors from Deep River, who relied on provincial booze to enliven their Saturday nights in the staff hotel.

Between Pembroke and Chalk River, along twenty miles of Highway 17, there lies a large army base, Petawawa, which in 1945 was home to a mixture of soldiers, prisoners of war, and civilian internees shipped north for the duration of the war. With the end of the war the latter two categories could go home and Petawawa could resume its function as one of Canada's larger military bases. The sand dunes of Petawawa make ideal artillery practice ranges; the sandy soil extended upriver.

Farming country extends north as far as Pembroke. Renfrew is not prime farmland, but south of Pembroke there is enough soil to support dairy farming and market gardens; north, the Precambrian Shield asserts itself, hemming in the Ottawa between granite cliffs. Between the cliff and the river there is a narrow, sandy ledge, up to a mile in width; there the government built its laboratory and its townsite.

There were recent examples of what such a site could look like. The Americans had built a town on a New Mexico mesa, and another in a Tennessee wilderness, near Knoxville. Neither site was a triumph of the industrial imagination, though at the last moment an architectural firm was called in to salvage house design. The Tennessee site would house, so its designers imagined, thirteen thousand workers, to run uranium separation plants and, along the way, to run a prototype plutonium reactor. It was officially known as Oak Ridge; its inhabitants called it Dogpatch.[71]

Oak Ridge was, of course, designed to be permanent. That was the nature of an atomic facility; no matter what the politicians thought, an atomic site is at least semi-permanent, as a prolonged clean-up at the Université de Montréal had shown during 1946. It was a point that Howe did not confront in 1944, when he urged the reuse of wartime buildings originally designed to be cheap, convenient, and temporary. Radioactive products have a habit of staying around and of inhibiting carefree recycling of places where they have been. Prudence dictated the long term.

Chalk River was permanent but the new settlement was grafted on the existing life of north Renfrew County. Apart from lumber and other forest products, and farming towards the south end of the county, Renfrew was a backwater. A recent study has concluded that the upper Ottawa Valley has managed to produce no fewer than ten dialects, of which something described as 'Hiberno-English Ulster Irish' was said to be the most important. 'Lack of economic progress,' another study argues, 'has protected [the Upper Ottawa] from dilution and change.' In 1941 the most unchanged part of the valley was Renfrew County; Pembroke had 11,159 people, 50.9 per cent female, and 54.68 per cent Protestant. (The rest were Catholic; non-Christians made up 0.26 per cent of Pembroke's population.)[72]

Pembroke did not think of itself as a backwater. It was one of the first towns in Canada to enjoy electric street lighting, thanks to hydro power from the Ottawa. It had O'Brien's Opera House which, it must be conceded, was not much used for opera. Its great families built large and comfortable houses which lined 'the treelined streets of the upper town and in the east end.' Most people in Pembroke were Irish or French in origin; most of them did not rise very high on the social scale. The Germans, however, were regarded as 'a very hardworking lot' who provided an example, whether desired or not, to the rest of the community. Hardworking or not, Pembroke's people earned less than the provincial average.[73]

In any economic development in north Renfrew, Pembroke was the natural source of labour as well as a service centre; it was not and could not be a supplier of exotic components and scientific skills. Its existence saved NRC and DIL from the complications that would have been involved in moving and housing hundreds of tradesmen; many if not most of these people could be housed in Pembroke. Further north there was little even of manpower. At the Chalk River site and at Deep River there was little to disturb; and that little – a few families with cabins by the river – was quickly moved off. One farmer who ran a lighthouse was hired to tend it; he nevertheless had to live elsewhere.

The old settlers were gone, but their names remained. The new scientific settlers wished to adopt one, Indian Point, for their town. That was ruled out by the Post Office; there was an Indian Point somewhere else. If the townsite could not be named after the land, why not water? The Ottawa River between the upstream rapids at Des Joachims and Petawawa was called Deep River, for obvious reasons. Father McElligott, the Catholic priest at Chalk River, made the suggestion, sometime around December 1944; it was subsequently formally adopted.[74]

No sooner were the old inhabitants off than the new ones started to move in. First came surveyors and then construction crews; these could be accommodated in temporary camps in the area.[75] A town planner was next, from McGill University. He drew up plans for a comfortable suburb with curving streets feeding into an administrative and shopping centre. His role had a larger purpose. The new reactor complex was to be permanent. Its attractions could not be purely intellectual, however exciting and novel the work. There must also be a tolerable community for the reactor staff to live in, and it was obvious that a location 225 miles up the Ottawa Valley from Montreal would not immediately strike them as plausible. But while the planner designed for the long haul, budgets were important in the short run. The result, inevitably, would be a compromise if not a paradox.

Early in 1945 regular staff began to appear, sometimes temporarily, often for the duration. Deep River was acquiring its first inhabitants. To accommodate them, DIL built thirty new houses, two staff houses (in effect dormitories), and brought 120 wartime houses from Parry Sound, from the Nobel site. These were four-room single houses, or larger multiple-unit variations, all frame constructions, built for speed rather than comfort. 'They are rather nicely designed little houses,' Cockcroft wrote, with 'a living room, two bedrooms, a kitchen and a bathroom.' As a recent arrival from temperate England, he did not mention that they had no basements. Land was cleared and flattened, and the wartime houses placed on top. Only the most senior scientists and administrators got more – basements, with all that they implied for keeping out the cold Ottawa Valley weather. Insulation was regarded as a dispensable luxury. In the first week of October 1945, Cockcroft and his family moved in.

The senior management were assigned houses on Beach Avenue, down by the river under a stand of great pines. (Their housing was somewhat delayed by drainage problems.) Cockcroft had a large family, plus nanny, and to accommodate them a specially large house was built. Beach remained a favourite locale for senior staff: Laurence, Lewis, and later Lorne Gray all lived there, along with the senior American and British representatives at Chalk River. The next echelon of scientists got 'C-6' houses, duplexes without basements. Les Cook, who moved up from Montreal at about the same time, recalled that when this fact sank in, 'a near-riot broke out, or at least as near a riot as ... individualists could muster.' The senior scientists immediately formed a 'housing committee' and set to work on Cockcroft. Cockcroft in turn worked on DIL. A compromise was struck. The wartime housing shipped from Nobel would stay as was: no basements. But the new C-6s would get basements. The 'housing committee,' made up of scientists, got what it wanted. The technicians below them languished without cellars.[76]

The results were predictable. A shop foreman, Bill Power, reported in January 1946 that 'his wife's mop had frozen to the floor.' Leo Yaffe, another chemist, verified the gusts of wind coming through Power's floor. B.W. Sargent, a physicist, had the opposite problem. His coal-fired furnace could turn water to steam. DIL crews kept houses heated if their owners were absent; returning from a vacation, Sargent found that his candles had melted from the heat.[77]

Cockcroft also had to cope with the problem of what the NRC staff considered to be excessive rents for their houses or apartments; this problem, along with inadequate heat, he took up with DIL in February 1946. DIL resisted any change in the rents; they had been approved by the minister, and that, apparently, was that. As for the heat, that would be investigated.[78]

Houses or apartments were for the married (Lewis, a bachelor throughout his career at Chalk River, was the exception). Single staff stayed in dormitories, one to a room. That, in Canada in 1945, posed a problem. The women had to be separated from the men, an absurd situation in the opinion of Bertrand Goldschmidt, who protested to Cockcroft. Surely, the Frenchman argued, if the staff were to be trusted with the secret of the atomic bomb, they could be trusted to run their private lives as they saw fit? Cockcroft solved the problem. He placed a curfew of 10 PM on male visitors in the women's rooms, but he would not seriously enforce it. In any case, if more men showed up for the night than expected, they were put 'over the line' in the women's section as a matter of administrative convenience.[79]

Morality was never a major problem in the staff hotel. Drinking, at least in the early years, was. Goldschmidt chose to spend his Saturday nights elsewhere: 'those hellish Saturdays,' he remembered, as the staff dashed into Pembroke to stock up on liquor, and spent the evening running around madly, singing and yelling. Perhaps time took its toll, or the more respectable 1950s imposed a different discipline. By 1948 most males in town, though still under thirty years old, were married. The dormitories reflected their absence. 'Life there has been compared to a goldfish bowl with a preponderance of female fish,' Freda Kinsey wrote in 1953. The fish were effectively confined to the bowl; bachelors or spinsters, they did not usually participate in the family life of the village. While at first this did not matter, by the late 1940s or early 1950s unmarried staff life was a charmless existence, unalleviated by restaurants or nightclubs, and with only the radio for comfort. (Pembroke's first radio station, CHOV, was established in 1942.)[80]

At first only the most senior administrators had telephones, connected to the plant switchboard. Ordinary communication around the village meant the sociable method of dropping by; contact with the outside world was through pay phones, if at all. (Not to have one's own phone was not uncommon in Canada in 1945, when there were 15.3 phones per 100 population, less than a third of today's figure.) It was uncommon to have other modern conveniences, such as electric refrigerators; refrigerator production was a casualty of the war, and was complicated in Ontario by the fact that the Ottawa Valley electrical system ran on 60-cycle current while Toronto and Niagara ran on 25-cycle. Deep River, like any Canadian settlement of any size, had its own ice-house from which the iceman set out daily to deliver portable refigeration units – blocks of the stuff – covered in insulating sawdust. Electric refrigerators took a while; Yaffe got his only in 1951. These conditions were variously interpreted. City people found them primitive, but to somebody fresh off the farm, or from a less modern and less favoured town, Deep River was in the vanguard of civilization.

Shopping was a problem. Deep River was laid out with an eye to convenient shopping; unfortunately there were, to begin with, no stores. When one opened, in mid-1945, it was regarded as essential, but not popular. Its proprietors were darkly suspected of favouring high-ranking and high-paying customers with scarce under-the-counter goods.

At first the real shopping centre was Pembroke. Some trade could be done in Chalk River, or in Petawawa, but Pembroke was the metropolis. It had, in the early 1940s, 163 stores, doing just over $5 million worth of business. The number of stores would decline to 152 by 1951, but the value of retail trade would rise – to over $15 million. Doubtless there were many factors in the change. Canada was prosperous in the late 1940s. There was inflation after the war. There was the Petawawa army base. But there was also Chalk River, and for some purposes it was a truly captive market, though less so after the Deep River shopping centre finally opened for business in 1946.

Three times a week a small convoy pulled out of Deep River for Pembroke. In the lead was 'the shopping bus' carrying wives and mothers and infants. In Les Cook's memory it was 'followed closely by a truck loaded with go-carts and baby carriages. The cavalcade soon became accustomed to a series of unscheduled roadside stops to enable small children and pregnant mothers to be sick.' Awaiting them were five branch banks, a hotel, even restaurants. On the main street, 'merchants rubbed their hands with glee as this money-bearing squadron swept into town.' The mothers bought what they needed, and stayed for lunch too, enhancing receipts at the Copeland Hotel. Then the process reversed itself. The groceries were labelled and packed into the truck for door-to-door delivery that evening, and mothers and children packed back into the bus. 'It was always interesting to see whose [groceries] you were lucky enough to get and what other people bought,' Cook wrote.[81]

Health was a further concern. Pembroke did not filter its water, so neither did Deep River. This circumstance gave rise to some suspicion when an epidemic of diarrhoea swept the townsite in the winter of 1945–6. As it turned out it was not the water supply but the milk; attempts by DIL's medical staff to reassure townspeople were unsupported by the known facts, which did not enhance the company's reputation with the scientists.[82]

That problem was at least temporary. More serious was the question of health care in an area where doctors were in short supply, where the nearest civilian hospital was twenty-five miles away, in Pembroke, and where most specialists were no closer than Ottawa, one hundred miles further on. These were normal conditions in Renfrew County, but they were not 'normal' for the highly educated scientists moving in, used to conditions in Montreal, Toronto, Vancouver, or, possibly, London or Cambridge. A hospital was eventually built, but because of the size of Deep River it was not designed to offer a full range of health care. As for doctors' services and hospital bills, Deep River's inhabitants had to make their own arrangements to pay.[83]

County and town officials in Pembroke were not pleased at the arrival of a highly demanding group of parents and patients. It was placing a strain on local facilities, NRC was told, and, under the circumstances, the council should make a contribution to Pembroke's schools and hospitals. Mackenzie was indignant. It was the local population that was leeching off Deep River's medical care and not the other way around, he wrote in April 1946. Any complaints from the county or the city were either unfounded or grossly exaggerated. (As evidence of local superstition, NRC personnel cited the belief in Pembroke that 1947's heavy rains were caused by the start-up of NRX.) The actual cost, by NRC's computation, was room and board for a single penniless job-seeker, a bit of road maintenance, and pay for a special constable.[84]

Deep River did not lack for either medicine or police. There were, by 1948, five doctors and fifteen nurses, not to mention a town hospital. The hospital was designed on standard lines, but standards turned out to date from the 1930s. Nor did they take age distribution into account. It failed to cope with the baby boom because the obstetrics ward was undersized. The birth rate was reputed to be double the Canadian average. In a three-month period in 1951–2, forty-one children were born in Deep River, out of a total population of approximately three thousand.[85]

After health, schooling. The baby boom was on, and Deep River was no exception. The population was young, newly married, and increasingly well-off if not as yet housed to its complete satisfaction. (The census of 1951 put the number of children under ten in the relevant combined townships of Rolph, Wylie, McKay, and Buchanan at 1336, or just over 35 per cent of the total population.) There had been, from the beginning, numbers of children to be educated, at first in a four-room school, then in an eight-room school, as well as, for Catholic students, in a school attached to Father McElligott's church. The primary school could be controlled locally, by an elected school-board – an early concession to home rule by NRC and DIL. High school education was another matter. Ontario's high school system was, in the 1940s, highly centralized as to curriculum, but decentralized as to salaries and expenditures, which were left to the judgement of the local school-boards. The salaries and qualifications of teachers varied accordingly. Ontario residents understood this phenomenon and expected to be unimpressed by the Pembroke high school, to which fifty Deep River students were bussed. As soon as enough students accumulated they would demand a high school of their own.

In building a high school, Deep River residents had their first, but not last, contact with Central Mortgage and Housing Corporation (CMHC), the federal government's agency for mortgage assistance, housing research, veterans' housing, and a variety of other housing-related functions. It also developed government housing projects, for example on military bases, and included in its mandate were ancillary services, such as schools. Like other government institutions in the postwar period, it was overburdened and understaffed.

At first it all seemed easy. The townspeople wanted a first-class brick school with all modern frills, and CMHC was happy to prepare the plans, or so it said. Time passed. The school year approached, and finally, at the last moment, the corporation revealed that because of the shortage of time remaining it could only produce plans appropriate to an emergency, wartime, school. The Deep River school committee agreed, but as it turned out even this reduced school was late; the year was well underway before students could occupy it.[86]

Mackenzie also entertained fantasies of handing over the actual management of Deep River to CMHC. The corporation, however, had no such ambition, and Deep River remained the property and responsibility of NRC.[87] The reason for Mackenzie's anxiety to rid NRC of Deep River was the fact that housing and maintenance were a daily headache. NRC or its agents ran cafeterias, drove buses, allotted housing, directed recreation, and set rents. They collected garbage and delivered coal and even sometimes shovelled it. From Mackenzie's point of view, his charges were not grateful. Perhaps it was too much to hope that they would be. But neither did they seem to want to run their own affairs. Attempts at establishing local representation and control were short-lived and ultimately futile.

Because the company was landlord as well as employer, rent relations took on an almost feudal cast, and the cost of housing became a direct and unpopular responsibility of the landlord. There were, at the end of 1947, 358 houses. Rents varied between $22 and $65 a month. Their homes were virtually free of maintenance costs, as well as municipal taxation, and some residents were extremely doubtful of the benefits of buying them. The same feeling applied to municipal government. A school-board was one thing: to a group of people who owed their careers to their education it was not unreasonable to dedicate considerable time and effort to schooling. A town council was another matter. An early attempt to establish a representative government, starting with a town meeting, failed. Although a town council would later be established, some of the scientists always considered its activities as at best a necessary evil.[88]

Inflation reduced the real cost of rents at the same time that salaries finally improved – moderately – for federal scientific and technical staff. This meant that income levels were 'slightly higher than in most single industry communities' (average scientific salary: $7900 in 1958). The council and later the company thought twice about raising rents especially after increases in the early

1950s caused the level of grumbling to rise. Since rent and electricity were deducted at source, there was never any question of a rent strike, merely of a lesser amount in the pay cheque once the ordained increase had gone through.

This might suggest that a program of house purchases was desirable: for the company, to be rid of the whole rental business, and for employees, to carve out greater freedom from a paternalistic regime. But this was not exactly the case.

Other communal pursuits were more spontaneously successful. The original DIL plans included a recreation hall; a stage was later added with the explanation that it was standard issue in war plantations. There was a recreation director. There would also be a community church. The recreation hall could serve a variety of functions, most obviously for drama and musical activities. The church soon became multiple rather than plural as originally planned, with separate Catholic and Anglican congregations; W.B. Lewis was a pillar of the latter.[89]

The climate was a controlling factor on activity. The Deep River area is cold in winter, November to March, with the temperature descending to between −6 and −40° Celsius; cool during the thaw (it hardly deserves the name spring); and hot in summer, May to September. The fall is a glorious season, because of the colourful forest; the view of the forest around Mount Martin, on the Quebec side, with the placid river between, is a spectacular bonus. To handle the organization of sports, NRC sponsored a Community Association, with its own building. Inside there was bowling, the most popular sport in the late 1940s. Outside, during the winter, skiing, hockey, and skating prevailed; inside, it was basketball, badminton, and curling. Summer was dominated by golf and sailing. Lorne Gray, general manager at Chalk River during the 1950s and later president of AECL, organized the golf and built the clubhouse; as an amateur stonemason, he left his mark in the club fireplace. Lewis was a sailor, but not a good one; others were, however, and Deep River became a factor to contend with in sailing competitions.

Lewis's true recreation was the public library, which even in the 1950s was considered outstanding; and for a town the size of Deep River, it was. Educational levels were high; in 1958 106 PhDs, almost all resident in Deep River, worked at the plant. Contemplation could not be easily organized, though it could be facilitated. Deep River was a town of easy and frequent meetings, from neighbourly chats over garden and children to earnestness on a larger scale; like other Canadian school districts, it had its Home and School Association, meeting at the school to glean the latest wisdom on child-rearing. It was supportive, but it could also be oppressive, especially for the families left behind when the male-bearing buses chugged out of town in the early morning, bound for the plant.

Isolation was a factor in other ways. Unless one was a senior scientist or administrator, travel was a scarce commodity. Weekend voyages to Ottawa were a major undertaking in an age of slow buses and increasingly rare and inconvenient trains; by car it was seldom less than three hours to thread the main streets of Pembroke, Cobden, Renfrew, and Carp before arriving at the national capital. And Ottawa in its turn was a city whose more privileged inhabitants preferred to do their shopping elsewhere. With vacation limited, in the early 1950s, to fifteen days or less a year, families had to ration their available travel time carefully.

The experience of going outside could be startling. One scientist found to his surprise that his children were frightened of escalators, and then elevators, in a Montreal department store; young as they were, he decided, it was time to leave. One family is reported to have taken its children to Toronto to show them what a real slum was. 'For incurably urban-minded people,' Freda Kinsey wrote, 'there can be no satisfactory life here.' It was not the large issues that counted in the decision to leave or stay. '[It] may well be,' she explained, 'that an accumulation of trivialities drives away more people than any of the larger complaints.'[90]

A physicist put it more succinctly in a letter to the *North Renfrew Times* in 1958:

Although the town is trim and neat
With cozy houses on every street
Though saying so is indiscreet
I hate it.

It was, a resident complained, depressing to know that company rules required that every bedroom be ivory, every kitchen white. It was even more depressing that the number of clubs to combat boredom had gone up to sixty-eight by 1958 – 'enough recreation to destroy family life,' Father McElligott complained, presumably meaning recreation of the organized and public as opposed to the occasional and private kind. To keep things ticking over, there were twelve 'recreational experts' available.[91]

NRX

When W.B. Lewis arrived in Chalk River in September 1946 the scientific laboratories were almost complete. Seen from the gate, the site occupied a gentle slope leading down to the Ottawa River. Most of the buildings were of a type familiar on any military base in Canada: frame clapboard buildings, painted white. Inside the gatehouse the military impression continued: drab hallways floored with linoleum. Just inside, and attached by a corridor, were the administrative offices, including space for the senior scientific staff. Cockcroft had worked here; now Lewis would.

Some distance inside the grounds, on the north side, there was an inner fence, with radiation monitors, and beyond it the various scientific laboratories – the 'operating area.' They were on the west side; when, eventually, the NRX pile became operational, its effluent gases would be discharged downstream –as would smoke from Chalk River's coal-fired boiler plant. Inside the fence were Building 105, a large shed, housing ZEEP, and, towering over ZEEP, a large steel frame and brick edifice, looking like any recent middling factory. This was Building 100; it housed the components for NRX, or most of them. It was the reason for the site's existence, and the justification for its continuation. What went on in Building 100 would be Lewis's principal concern.

It was not, however, his only concern. Down the road from Building 100, and connected to it by a water trench out its southeast corner, were the 200-series buildings, the chemical section: the '23 Plant' and the '49 Plant.' They too embodied much of the raison d'être of Chalk River, for NRX was to be, among other things, a prototype for the extraction of plutonium from irradiated uranium. The 200 buildings were late additions to the site plan, but they were adjudged necessary. If uranium proved to be as scarce as some authorities feared, the chemical plant was required for the extraction of uranium from thorium. The state of NRX, however, was the most pressing problem.

The first pile in Chicago (CP-1, reincarnated elsewhere as CP-2) had consisted of blocks of graphite, hollowed out in spots to contain balls of uranium oxide or metal. (The arrangement of uranium is called a *lattice*, a term coined by Leo Szilard.) Shielding was minimal in the first reactor, because it operated at very low power, but it was greater in its successors, especially as its descendants were modified to provide for plutonium production. Control rods made from a highly neutron-absorbent material, such as cadmium or boron, were used to control both the start-up and subsequent operation of the pile.[92]

The first piles were literally that; to work on them or to dismantle them it was necessary to unpile the pile, and that, the Americans decided, would not do for a permanent, and expensive, plutonium factory. The first prototype at Oak Ridge, Tennessee, continued to use graphite both as a moderator for neutrons and as a means of sustaining the uranium lattice, but instead of balls or slugs of uranium embedded in the graphite, the Oak Ridge unit had horizontal tunnels, big enough to hold the uranium fuel canned in the form of slugs. These slugs could be pushed through the tunnel following irradiation, to drop into a water tank on the other side. There they would sit until the most intense radiation had dissipated. At the same time, because the heat generated by the reaction of the uranium was sufficient to melt the pile, some form of heat removal became necessary. This was done by making the passages inside the pile large enough to permit air-cooling. At Hanford, gas-cooling was first considered and then abandoned; cooling was by water piped in from the nearby Columbia River. The utility of a large, freshwater supply was thereby underlined.

This information was passed on to the Anglo-Canadian team, though not always without a struggle. As Howe later put it, the Americans 'would permit us to make mistakes if they thought that our people would be able by trial and error to rectify the mistakes.' Du Pont in the United States – a distant parent of DIL, via CIL – refused to provide any direct help although it had the most experience in heavy water pile work in the United States. Instead, questions for du Pont had to be routed through Groves's representatives, known as the 'Evergreen Area.'[93]

As design got underway, Groves's representatives, military and scientific, were expected to sit in on any and all important discussions, without necessarily being able to contribute equivalent information in return. Thus they attended a meeting in June 1944 when Montreal scientists and engineers debated the relative merits of stainless steel versus aluminum, and of a water-cooled as against a heavy-water-cooled pile (the former prevailed). Interestingly, the Americans reported that the Montreal team made 'good' decisions, because they were manifesting a desire to explore 'new fields rather than [duplicate] certain experiments carried on in the United States.'[94]

The first new field was that of size. The NRX pile was to produce 20 megawatts (thermal).* Size introduced, as we have seen, complications in terms of heat removal. Any heat removal system must operate, as Les Cook put it, 'continuously and reliably, with absolutely reliable shut-off mechanisms in case the system failed for any reason.' Next, the heavy-water moderator must be kept 'clean and pure'; it must not decompose into its constituents, hydrogen and oxygen – 'to prevent the formation of an explosive mixture,' as well as to prevent corrosion or other kinds of contamination. But how would it behave under radiation conditions? Last and most important, there must be conservation of neutrons, for without the steady replacement of one neutron by another during the chain reaction there would be no reaction at all, and no reactor.[95]

Neutron economy was the business of the physicists, Placzek, Volkoff, and company. 'I vividly recall Volkoff remarking with a smile,' Les Cook wrote, '"We don't know whether it will ever be able to get over 10 megawatts; in fact, we don't know whether it will ever be able to react at all!"'

The chemists, Goldschmidt, Guéron, Cook, and Yaffe, tackled

* The reader should know at this point that megawatts (1 million watts) come in two varieties. Megawatts (thermal), usually abbreviated as MWt, describe the heat produced by a reactor or pile. There are also megawatts (electric), written as MWe, which measure the output of electricity.

data from Chicago which suggested that heavy water would decompose under radiation, with an attendant risk of explosion. Elaborate measures were taken to ventilate the system, and a catalytic converter was installed to recombine the separated elements. The problem failed to materialize. As Cook concluded, 'with really pure water in a reactor, net radiation decomposition is essentially zero.'[96]

The next conundrum, also derived from the Americans, was xenon poisoning. At Hanford a unit had gone critical and then declined, shutting itself down within a few hours. Consternation ensued. Pile data showed that xenon-135, a neutron-gobbling substance, had been created. Hanford was graphite moderated, but a quick check with the just completed CP-3 reactor in Chicago, a heavy water pile, showed that the reaction was consistent. By adding extra uranium slugs the pile engineers could restart the reaction; fortunately there was space available, built in as a precaution. It took money and time – three months – but it could be done, and it was. All this was reported to Cockcroft, and his staff took note; one, Alan Nunn May, was reported as 'kind enough to offer services for some of the work.'[97]

The specific design of the future pile was the responsibility of D.W. Ginns, an ICI engineer seconded to the Montreal laboratory. Although he worked under the unpopular Newell, Ginns was generally liked, impressing all those who dealt with him. This, despite his ICI past. Eugene Wigner, who was ultimately responsible for the design at Hanford, visited Montreal in October 1944 and was struck not merely by Ginns's redesign of the NRX cooling circuit and his conception of a completely enclosed system for heavy water, but also by the fact that he had taken only a few days to do it. What emerged was, in Wigner's opinion, 'a very substantial and important improvement' in pile design.[98]

It helped to have Wigner's approval, for in General Groves's view the Montreal project was lamentably slow in the design of NRX, partly because the design team was too small (Groves believed that it had only two suitably experienced engineers) and also because it had failed to make proper use of his Evergreen Area staff.[99] It was also decided to cool the reactor by water rather than gas: a decision that incurred further delay and more expense too, for the water filtration plant could have serviced a small city and the standards enforced were high – too high in the opinion of some. Given the necessity of pure water so as to avoid build-ups of sediment as well as the equal necessity of preventing radioactive contamination of the river, caution was indicated. As Groves noted, there was a price exacted, in time and money.[100]

With the method of operation of the NRX reactor under active consideration, attention turned to its products. Its fuel rods would eventually have to be replaced after use; when they were it would be found that the reactor had produced roughly 20 grams of plutonium every day. 'Then,' as Cook wrote, 'what would we want to do with the highly radioactive fuel rods and with the plutonium?'

The question appears to have exercised Steacie and Mackenzie. The original estimates for the Chalk River site excluded the cost of a plutonium separation plant;[101] this may have been because the process for separating plutonium from uranium was still uncertain, forcing delays in the design of the relevant chemical facilities. (This was an area where the Americans were not co-operating.) Mackenzie's desire for further collaboration with the British could not have implied any objection to making plutonium, and the pursuit of suitable methods showed that the laboratory was still aiming at plutonium extraction as an essential goal. As far as the British were concerned, Chadwick wrote in April 1946 that 'the extraction work in Canada must be pushed ahead and transferred to the pilot plant stage as soon as possible so that we may gain experience in time for our use in England.'[102]

Bertrand Goldschmidt remembered the summer of 1944 as one of endless vats of solvents. The chemists decided on a solvent extraction process of their own: 'trigly.' Triglycoldichloride – trigly for obvious reasons – would 'selectively extract plutonium nitrate (in its stable IV valent form) and leave behind not only the fission products ... but also the uranium nitrate (VI valent), which was certainly most unexpected.' Unexpected or not, it worked. It worked, however, on a relatively small scale – 'in a batch extraction fashion' as Les Cook explained.

There followed a dispute of an apparently (but only apparently) academic kind between the chemists on the one side and Steacie on the other. Steacie did not want a large extraction plant; it was far preferable to incorporate the trigly batch process and to place any excess radioactive compounds in storage tanks. With the matter partly resolved, the plutonium extraction plant was scheduled for completion in the fall of 1945. The records of the project suggest that the precise method of extraction was nevertheless still somewhat uncertain; a November report stated that it was 'proposed to adopt the trigly process with a possibility of changing to trigly-ether later if necesssary.' Goldschmidt and the chemists had moved up in the summer, to assist in design and to prepare for the arrival of 'hot slugs' from Chicago.[103]

That was still not the end. When Goldschmidt left the project for home in January 1946, the future of the plutonium plant was unclear. The division of labour in the chemistry section may have been partly responsible. The British admired, but did not want, the Goldschmidt batch process for plutonium. They needed something that would work on a large scale, and continuously. Their work would be crowned with eventual success; meanwhile it was virtually a separate entity at Chalk River. On the Canadian side, progress was made on the design of the 200-series buildings that would house the plutonium separation unit. The chemical plant was designed to handle all the rods in the NRX pile, which was connected to the chemistry building by a trench filled with water, at the bottom of which the irradiated rods could be moved with safety – water being a good insulator against radiation. In case of emergencies, rods could be removed from the pile and stored.

Despite the increasing sophistication of the chemistry section, there were discontents. 'The pile is still lagging,' Mackenzie wrote in December 1946, 'and the chemistry end is not going well.' This, Mackenzie decided, was because the chemists' – in his view not just the British ones – attention was fixed on developments in England, where there was going to be a large-scale plutonium extraction plant; in January 1947 he convened a meeting at Chalk River to put things right. Chemistry nevertheless remained a sore point; chemistry staff and the leadership of NRC lived uneasily together.[104]

If plutonium enjoyed an unstable present and an uncertain future, the '23 Plant' was Chalk River's Cinderella project. Cockcroft believed that '23,' the thorium into uranium process, held desirable possibilities. NRX's design incorporated a number of channels specifically for the irradiation of thorium into uranium-233, a fissile isotope which, it was hoped, might hold part of the answer for what was believed to be a world-wide shortage of uranium.

The chemistry for extracting uranium-233 from its parent, thorium-233, was complicated by an intermediate stage, in which the isotope took the form of protoactinium (or protactinium: atomic number 91), requiring the removal of traces of the latter element from the final product. There was, once again, a conflict between two processes. Design went ahead on one that used manganese dioxide, and the alternative, which had not found favour, was put to the side. In the fall of 1945, with Goldschmidt in charge, the process was reconsidered. As Cook remembered, 'A rapid calculation showed that the plant being designed would need more manganese dioxide than the entire continental output. No one had thought to do such a calculation.' It was obviously time to re-think, to the annoyance of the plant's designers, and in the event it was redesigned to incorporate a different method of extraction.[105]

Situated across the road from the plutonium plant, the '23' facility never had as prominent a part in Chalk River's activities. It began operations only in January 1951, and thereafter worked in starts and stops until it was finally closed down in 1958. The uranium shortage that had called it into being had, in the meantime, evaporated.[106]

The uncertainties of plutonium production were matched by larger hesitations about the project as a whole. Its deadlines receded on a regular basis, while its costs mounted. The NRX plant, with its auxiliary buildings, had been projected at $1.2 million in August 1944; by January 1945 it was $2 million. The pile's cost rose in regular steps over the next two years, reflecting unexpected problems, higher labour costs, and delays, and it would end up at about $10 million, when all was complete.[107] DIL and Fraser Brace, the operator and the contractor, were not especially to blame. Materials were scarce and labour even scarcer. Draftsmen were in short supply throughout, and electricians at the end, while some parts of the work required a degree of precision unprecedented in the experience of most of those who worked at the site.

The fuel itself, uranium metal rods ten feet long, had been made in Britain or the United States. Sheathing, or canning or cladding, could be made in Canada, but the larger aluminum calandria could not. It would come from Pittsburgh. The heavy water, at least, was now available, but each drum had to be tested as it arrived to make sure that it was sufficiently free from impurities and from an excessive residue of ordinary (light) water.

The weight of heavy water, of a graphite reflector, and of the machinery for lifting fuel elements out of the pile made a great deal of concrete, on exceptionally firm foundations, essential. There was, inevitably, some difficulty in finding bedrock at the site of the 100 Building; excavators had to be taken off all other projects and set to work in shifts before rock was struck, some thirty feet down.[108]

Great care was necessary in fitting together the pile's components. Leakage in, of impurities, and out, of radioactivity, had to be prevented. Nor could materials be tested only in their pristine, pre-radioactive state; it was necessary to ascertain the effect radiation would have on them. An unexpected expansion or contraction could break a seal or open a plug and cause a hazard. If the materials were satisfactory, so must be the labour. The chief DIL engineer on the site compared the process, not unreasonably, to 'a watchmaker's job.'[109]

Safety during operation was the next concern. NRX was designed so that 'tripping any one of a good many hundred individual units will shut down the pile.' Monitors and measuring devices were distributed around Building 100 and wherever else radioactivity might be found. In the chemistry division, thick concrete walls were required. The staff worked on a home-made control board on one side, and the used fuel rods were manipulated on the other. Between them was an array of levers, circuits, and mechanical arms, many of them of necessity locally designed to handle the job of stripping the cladding off the fuel rods, prior to dropping the fuel into solvents for plutonium separation.[110]

By early 1945 costs had begun their upward spiral. This caused a certain amount of annoyance at the time, and memoranda flew back and forth among contractor, DIL, and NRC bearing laments and complaints. The time consumed from inception of design in

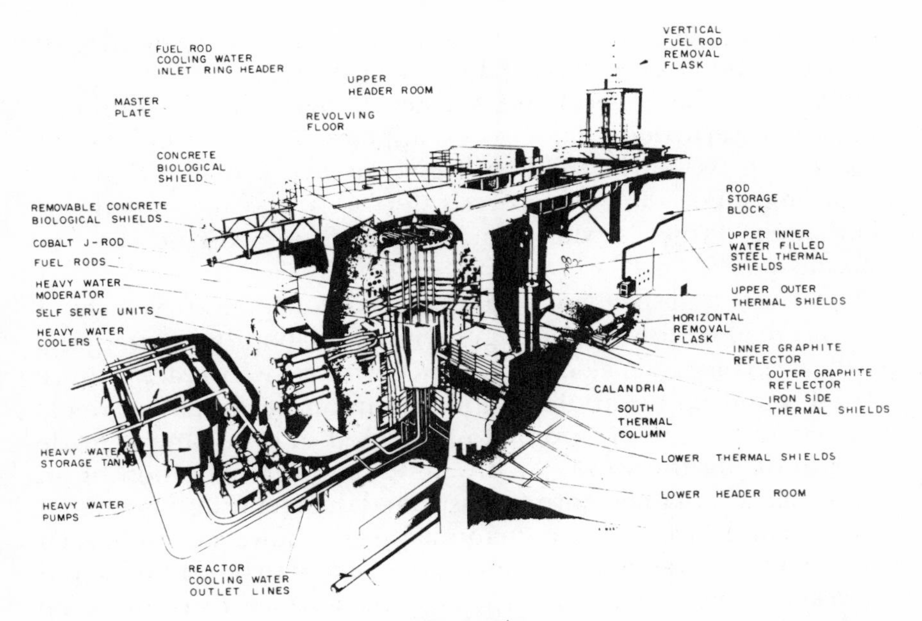

NRX reactor

the late spring of 1944 to the final criticality of the pile was just over three years. Design and construction took place at a time of material and labour shortages (in September 1944 the project was losing an average of seventeen labourers a day); nor must the novelty of the experience be underestimated as a factor in delay.[111] Mackenzie felt the delay keenly. 'We have had our scientific and technical troubles at Chalk River,' he wrote Cockcroft in January 1947. 'The pile is not operating yet and probably won't operate until the end of March, all of which is most discouraging.'[112]

Without uranium metal from the United States there would be no start-up. Groves, in his last months as head of the Manhattan Project, could or would do little. A visit to his Canadian stepchild in October 1946 did nothing to ease matters. Chalk River laid on a party at the only acceptable local eating place, the Byeways. Liquor flowed, and the general rose to speak. He spoke for about twenty minutes. He had little or nothing to say about the atomic project, its significance or its future. But he disliked England and Englishmen, and he wanted his audience to know it. The audience, consisting largely of Englishmen, was politely stunned. The general's car drove up, and Groves departed for Ottawa.[113]

In Ottawa, the general and Mackenzie held a last conference. Mackenzie recorded that Groves appeared to be 'very kindly disposed to us and I think will give us every help.' This was promising, since some of Groves's staff were staying on with the Manhattan Project's successor, the civilian Atomic Energy Commission.[114]

The winter and spring of 1946–7 passed slowly and anxiously, as NRX was readied for criticality. A division of labour was emerging, partly as a consequence of experience with ZEEP. A reactor, once started, required twenty-four-hour attention, seven days a week. The scientists were not trained or inclined to do it, and it made sense to divide the work. At the same time the operation of an atomic pile was plainly more sophisticated than tending a basement boiler in a factory. Some manufacturers, however, produced equipment that was at best boiler quality; it had to be sent back and reworked. Finally, in the late spring, K.F. Tupper, director of the engineering division, made the final dispositions. The operators would tend the reactor as they had already done with ZEEP. Scientists would use it for experiments, but under the authority of the operators. So at least went the theory; the practice, over the years, was not always so clear-cut.[115]

The last stages were dominated by worries over the integrity of the heavy-water circuit; unexplained amounts of ordinary (light) water were showing up. Mackenzie, whose visits to Chalk River increased as the crucial day approached, had Tupper set up a committee of physicists, engineers, and operators to worry over the problem – and anything else that might now arise. W.H. Watson, the head of theoretical physics, took up station at a desk on the operating floor to expedite matters. A tube in the circuit had to be drilled to remove fifteen pounds of unwanted water that could not be flushed out. Tupper informed the scientific staff that very few scientists would be present when the pile went critical. The rest, including Lewis, were kept at a distance. The scientists, it was believed, would each want to see something different as the chain reaction proceeded; each might interfere in a different way and the result would be chaos.

B.W. Sargent and Bruno Pontecorvo were among the four chosen physicists who joined the operators on the night of 21–22 July 1947. The start-up consisted of pumping in heavy-water moderator; when enough moderator was present to slow down neutrons in the uranium fuel, a divergent reaction would begin. The process was measured with clicking neutron counters. 'The clicking speeded up,' Sargent recalled, 'slowly at first, then more and more rapidly, culminating in a rattle as if a shackled giant had been awakened from his sleep and was struggling to be free of his chains. The nuclear reaction was self-sustaining at very nearly the expected moderator level.' Mackenzie, who was notified by phone in Ottawa, wrote prosaically that this was the 'Big day for which we have waited for so long.'

"It was thirteen minutes past six by the clock in the control room, and a beautiful July morning in the Ottawa Valley,' Sargent wrote. A few metres away, in front of his blackboard, W.B. Lewis contemplated the results of his calculations. They showed that the reaction would start when the level of heavy water in the pile reached 168 centimetres. 'Actually,' Mackenzie modestly wrote, 'the chain reaction started at 168 cm....'[116]

3

Atomic Energy?

A scientific community is by its nature isolated, insulated, as Thomas Kuhn put it, 'from the demands of the laity and of everyday life.' Kuhn added that the isolation was never total – was, of necessity, a matter of degree. Any profession has its unique store of learning, arcane and difficult of access by those outside its boundaries. 'Nevertheless,' he continued, 'there are no other professional communities in which individual creative work is so exclusively addressed to and evaluated by other members of the profession.'[1]

What was true and meritorious in individuals was not necessarily true, or even possible, on the grand scale. Chalk River had a single paymaster, the government of Canada. The government relied on C.D. Howe, and C.D. Howe depended on C.J. Mackenzie. Mackenzie sat, most of the year, in an office over a hundred miles away, at NRC. Yet Chalk River was NRC's largest and biggest-spending division. For knowledge of its activities, Mackenzie relied on two men. The first was W.B. Lewis, who would run the scientific side. But for the second, who would manage everything else, Mackenzie needed somebody with local experience. The identity of the second was, for some years, a matter of considerable doubt.

Chalk River was an unusual community; rare in the world (its closest relatives were Oak Ridge and Los Alamos), it was unique in Canada. Nowhere else in Canada was there a self-contained community of upwards of a thousand scientific workers. In

Ottawa, Montreal, Vancouver, or Toronto, scientists dispersed after work like anyone else and went home to a neighbourhood of businessmen or lawyers. While there could be visitors to Deep River, there were few to the plant, and those few were previously screened by security. Scientists or engineers dropped by from the United States or the United Kingdom, or from elsewhere in Canada. But no laymen crossed the threshold of the plant without an official reason until 1950.[2]

W.B. Lewis

Wilfrid Bennett Lewis was thirty-eight years old when he arrived in Canada in September 1946. Born into a professional family (he later described his father and grandfather as 'both distinguished civil engineers'), he attended preparatory and public schools before enrolling at Cambridge in 1927. Like Cockcroft, he had some background in industry; like Cockcroft he came to the attention of Lord Rutherford, the head of the Cavendish Laboratory. Lewis's particular skill, it seemed, lay in electronics. 'In modern words,' Lewis wrote in 1982, 'electronic circuit engineering or electronics was my qualification of special value.' He felt no strong attraction to the radioactive side of physics, where Rutherford had made his and his laboratory's reputation.[3]

Lewis had looked forward to graduate work in physics, but 'in anything other than radioactivity.' He soon veered into radiation, however, working for Rutherford on accurate analysis of alpha particle groups from radioactive substances. On acquiring his PhD in 1934 he took a research fellowship from his old college, Gonville and Caius, and successively became a demonstrator and then a lecturer in physics at Cambridge. He worked on artificial disintegration with Cockcroft in the mid-1930s, and between 1937 and 1939 he worked on the design, construction, and operation of Cambridge's cyclotron.

His electronics specialty drew Lewis off into radar as war approached; his one book, *Electrical Counting*, was in electronics and would be published in 1943 by Cambridge University Press. At the end of the war Lewis had not been drawn back into nuclear research; instead, he was chief superintendent at the Telecommunications Research Establishment (TRE) at Malvern – the principal British nuclear scientist not to have become involved in the Manhattan Project and its many ramifications. It was in that capacity that he first visited Canada and met Mackenzie.

Lewis said little of the motives that caused him to come to Canada in 1946. TRE was winding down, though it did not disappear. Lewis did say, and there is no reason to contradict him, that he was moved by the idea that atomic energy held the key to mankind's betterment; Lewis held strongly to his Christian faith. Faith, he may have believed, was expressed through works, and what better work could be required of a believer than the spread of abundant energy to the impoverished – because energy-scarce – parts of the globe?

We have also seen the context in which Lewis accepted the post of scientific director at Chalk River. He was almost certainly asked to take the position by Cockcroft, a trusted colleague at Cambridge, at a time when Chalk River still held great importance for the British nuclear program. He was sponsored by no less a person than Sir John Anderson. Acceptance of Chalk River could be (and probably was) construed as a patriotic as well as a humanitarian act.

Although he may not have intended it, Lewis thereby became a man with two countries. The year he came to Canada was the year that Canadian citizenship was differentiated from British, although for most of his remaining life the distinction between the two was easily blurred. (Until the 1970s British subjects resident in Canada could vote in Canadian elections, and vice-versa.) In the United Kingdom, Lewis was regarded as a great Englishman, and his Chalk River laboratory was considered, at least in an intellectual sense and certainly in a sentimental one, to be an extension of British science. Some British scientists idly wondered whether Lewis would not have been a better choice than Cockcroft to head the British research program; some, like Vivian Bowden, another Cavendish veteran, became convinced that Lewis's and not Cockcroft's course of atomic research was the true one.[4]

Lewis took years to put down roots. He was not an unsocial man, and could even be convivial when the occasion demanded, but he put his work first. A bachelor at thirty-eight, he joined the other bachelors in the staff hotel and only later moved to 13 Beach

Avenue, across from Cockcroft's former house on the riverfront road. His mother came out from England to take care of 'Benny'; later his sister acted as chatelaine. In the 1940s and 1950s Lewis sailed at the Yacht Club – he was not a particularly accomplished sailor, and his boat usually lost – and presided over the creation of a public library, later named in his honour.

For years Lewis rented his house; as the town was gradually converted to house purchase he was one of the final hold-outs. In exasperation, Lorne Gray, AECL's president, asked Lewis whether he intended to stay or not. Lewis, surprised that anyone should ask, bought; but as Gray remembered the occasion, he could not be sure – never knew – whether Lewis also took the political step of becoming a Canadian citizen.[5]

Lewis had the responsibility of allocating resources and personnel. In the final resort, his was the prevailing judgement that could speed the course of a project or cancel it. But the kind of work that was done was limited, and determined by other factors besides the director's will. Chalk River's agenda of work was set in the first place by the existence of two reactors, ZEEP and NRX, and in the second by the special attributes of NRX. It was limited by the availability of personnel, and that factor in turn was determined in large part by the salaries that NRC was willing to pay and by the relative scarcity of expertise in certain areas, such as radiochemistry. It was determined, too, by the ability of the scientific staff to establish, and defend, their own work.

Lewis had a clear notion of how a government laboratory should be run, and he was prepared to enforce it. There would be time clocks and regular hours, and summonses to the director's office to account for unauthorized absences. Those who worked late might see the director's light burning too, but there were few exemptions for those who worked best after dark. For Lewis, like Sir John Anderson, believed that problems were best resolved when confronted vigorously, head-on. He was not, his colleagues remembered, unkind; he was merely unconcerned by others' reactions and indifferent to the cosmetic forms of politeness. Those who could stand it, took it; those who could not sought refuge in timidity or obscurity. Others, whether relieved or embittered, simply left.

Lewis's seminars and staff presentations had a schoolmasterish tone. The director himself was not a natural speaker: set-pieces were not his forte, and even though he tried, he would not be remembered for his innate eloquence. But he indubitably had 'presence.' Once Chalk River acquired a lecture theatre, the drama was increased. The critic always sat in the front row, always in the same seat. Interruptions were possible, and frequently deadly. The critic was well prepared, for he had a library at his beck and call, and the time to use it. Unmarried, Lewis had few if any distractions. If a subject were unfamiliar, he might take a weekend to master it, and return with an uncannily accurate knowledge of an area that until a few days before had been a blank. It was a valuable attribute, and one in which Lewis trusted, but it had its limits. The limits were, to some extent, those of Chalk River itself.

Lewis's institutional characteristics were important, but they were not all. The laboratory director believed he had a personal mission. Deeply religious, an active lay member of the Anglican church, Lewis like many devout churchmen sought a partial justification for his faith in his work. Science was the instrument: it held, he believed, the possibility to create a better secular future for mankind. In his briefcase when he arrived in Canada was a blueprint: a prediction in the form of an economic study of the superiority and economic inevitability of atomic power, devised by one of the Harwell staff. It would take time; first things must after all come first, but he would not forget it.

For twenty-five years Lewis, through character and intellect, dominated Chalk River. When he was absent, he yearned to return, especially if a conference or a seminar was impending. By the late 1950s, his staff were known to rehearse what they had to say. If a new boy (for in Lewis's shop no women were scientists) had the jitters, he could run through his lines with his friends taking the part of Lewis from the floor. Success was measured in accurate anticipation of the director's comments and questions; failure was not to be thought of. The fact that Chalk River was an isolated society, remote even from science departments in universities, added to the psychological pressure. To many, the pressure meant excellence; to not a few others, though a minority, it meant intolerable egotism.

Some of his staff suspected that physics was closest to Lewis's heart. Biology and health were not. Though not neglected, they were distant from Lewis's centre of interest. And so, some suspected, were the chemists. Chemistry was Chalk River's problem child. Physicists proverbially regard chemists as a lower order of scientific being; however exaggerated the proverb, it has the occasional bite of truth. At Chalk River, chemistry was the last division to find a permanent home, and even when its buildings were complete it was never clear what its future (and that of its twin, chemical engineering) would be. Would there be a real extraction program to produce plutonium? If there were, would it be in batch lots or would it be set up as an industrial process? What would be done with the plutonium once it was produced?

T.W. Boyer, a senior chemical engineer, concluded as early as March 1947 that 'only a decision by the Canadian government to proceed with the operation of a large production pile' could justify a large-scale chemical effort. This he thought unlikely. 'Such a programme,' he wrote, 'would involve tremendous expenditures of public funds for what in the final analysis is probably only prestige in the eyes of the world.' Even a commitment to nuclear power was doubtful for an energy-rich country like Canada, with cheap and abundant coal, oil, and water power. Under the circumstances, only a small research group, pure and applied, would be necessary. Such a group, however, must have a higher quotient of PhDs than the existing Chalk River staff, with a higher research capability.[6]

Plutonium remained a priority for Chalk River, but only for a time. As long as any possibility for military use remained, it was practical to keep a production capacity in being. The British might need a back-up facility for their own plants in case of war, and during 1948 and 1949 the possibility of a conflagration in Europe was never far from the surface in discussions among the Canadians, Americans, and British.[7] War was never more than a possibility, and without a firm commitment there was no disposition to disrupt plans and budgets at Chalk River. As time passed, war scares dwindled. That Chalk River did no work of its own for warlike purposes became less an accident and more of a tradition. Plutonium production was something that would largely take place at another location.

On his arrival Lewis had to take on the problems of morale at Chalk River. There were complaints about housing, services, and pay. DIL staff got more money and were, in the eyes of the scientists, less qualified. Mackenzie had tried to hold the line; failing that, he told the public that 'the rumours in Chalk River' that scientists were underpaid by comparison with their peers elsewhere were utterly unfounded. He deplored 'the absolute juvenile thinking that we get from our ablest scientific people,' equivalent to 'the sort of thing you would expect from an undergraduate debating society.' He advised Lewis that 'unless you are firm you will find that there will be no solution.'[8]

Lewis, Mackenzie soon discovered, would be firm. Firm, to Lewis, meant discipline in the laboratory, but it also meant firmness in defending the laboratory and its interests, be they pay or the most modern equipment. Morale was best ensured by a sense of participating in a world-class institution. For that there was, and would always be, a price, which he expected NRC and its successors to be ready to pay. In return, he would keep the laboratory on track, and prevent it from exploding in unrestrained research. Being firm left time for other activities: discovery, research, and debate. Above all, it left time to consider where the Chalk River laboratory should go; that, more than anything else, would be Lewis's responsibility.

'Are You DIL or NRC?'

In a cartoon circulated around Chalk River in 1946, two men were out for a day on the river. One, drowning, is screaming for help. The other, secure in his canoe, is debating the issue: 'I say, old chap: are you DIL or NRC?' As good cartoons should, it had a number of meanings. Was Chalk River an industrial site or a scientific lab? Who was best equipped to run it? And was it even a Canadian lab, or was it an extension, still, of British science and government?

It was not easy to know. Chalk River might have passed under exclusively Canadian control when Cockcroft left in September 1946, but it was still part of a larger Anglo-American atomic universe, whose priorities were still largely military. The possibility of war was never far away. Relations worsened between the western allies, including Canada, and the Soviet Union. Canadians could not fail to remember that the Montreal laboratory had been a focus for Soviet espionage during the war; during 1946 and 1947 the spy trials resulting from Igor Gouzenko's revelations held the front pages of the country's newspapers. In New York the United Nations Atomic Energy Commission, which included Canada, struggled to find some way of reserving atomic energy for peaceful purposes. But the Americans refused to dismantle their nuclear weapons – of which they had very few – without a proper international inspection system, while the Soviet Union demanded that all atomic bombs be dismantled before any further steps were taken. The Soviets had in any case launched a crash program for nuclear development; making up for time lost during the war, it produced an atomic explosion in September 1949.[9]

The hope of a resumed collaboration between Britain and the United States directly shaped the Canadian atomic project. Indirectly the hope helped define the purpose and future of Chalk River, if only in a negative way. Throughout the first five postwar years, while hope remained, the Canadians acted the part of concerned bystanders, friends to both parties, offering very occasional advice and encouragement, but determined to stay out of the way in case British insistence and American resistance led to a complete breakdown reminiscent of 1942–3.

What Britain insisted on was the integrity of its own atomic program. As the British advanced from pilot plants to piles towards the manufacture of actual weapons, the Americans sought ways of diverting them, offering instead limited co-operation, even over weapons, provided the British abstained from, or deferred, the final manufacture of weapons in the United Kingdom. In American eyes, Britain was too close to the Red Army, should the Communists decide to move west. The British, for their part, from time to time toyed with the idea of locating 'production plants' – the euphemism for plutonium-producing piles – in Canada, a proposition which the United States would have strenuously supported. British interest in Chalk River remained proportionately acute.[10]

The American interest was, at first, more speculative. The

United States maintained a representative at the site to keep an eye on things; Howe had promised that anything he wanted to know was his for the asking. The colonel's ability to reciprocate was, however, quite limited, thanks to the American McMahon Act which, in addition to establishing the Atomic Energy Commission, strictly limited the atomic information that Americans might divulge to foreigners.

Howe regretted the inconvenience, and he discussed with Mackenzie 'the desirability of a bilateral approach to the United States, proposing,' as he explained in September 1947, 'that Chalk River be incorporated into the U.S. program for developing peacetime uses for the product, provided complete exchange of information can be arranged.' If the tripartite arrangement collapsed, he added, Canada's best option would be to go with the Americans.[11]

Others were inclined to complain directly, and their feelings touched General McNaughton, still at the United Nations in New York. 'He considers the McMahon Act very badly drawn,' an American diplomat reported in October 1947, 'and while he has been able to sit on the lid effectively, the situation is now causing him deep anxiety.' The British, for their part, drew up an itemized list of subjects on which information had been denied to themselves or to the Canadians, which 'might delay specific operations [at Chalk River] six months or more.'[12]

The imbalance was redressed to some extent by the quiet operation of an agreement made under Groves. The Canadians were quite aware that almost all the heavy water at Chalk River came from the United States. It represented Groves's *quid* for the informational *quo*, for without it the entire site would grind to a halt. Groves had assured George Bateman 'that as long as the water was used for purposes for which it was intended when loaned to Canada there would be no possibility of its being recalled.'[13]

All this took years to work out. Until the mid-1950s the central forum for atomic diplomacy was still the Combined Policy Committee, on which Howe still sat. When he was unable to attend, he sent Mackenzie or the ambassador in Washington, Hume Wrong, in his place. In between meetings, staff from the Canadian embassy handled whatever problems there were, and saw to housekeeping.

The CPC sponsored declassification conferences to downgrade superannuated secrets; and it uneasily discussed what to do with the useful products of allied reactors, radioactive isotopes being the most prominent example. Membership on the CPC allowed Canada to keep abreast in a general way of developments in atomic energy; to learn, for example, that uranium was still in desperately short supply, too valuable, according to one of McNaughton's staff, to 'squander' on industrial power plants.[14]

The CPC was the business of other committees, political and official, which dealt with the larger meaning of atomic energy. The cabinet itself seldom discussed the topic. It was a technical matter reserved for a cabinet committee, in this case the ponderously named Privy Council Committee on Scientific and Industrial Research. That committee seldom met, and when it did it decided little; both these phenomena were not unrelated to the fact that C.D. Howe chaired the committee.

Below the ministers there were the officials. The Atomic Energy Control Board enjoyed a constricted existence, its powers for the most part confided to the NRC. The AECB's president, General McNaughton, found he had other affairs to occupy his attention, and in March 1948 he resigned. Howe set out to find a chairman more to his personal taste. His first choice was George Bateman, and when Bateman declined, Mackenzie stepped into the breach. He remained president from the beginning of 1948 until 1961; as president his first action was formally to recommend that Bill Bennett, Howe's former executive assistant and, since 1947, president of Eldorado, be placed on the board. Howe commented, 'it will probably save trouble if we keep our atomic energy affairs under a single direction.'[15]

From 1948 until 1957 they remained 'under a single direction' – Howe's, executed through Mackenzie and later Bennett. It did not please everybody. The British thought Howe and Mackenzie were prey to nerves and imaginary alarms, including their persistent fear that Chalk River was 'too small a unit to proceed independently on a development programme.' This preoccupation became increasingly occasional as time passed, but it showed itself in periodic discussions of larger co-operative projects with the British, and in Mackenzie's tendency to fuss over relatively small details in his overall administration of the atomic project. After all, NRC had never handled anything as big as Chalk River before, and Canada was a comparative stranger to large investments in research.[16] But having started on the road, Howe did not shrink from the financial demands that an atomic program entailed. If Canada was to have an atomic program, it would have to find the money to fund it. All that remained was to discover where the atomic project ought to go.

That was Howe's business. In the postwar years his power waxed; those who commanded his confidence had little need to go through committees of their peers. Mackenzie had decidedly mixed feelings about it. As early as March 1947 he complained to Cockcroft that he 'would give a great deal to be free from all atomic energy activities as I find it very uncongenial to be always hounded ... and I fancy you feel the same way.' To Howe he fretted that he was overburdened – 'overworked,' in the minister's terse summary.[17] Mackenzie intended to parcel out his burden – not horizontally to the AECB, but vertically, to carefully selected lieutenants. Cockcroft was his first choice to run atomic energy. When he refused, Mackenzie settled for a scientific director, Lewis, without finally deciding how to run the operations side of Chalk River.

At first that did not matter, because DIL and the contractor, Fraser Brace, managed Chalk River. In return for its corporate services, from design through management, DIL was paid its expenses, but no fee. DIL was a purely wartime creation, a convenience for its parent, CIL. Its 'no fee' policy warded off criticism that a company specializing in chemicals and armaments was plundering the public purse at a time of national crisis. By 1946 DIL was ready to wind itself up. Only Chalk River stood in the way.[18]

DIL would eventually have extracted itself from Chalk River come what might; the excuse employed, however, was that Chalk River was really a 'pilot plant' and not 'a normal industrial unit.' More precisely, and more accurately, Greville Smith, DIL's general manager, told the AECB that 'no clear line can be drawn between the spheres of DIL and NRC,' a situation which was productive of certain 'difficulties.' Mackenzie then wrote to members of the NRC in October 1946 that 'it would be better if some other organization were to assume the overall operating responsibilities.' CIL was considered for the role, but showed no eagerness to undertake it. The AECB resolved against running Chalk River itself. Howe did not wish to establish a new crown company to do it. Under the circumstances, the NRC had no option but to run Chalk River. But how?[19]

If Chalk River was to be a division of NRC – the largest and most expensive by far – it should not be run in the ordinary way. Mackenzie held as an article of faith that capable research scientists should not be wasted on administration; the doctrine may, of course, be inverted to hold that research scientists make bad administrators. In any case, valuable talent and time, if not more, would be lost. Lewis, the newly appointed director, might otherwise have seemed an obvious choice, but besides the burden of his credentials as a scientist he had been in Canada less than two months. Lewis's arrival seems only to have increased Mackenzie's burdens, for the NRC president felt it appropriate to shepherd Lewis in an unfamiliar field.

Mackenzie then turned to the approaching departure of DIL. An administrator was required who could cope with the special conditions at Chalk River. As for scientific personnel already there, Mackenzie was cool. They were mostly under thirty and had 'no real appreciation ... what normal, peacetime operations mean.' Normal peacetime operations had last occurred during the Depression, when the cost of living was actually falling, not rising, and when prudent people postponed such luxurious activities as having a family or buying a house. Younger Canadians were no longer prepared to put up with that. Indeed, as the NRC president wrote in February 1946, 'the young Canadian group of theoreticians give me the greatest trouble.'[20]

Although he did not yet know that Lewis would be a strong director, the NRC president had to select someone who could balance his scientific chief at Chalk River; indeed, much of the administrative history of Chalk River over the next twenty years would turn on just that criterion. Too strong or assertive, and there would be clashes; too weak, and he would be eaten alive. Mackenzie first lighted upon W.H. Cook, the head of the division of applied biology at NRC since 1941. Cook refused the offer of the director generalship of the Chalk River plant at the end of October 1946. Mackenzie continued his anxious searching.[21]

The details need not concern us. He consulted widely, at Chalk River and among his friends, scientific and business. He engaged a consulting firm, Wallace Clark, to advise what to do. They urged that he hire a proper vice-president, 'a man who is married, knowledgeable on the scientific end of things and who would carry the respect of everyone there rather than a senior business executive.' Mackenzie rejected this advice. He wanted a businessman, a practical counterpart to Lewis: a fatherly type, relatively senior, to cope with the 'junior' staff at the laboratory. Failing a businessman – he tried to land one, without success – he turned to a proven scientific administrator. This time his choice was Professor David Keys.[22]

Keys was apparently well qualified. He was senior – fifty-six years old, two years younger than Mackenzie. He had got his start in life as head boy at the exclusive Upper Canada College; since then he had not looked back. A gold medallist in physics at the University of Toronto, he had studied with the eminent Roentgen in Munich. He then earned successive PhDs from Harvard (1920) and Cambridge (1922). While teaching at McGill, he helped edit a German geophysical journal. A successful textbook in geophysics spread his name between the wars; and in 1929 Keys became McGill's Macdonald Professor of Physics, Rutherford's old job. He made a good reputation as a professor ('a wonderful teacher' according to one of his students), drawing on an innate gift for exposition as well as a real natural humility. During the Second World War Keys set up the Royal Canadian Air Force's training program for electronic technicians; it was a resounding success. His eminence was recognized by his appointment in 1945 to the council of NRC. When he got a phone call from Mackenzie at the beginning of December 1946 it must have seemed just another item of council business.

So it was, but not quite what Keys expected. 'David, how'd you like to go to Chalk River?' Keys hesitated. Would it be temporary? No, said Mackenzie. It would be permanent; but of course it would be a challenge. The first challenge seems to have been the specifications of the position. It appears that at first Mackenzie held out the title of director general. When Keys hesitated, Mackenzie upgraded the offer. He would make him vice-president (scientific) of the NRC. That title had its uses, Mackenzie reflected; if Keys didn't work out at Chalk River he could always bring him down to Ottawa to work at NRC. Mackenzie squared things with McGill, while Keys and his wife toured Chalk River and the Deep River townsite. On 17 January 1947 he phoned Mackenzie to tell him he would take the job. There were just two weeks to go before DIL handed over Chalk River.[23]

Mackenzie had been working through that too. It would be difficult to replace the DIL personnel, especially in view of a nationwide shortage of engineers. They had to be offered positions with NRC, and to Mackenzie's relief virtually all of DIL's senior staff agreed to the transfer; only five, including H.J. Desbarats, the works manager, returned to CIL.[24]

At the end of January 1947 Mackenzie travelled to Chalk River to welcome Keys and supervise the transfer of authority in person. DIL briefly continued certain services for NRC, including liaison with Fraser Brace. In Keys, Mackenzie had a buffer between himself and the Chalk River scientists as well as a proper channel of communications between Ottawa and the directing staff at the project site. His problems were not, however, quite at an end. Within weeks, one of the most senior administrative staff had departed, leaving a further vacuum at the top.

Keys did what he could. He applied his emollient personality to the rift that had developed between the old, practical DIL engineers and the scientists. He could not solve every problem, but he tried. And he relied on the proven fact that virtually nobody who met him could dislike him. Tensions abated, though they did not disappear. Keys's talents were pre-eminently those of persuasion rather than command. He could master a brief, and project it to any audience: a reporter, a politician, or a high school class.

His views on research at Chalk River were neither proffered nor sought. He had now to start over, painfully boning up the last twenty years in nuclear physics and radio-chemistry. Keys made the best of it. He began regular prowls through the laboratories, gently inquiring what was up and asking what it might lead to, and reporting the results to Ottawa. His efforts evoked a mixed reaction. Some appreciated the chance to explain things. Others were exasperated by the elementary questions. 'Many times he'd say to me, "I really don't understand this,"' one of the younger scientists recalled. But Keys did not complain, even if his work at Chalk River fell far short of what he might have hoped for.

The role that Keys ended up playing was very different from the one Mackenzie had contemplated. At Chalk River the old professor learned rather than taught; he then used his knowledge to teach others, outside the precincts of the nuclear fortress. He had no great impact on the operational side, where the DIL structure remained intact. It would take years for the inherited conflicts to work themselves out of the system, and there Keys could help by smoothing over differences that he was unable to resolve. Universally admired for his human qualities, Keys imparted compassion and civility to Chalk River. He expressed the best of the laboratory's character, and added a civilizing influence. He did not, however, administer it or shape its basic direction. 'Keys was done a very grave disservice,' Leo Yaffe reflected, 'by taking him from McGill.'[25]

The existence of a vacuum at the top was noticed. Though Keys stayed, Mackenzie next pinned his hopes on Ken Tupper, the director of engineering, who had, he hoped, the right mix of administrative skills, NRC background, and scientific achievement. But in 1949 Tupper was lost to the University of Toronto, where he became dean of engineering. Two men replaced him, Gordon Hatfield, a DIL veteran, and Lorne Gray. Hatfield was a popular figure, but not as effective as Tupper. Gray, however, was a very different animal.[26]

Gray was a late arrival at Chalk River. He was an engineer, who had taught briefly at the University of Saskatchewan. He had spent the war in the RCAF, and afterwards turned up in the Department of Reconstruction in Ottawa, where Mackenzie, who knew him and who in any case always kept a soft spot in his heart for Saskatchewanians, had spotted him. 'Lorne Gray in to see me for a few minutes,' he wrote in his diary in February 1947. 'I think we might entice him to go to Chalk River if we tried hard enough.' Gray had other ideas, including a spell in business in Montreal; but in the long run he responded to Mackenzie's enticements.[27]

His first job was working for Mackenzie in Ottawa. He liked Mackenzie, and defended him, but he had his reservations about NRC. The council, Gray remembered, was 'like a badly run university,' and his own job was close to its centre. When Mackenzie asked if he would like to try his hand at straightening out Chalk River, his problem child, Gray agreed. It was in all ways a happy choice. Gray loved Chalk River. It was a challenge to help build a new town, some of it with his own massive hands: Gray was a skilled amateur stonemason. He liked golf; and, with assistance, he built a golf club. Employees, even scientists, pitched in, some in the naive belief that what was expected was also in some form compulsory. Gray was not the man to shatter such a convenient illusion, nor was he beyond using idle equipment from the Chalk River plant to facilitate his dream.

Gray's administrative skills matched the project's needs. He preferred not to confront an immovable object head on; he moved around obstacles, usually at speed. He liked, for example, to re-create expense accounts. One of Chalk River's scientists, travelling abroad, found himself confronting a formal dinner. He rented an appropriate suit and attended in proper form, a credit to himself and his institution. Gray agreed, but the NRC expense guidelines left him no leeway to repay the money. If, however, the item were re-jigged as something else, taxis perhaps, he could meet the cost. Pleased with his expedient, Gray explained it to his scientist-client in some detail. The reaction was not what he had hoped.

'Do you expect me to lie?' the scientist demanded.

'Oh, go to Hell,' Gray said wearily.

But one way or the other, the scientist's and NRC's sense of propriety would be satisfied. Gray understood that propriety had a cost. It was a commodity, like any other. Gray was expert at obeying the letter of a foolish regulation, while transcending its spirit. Mackenzie had offered him the opportunity to put his expertise to work and, when he saw the field in which he was expected to work, he was satisfied. It was the boneyard of other people's reputations, but it would not be Lorne Gray's. In effect, he put his expertise in the service of a dream, a dream of construction, of accomplishment, and, of course, of science. Science had such tremendous possibilities, and atomic science, obviously, was where the action was. For Gray, like many engineers, was a romantic, with a vision.

The dream was Chalk River, in its beautiful valley, with its neat scientific suburb and its bright young population. Chalk River represented something of value to Gray. He was determined to make it work, to expand it, and to protect it. For Gray, as for most of those who chose to stay, it was the centre of their universe. He was determined that it would also be at the centre of Canada's. It was hard to believe that anybody would not share his excitement, or his purpose. But there were always some, like the scientist, who doubted Gray's talent, who failed to see his vision, or who refused to follow his call. At Chalk River, they left.

Light in the Forest

Lewis's forest kingdom was for many years Canada's principal centre of nuclear research. There had been interest in the universities, but no university had or could afford a nuclear reactor. McMaster had Harry Thode and his laboratory, and McGill was building a cyclotron. Cockcroft, before he left, recommended that Chalk River make sources of fission products available, and drew attention to an American idea of establishing regional laboratories. That might eventually be tried in Canada, where the distant western provinces could use a small, 100-kilowatt pile – 'in due course.' Until then, Chalk River and its experiments would define atomic research as far as Canada was concerned. The definition was Chalk River's principal line of business, and scientists and technicians mastered the marvellous machines the war had bequeathed to Canada.[28]

Inside the plant, jurisdiction was divided among several divisions. Broadly, there was Administration, whose 'chief' by 1950 was Lorne Gray, and there was Research, with W.B. Lewis as director. (The nomenclature, which may seem petty if not actually byzantine, nevertheless reflects the actual lines of authority in the laboratory; it was no small matter to those directly concerned at the time, and it therefore bears a certain amount of historical scrutiny.) Operations, which in DIL days was strictly separate, was subordinated to Research, and to Lewis, on Tupper's departure in 1949;

Medical Services, which remained a separate division, was restricted in its responsibilities and placed under a 'chief' rather than a 'director' after 1948. As a consequence there was only one *director* on site: Lewis.

Geoff Hanna, who arrived at Chalk River to work as a junior physicist in 1946, discovered that he had no problems in picking his own research subject. It was, in fact, a collaborative enterprise with one of the chemists, Bernard Harvey. 'I could set up my own programs, do my own work,' Leo Yaffe, a chemist, remembered. If it seemed promising, it could be done, better than in any university of the period because of the resources, material and human, available at Chalk River. 'In a rough and ready way,' Hanna reflected, '[Chalk River] was set up to do what it bloody well liked in the field of atomic energy.'[29]

That was not precisely the view from Ottawa. The AECB, while McNaughton was there, still had a role to play, but it confined itself to registering reports from Chalk River and approving Mackenzie's recommendation that NRX be directed towards the production of isotopes and what the AECB minutes mysteriously labelled 'research.' The pile would take a while to work up to its full 20 megawatt (thermal) capacity; but the production of isotopes showed that Chalk River was doing useful work.[30] That kept Canada's hand in the nuclear game, and permitted Canadian representatives a certain weight in conclaves with the Americans and the British, or at the United Nations.

It remained to be seen what these considerations were worth when it came time to prepare Chalk River's budget. Chalk River wanted $8 million for the fiscal year 1947–8. Mackenzie, in his diary, recorded the opinion that they would never get it. After discussion with the deputy minister, the estimates went forward to Howe, and back to the AECB. With Howe's support, NRC got most of what it wanted.[31]

The pile did not function smoothly, at least at first. Only experience of operation would show up flaws in design and construction; and only experience could compensate for the inexperience of staff. Keys's reports showed there was much to learn. In August and September 1947, he reported that there had been problems with leaky valves, with a consequent loss of deuterium; and on 7 October there was a disquieting accident in the pile, involving the loss or contamination of 5000 pounds of expensive heavy water. Important as the accident was in itself, it was also indicative of Chalk River's dependence on a single very large and expensive machine.

The reactor control room featured an 'annunciator board.' At 11 PM on 6 October the board sounded an alarm for valve number 370. An operator was sent down to take a look, to be greeted by the sight of heavy water, costing $200 a pound, 'gushing over the storage tank.' The heavy water was collected and the valve was isolated. The next day three leaky valves were repaired. So far, so good.

Testing the pile could go forward, so as to permit the irradiation of certain samples. But while testing, valve 370 was left open. The operators were puzzled at the readings of their gauges: either the instruments were wrong or a large amount of heavy water was *missing*. Another leak was discovered, the calandria was drained, and the operators began to look for the missing heavy water. A glance into the 'helium gas holder room' showed where it was: eight inches deep on the floor with hundreds of gallons of oil on top.

It was by now close to midnight. At 12:35 AM staff, roused from their beds, began to arrive. Gordon Hatfield, chief of operations, sacrificed his pajama jacket to plug the leak. The pumps were manned, and the heavy water, much of it of very doubtful purity, was drained into drums. The pile was shut down for three weeks, while the cost in heavy water could only be guessed at. According to Tupper, the mishap meant that ZEEP and NRX could no longer be operated simultaneously. It was, Tupper wrote, a good reason to contemplate the mortality of reactors, and to consider supplementing Chalk River's reactor capacity by adding another unit.

An investigation showed operator error, as well as evidence of tampering. The operator responsible was fired, and the incident closed. The pile was out of service for six weeks, pending the arrival of more expensive heavy water.[32]

It would not be the last time the pile was 'down'; it was out again for leakage in May 1948, and briefly for other reasons in September. Time 'up' was improving, however, until by May 1949

The Physics Division, and guests, of the National Research Council's Montreal Laboratory, 1943. Front row, left to right: Warwick Knowles, Pierre Demers, James Leicester, Henry Seligmann, Ernest Courant, Ted Hinks, Fred Fenning, Bruno Pontecorvo, George Volkoff, US liaison officer. Back row, left to right: Alan Munn, Bertrand Goldschmidt, John Ozeroff, B.W. Sargent, Gordon Graham, Jules Guéron, H.F. Freundlich, Hans von Halban, R.E. Newell, Frank Jackson, John Cockcroft, Pierre Auger, George Laurence, Stefan Bauer, Quenton Lawrence, Allan Nunn May

'Les canadiens' – a reunion of the founders of the Montreal laboratory. From the left: Jules Guéron, Bertrand Goldschmidt, Sir John Cockcroft, Lew Kowarski, and Pierre Auger

Cockcroft among the constructors: Chalk River, 1945. From the left: J.H. Brace (president of Fraser Brace), A.N. Budden, unidentified, C.J. Mackenzie, F.J. Palmer (Fraser Brace), H. Greville Smith (vice-president, CIL), David Kirkbride (general superintendent of DIL at Chalk River), unidentified, A.B. MacEwen (DIL), H.J. Desbarats (DIL), J.D. Cockcroft, C.H. Jackson (DIL), G.R. Stephens (Fraser Brace), H.S. Milne (DIL), and M. Green (Fraser Brace)

ZEEP

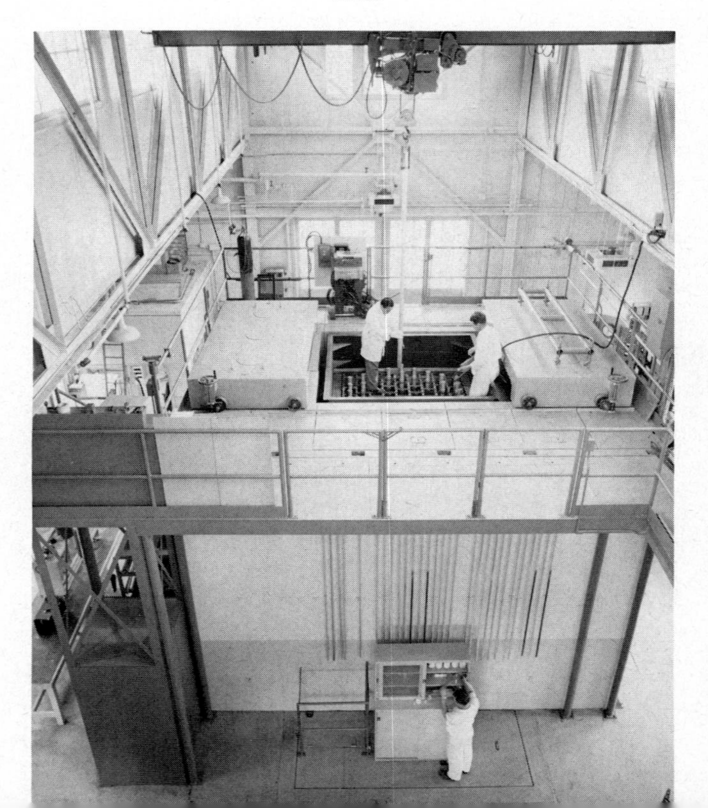

Deep River townsite

Wartime housing at Deep River

C.J. Mackenzie and W.B. Lewis commemorate ZEEP

W.B. Lewis's calculation of the start-up of NRX

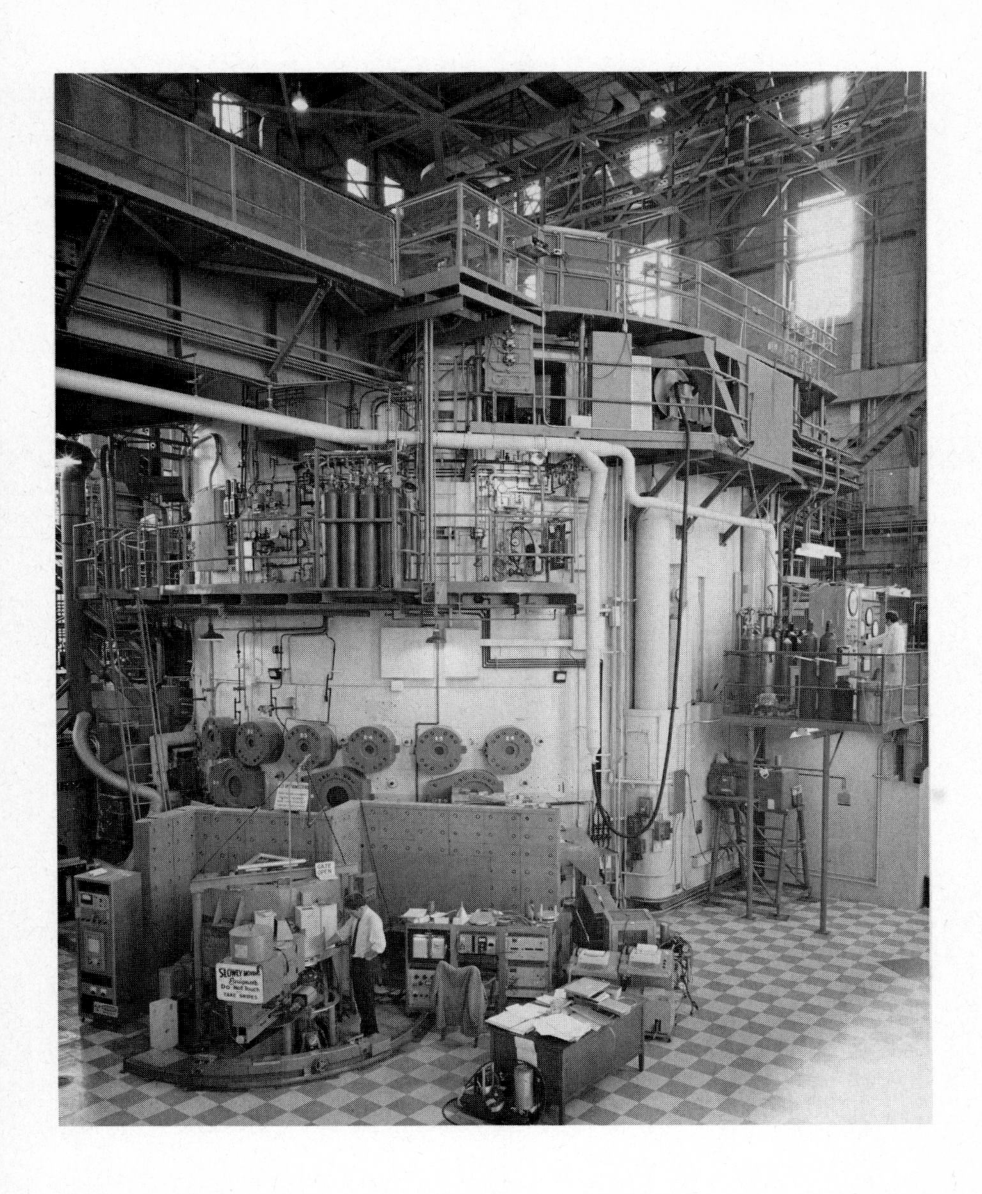

NRX

After the accident: cleaning up NRX, 1952

Aid to India. From the left: W.J. Bennett, Homi Bhabha, W.B. Lewis, and Nik Cavell

Ceremonial curling at Deep River. In the front row, from the left, Lorne Gray, W.J. Bennett, and David Keys

NRU

Founder's visit: Gray, Cockcroft, Lewis, and Sir William Cook
at Chalk River

Chalk River: the tall buildings in the centre rear house NRX and NRU.

John Foster and Harold Smith at NPPD, 1958. Ontario Hydro

CIRUS: Trombay, India

Whiteshell Nuclear Research Establishment

W.J. Bennett (left) and Richard Hearn (right) smile for the cameras as Premier Leslie Frost and C.D. Howe turn a ceremonial shovelful of sod at NPD.

NPD, Rolphton

The fuelling machine at NPD was the first demonstration of the value of a system which enabled operators to refuel the reactor without shutting down.

NPPD's eventual home at Sheridan Park

Starting up NPD: Lorne McConnell in the centre

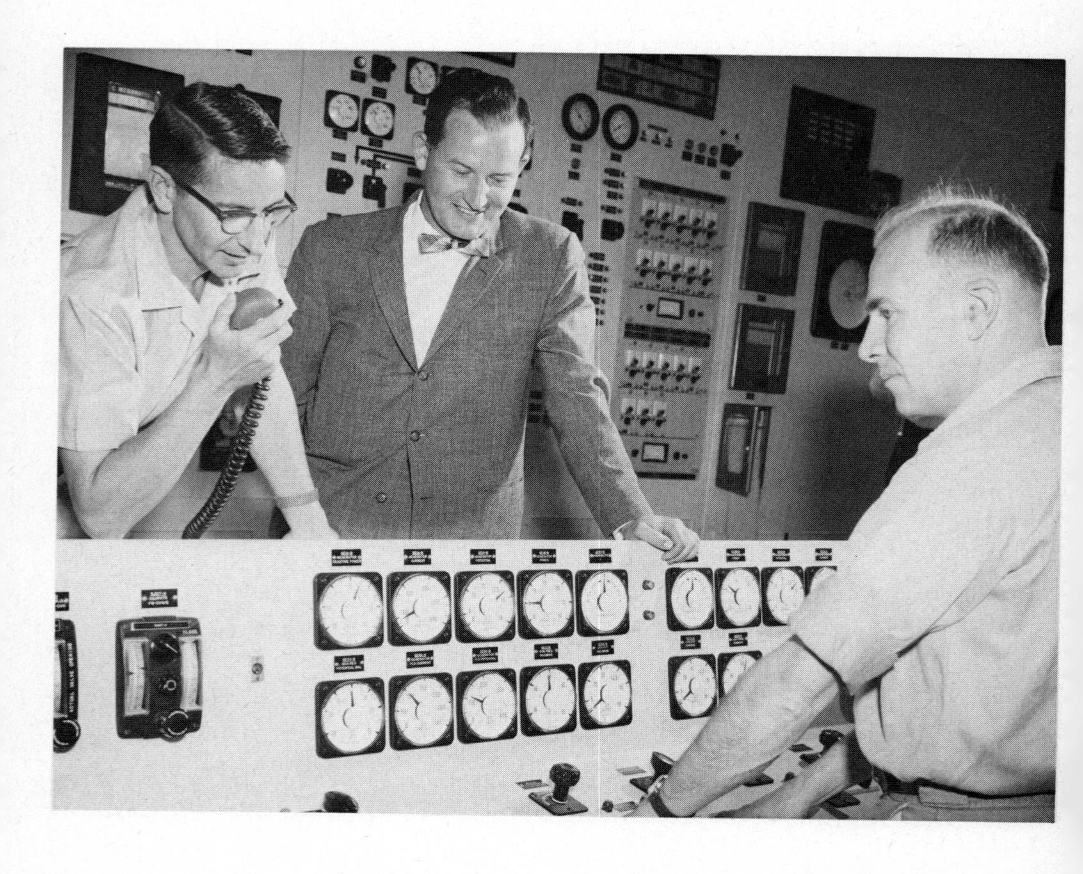

it was working at 99.3 per cent capacity. Down time by then reflected only refuelling and maintenance.[33]

As NRX's operations were routinized, the physicists moved in to learn more about their marvellous machine. They wanted to know what they had, and how it operated; how, for example, it achieved different levels of neutron flux in different parts of the reactor. In the spring of 1948 they notified Keys that NRX's neutron flux was five times that of any other reactor in existence; NRX was functioning well, and its high neutron flux made it, in Cockcroft's opinion, 'the most important instrument in the three countries for the investigation of many problems of importance to the future development of atomic energy.' Cockcroft found time to visit as, later, would Christopher Hinton, in charge of the British program for industrializing atomic energy.[34]

That summer the space around the reactor was jammed with experiments as physicists, chemists, and metallurgists competed for space and time. Great prestige and international notice came their way when they modified Chadwick's conclusions about the mass of the neutron. Meanwhile, Bruno Pontecorvo and Ted Hinks of the nuclear physics division made a name for themselves through the study of cosmic rays. (Cosmic rays also produce neutrons and can cause fission.)

Others among the scientists and engineers were busy with instrumentation; in one case, to study neutron diffraction they put together an automatic neutron spectrometer. Another group worked on microbalances and an evacuated torsion electroscope. Meters, counters, and amplifiers were designed and, if possible, farmed out to Canadian manufacturers. Work was hampered, especially in theoretical physics, by a lack of sophisticated computational equipment: 'keyboard machines' (mechanical adding and subtraction items) were the best they could do, and time passed slowly.[35]

As Harwell took shape, interaction among the British and Canadians flared, and then dimmed. Pontecorvo departed, and with him other British-paid staff. Some stayed: Geoff Hanna and Don Watson, for example, or the Canadian-born Arthur Ward. Relations did not cease, but became gradually more distant. It was thought wise to supplement the dwindling daily contacts with formal conferences, held alternately at Harwell and Chalk River.

The British had certain requirements. They wanted information based on Chalk River's superior experience. They needed to know what André Cipriani, Chalk River's health physics chief, had learned about detecting radiation, and how best to cope with problems like effluent disposal. In 1948, for example, Cipriani told them that he had concluded that fission products would have to be concentrated and stored rather than diluted and passed on to the river, as originally envisaged. It was the beginning, but not the end, of Chalk River's efforts towards the containment and storage of nuclear wastes.[36]

Mackenzie had no difficulty in supporting, not to say originating, proposals looking to military production at Chalk River. Those who knew him also understood that the NRC president might look to the military to secure an increased budget for Chalk River; as a pure research venture Chalk River was still, in Mackenzie's eyes, far too costly to be indefinitely sustained.[37] His qualms about the future were concerned not so much with the danger of war but with a new project for Chalk River: a new pile to supplement NRX.

Until July 1947 his mission or mandate, as he understood it, was to see NRX to successful completion, and to put Chalk River on a firm administrative and scientific foundation. With NRX operating, the goal shifted. At a meeting of the AECB in September 1947, Mackenzie, after reporting on the success of NRX, turned to the possibility of a second pile. The board did not, apparently, blink. 'It was agreed,' its minutes state, 'that the construction of a second Pile should receive early consideration and that the possibility of using oxides or carbides rather than metals for Pile construction should be explored.'[38]

The idea of a second pile received unexpected but persuasive reinforcement with the accident and consequent disabling of the NRX reactor the next month. To Lewis a second pile was an obvious necessity, and had been so for the past year. His reasons were complex but convincing. He began with the prospects for world energy supplies and what he took to be the mathematical certainty that world demand would use up most available non-nuclear supplies. Unfortunately, given the apparent short supply of uranium, they would very soon exhaust that too, leaving two alternative sources: plutonium, as produced by piles such as NRX or Hanford, and thorium.

> **THORIUM**
> Thorium (Th), atomic number 90, is a fertile radioactive metal, one that can be converted by neutron irradiation into a fissile isotope of uranium. Because of its abundance in nature it was the subject of intense study when it was believed that uranium was scarcer than it proved to be.

Plutonium was an unavoidable short-term necessity. In Lewis's view the probable cost of extracting it meant that Canada should not switch over to large-scale extraction plants, but should rely instead on British facilities. This was especially so because Lewis expected that processing thorium into uranium-233 would eventually supersede plutonium. But how to secure enough of either without supplementing the capacity of NRX? In any case, how long would NRX last? Two years? Ten at the outside? No one, of course, knew; no reactor had as yet expired through old age. Assuming four years as most likely, and assuming that Canada would need a largish quantity of both plutonium and thorium to build the next generation of piles, Lewis argued that planning should start immediately for an 'NRX replacement pile.'

Then, and only then, could serious work start on the next stage, which would be intermediate between experimental piles and power plants. One possibility entailed using beryllium as a moderator for enriched uranium, the whole to be gas-cooled. Breeder reactors – using plutonium to make more plutonium from uranium-238 or uranium-233, so that a net gain resulted – were another possibility, holding out the temptation to transcend the shortage of uranium. All this research would take time, but as Lewis concluded his discussion he reminded his readers that one of the reasons for Chalk River's existence was fundamental research.[39]

Once Lewis had grasped the proposition that nuclear power held the best promise of the future, he did not let it go. The route chosen was thorium, the process 'breeding.' His early analyses were concerned with the manipulation of figures so as to justify the expense of recovering uranium-233 from thorium. To do it Canada would have to operate on a very large scale, and resign itself to large expenditures on the chemical reprocessing of thorium. Initially, however, the reprocessing of plutonium and thorium must be done on a small scale: pilot plants. Thus, while the physicists went forward with their fundamental research, the chemists would for the moment embody the hopes of the future.[40]

While Lewis was thinking large thoughts, Mackenzie was coping with large budgets. He did not need much persuasion that a supplement to NRX was desirable; not to have it entailed the risk that Canada would lose ground in nuclear research, and possibly waste what had already been invested. Conceding that much did not mean corralling immediate funds. Possibly, Lewis mused, it would be necessary to invoke British co-operation for the sake of 'larger resources,' especially of personnel.[41]

It took three years, and a large number of extraneous developments, before a new pile received official authorization. Some of those developments were not even in prospect in mid-1947. The method of financing a new pile, and the date on which such a pile might be ready, therefore remained unknowable.

Lewis was not deterred from a further step. Late in 1947 or early in 1948 he established a Future Systems Group under George Laurence, who was also head of technical physics. Laurence had been on detached duty working with McNaughton at the United Nations in New York. The futility of the exercise as well as McNaughton's temper took their toll, and the two men were not sorry to part. The time spent in New York kept Laurence out of Canada while the command changed at Chalk River. It had never been Mackenzie's intention to appoint Laurence to head the project, but Laurence was the senior Canadian-born nuclear physicist. For that reason, if for no other, he was doomed to be perceived as a perennial second to Lewis; and Laurence himself occasionally reflected on the irony of his situation. It added spice to his relationship with Lewis, for Laurence was as opinionated as the director, and equally stubborn.

In 1949, in the aftermath of Tupper's departure, Laurence's responsibilities were expanded to include all of chemistry and engineering. A 'plant design branch' was created under Ian

MacKay, a young McGill-trained engineer who had moved to Chalk River at Mackenzie's behest (Mackenzie had known his father, McGill's engineering dean).

The Future Systems idea was an advance on the contemplated second pile; from its inception it worked on a somewhat divergent track, trying to conceive and if possible design a breeder pile that would replenish its own fuel. Its initial conclusions proved to be of considerable interest, and are of interest today because of the directions they indicated for Chalk River's future reactor designs.

They suggested, first of all, that uranium-235 enrichment was a false start; it cost too much to process, and the shortage of uranium had also to be a consideration. The thorium-uranium-233 cycle seemed to be another dead end; not enough was known about it, and its prospects were correspondingly uncertain. It made more sense to concentrate on the better-known uranium-238–plutonium cycle. All this was with an eye to breeding. One pile idea incorporated a plutonium-enriched core surrounded by a uranium-238 blanket; another envisaged a pile uniformly enriched with plutonium. Cooling systems were next on the agenda. A molten metal coolant was rejected because no suitable metal could be found that did not also have a high neutron absorption rate. That left gas and liquid coolants. If the latter, it would be best to learn from NRX and opt for a closed cooling system that did not have to be disconnected every time a fuel rod had to be changed.[42]

Cockcroft had established a separate health division in 1945, which both that year and the next took on a number of medical personnel. Technically the division was coequal, at least for a time, with Cockcroft's own responsibilities, and it was from that division that Mackenzie briefly considered extracting J.B. Mayneord to replace Cockcroft himself. Mayneord eventually returned to England, and it was A.J. Cipriani who took the lead when the division was transferred to Chalk River. The division had to concern itself with advising the reactor designers on the proper thicknesses of shielding around the NRX reactor, and then, when the pile went critical, to test its hypotheses.[43]

The health division confronted, by the end of 1947, the question of what to do with radioactive products of the pile. Argon-41, for example, was a short-lived gas expelled by the plant smokestack. It could be coped with because it did not remain radioactive for long but, as Lewis recognized, the longer-lived radioactive substances were a very different matter. There was no quick or easy solution to the problem, any more than there was to the question of how much radiation was to be adjudged 'safe' for Chalk River employees. Working in unknown territory, the health protection staff ultimately controlled the conditions of work for those engaged in and around the piles – and any future piles that might be built.[44]

Isotopes

As consideration was given to a possible second research pile, NRX was engaged in its first industrial production. The high neutron flux permitted the manufacture of radio-isotopes, radioactive variants of substances like cobalt.

> RADIO-ISOTOPES
> The term radio-isotope is unusually descriptive: it means any isotope which is radioactive. Radio-isotopes are usually produced in a nuclear reactor.

The idea was not new. Radioactive substances can alter the cell structure of the human body. Rashes, skin tumours, and radium burns on early researchers showed the possibility for harm. By the same token, radioactivity could be used to consume cancerous cells. The most powerful substance available was radium, Madame Curie's discovery, and by the 1920s radium was well-established in cancer therapy.[45]

Radium was difficult and time-consuming to produce. It required a very high grade of uranium to be possible at all, and high-grade uranium was hard to find. The mining company with the highest grade, the Belgian Union Minière du Haut-Katanga, dominated the market and controlled its competitors' prices through a cartel based in Brussels. The outbreak of war found the price of radium for therapy well above $20,000 a gram, out of reach of all but the richest hospitals.

That such a vital product should cost so much had been a public scandal for years. It was in the mid-1930s that an alternative first appeared, as a by-product of the discovery that it might be possible to induce radioactivity in non-radioactive elements. Lewis Strauss, a New York investment banker whose wife had recently died of cancer, had just become aware of the tragic shortage of radium in American hospitals. But if other elements, more easily obtained, could be irradiated, then quantities would rise and prices would drop. 'An isotope of cobalt thus produced,' Strauss wrote, 'would be radioactive and would emit gamma rays similar to the radiation produced by radium ... at a cost of a few dollars a gram.'[46]

The Hungarian emigré scientist Leo Szilard scented opportunity. In 1937 he wrote to Strauss offering his services in the struggle, as soon as he could patent a feasible process. In the meantime, after moving to the United States, he experimented with the irradiation of first cigars and then pork to eradicate impurities. There was no reason, Szilard thought, why irradiation should not work as a kind of universal purifier. The nuclear chain reactions which seemed feasible following the discoveries of 1938–9 could in principle provide the irradiation.[47]

Others had the same idea. The investigation of radiation's damage to cell tissue led naturally to its therapeutic properties. J.S. Mitchell, professor of radiotherapeutics at Cambridge, and in 1944–5 the Montreal laboratory's chief medical adviser, reported in 1945 that 'the recent developments in nuclear physics will exert a far-reaching influence on medical research and practice,' among them 'a new and extremely powerful method of preparation of large amounts of many radioactive isotopes.' These could be used as radioactive sources, like radium, to which the body would be exposed, or they could be allowed to accumulate in certain affected organs. A characteristic of radioactive iodine-131 is that it accumulates in the thyroid gland, with obvious implications for the treatment of thyroid cancer. Isotopes could also be used as a trace element in body scans, or for industrial scanning for imperfections, such as cracks, in a product.[48]

Isotopes were not the purpose of the earliest reactors. They were, however, a convenient by-product of the AEC's X-10 reactor at Oak Ridge, Tennessee; in the first postwar years all American domestic needs could be met from this source alone. The uses, as predicted, were both medical and industrial; by 1948 the isotope manufacturers were looking to produce them on a regular industrial basis.[49]

Isotope production was not NRX's prime order of business. The first year of reactor use was devoted to the study of reactor behaviour and particularly to the effect of radiation on fuel rods. The physicists who were interested in neutron beams, as Donald Hurst (then a physicist at Chalk River and later chairman of the AECB) explained, worked in and around holes in the reflector and shield leading out from the calandria wall. 'As long as the research was purely physics,' Hurst continued, 'a weak link between the operating timetable and the researcher sufficed.'

Isotope work, like neutron beams, originally depended on holes built into the reactor wall; later, with the reduction in the number of control rods, isotopes could be produced inside the core, in vertical tubes. Since chemists were also interested in the irradiation of research materials, and in order to facilitate the study of short-lived radio-isotopes, a pneumatic tube arrangement was established to carry samples direct from the chemistry building into the reactor.[50]

The effect of production schedules and chemical priorities was to bring about a considerable change in the way the reactor was run. The casual connection between scientists and operators typical of 1947–8 was superseded by a much greater attention to schedules; not coincidentally the number of supervisors (all graduate engineers) rose from fifteen in 1948 to twenty-two in the mid-1950s, while the number of operators grew from thirty to forty-four. A major ingredient in the change was the arrival of industrial isotope production.[51]

The advent of isotopes in science and medicine in 1946 occurred just as the radium industry was struggling to recover from the disruption of its markets in the Second World War. The cost of production had not changed, and neither had the price. Stockpiles had increased while hospitals' ability to buy had diminished. The European market had virtually disappeared, because of the scarcity of dollar exchange.

Eldorado Mining and Refining had been convulsed by internal scandals during 1944–6. Its new management, under Bill Bennett, scrambled to find markets. For though the company was government-owned, Howe and Bennett understood that unless Eldorado could be made profitable it would be wound up. Profits depended on establishing a regular, reliable market for Eldorado's high-cost uranium, and on selling the company's radium product. A first step was to establish a marketing organization. 'Shortly after I joined the company,' Bennett wrote, 'I borrowed the services from NRC of a physicist named Morrison. With this arrangement I hoped to get a professional assessment of the post war prospects for the use of radium. The experiment was not too helpful, and I therefore decided we needed a full time and highly qualified man for the job.' He turned to his directors for names, and came up with Roy Errington.[52]

URANIUM

Uranium is the heaviest naturally occurring element. Though discovered as early as 1789 by the scientist Martin Klaproth, and named after the planet Uranus, it was not until 1896 that its radioactive properties were ascertained by Becquerel. Natural uranium (whose chemical symbol is U) consists of two prime isotopes, 238 and 235. Only 0.7 per cent is 235, but this is the portion which readily fissions or splits and was the isotope of interest in the wartime search for the bomb. To separate 235 from natural uranium requires special plants. Natural uranium can be used in reactors or piles for the controlled production of nuclear energy.

Uranium was much sought after as a strategic mineral in the years after the war, because it was widely though not universally assumed that it was very rare in recoverable quantities. This proved not to be the case, though not until 1952 or 1953 was it obvious that uranium was a metal of abundance, not scarcity. Other radioactive substances that had until then been considered likely substitutes, such as thorium, precipitately declined in interest to reactor designers. Because of the large amounts of uranium discovered in Canada, the United States, Australia, South Africa, and elsewhere, its price has fluctuated downwards over time; because of the specific national interests involved, trade in uranium has frequently been the subject of political dispute and economic retaliation.

Uranium-bearing ores require varying degrees of milling and refining. In the 1940s and 1950s uranium reserves were estimated in terms of embodied U_3O_8 (one of the oxides of uranium, and the one most susceptible to further processing); refined uranium was produced in Canada by Eldorado Mining and Refining (later Eldorado Nuclear), a federal crown corporation with interests in the Northwest Territories and Saskatchewan and a refinery in Port Hope, Ontario.

Errington was a physicist by training and an executive in one of Howe's wartime companies, Research Enterprises Limited. REL's future was, in 1946, distinctly limited, for Howe was selling off the company as fast as he could find buyers. Errington was supervising an idle production line when he got a phone call from someone named Bennett. Bennett explained that he had a job marketing radium; if Errington took it, he could expect a 20 per cent raise in salary. Errington took it.[53]

The next question was where Errington's sales division, dubbed Commercial Products, would reside. Port Hope, site of the Eldorado refinery, was considered and discarded. Errington preferred a city, and settled on Ottawa, with Bennett. He and his staff, a salesman named Don Green and their secretary, May McGoey, moved into an office in Number Three Temporary Building; later, as sales increased and their number grew, they took larger quarters in the basement of a store in Eastview (now Vanier), in Ottawa's suburbs. They peddled radium in all its forms: needles, tubes, and radium-beryllium sources for oil exploration. Once ensconced in Eastview, they rented an adjacent garage for $5 a month and filled tubes on the spot; but that, of course, was after they found a market.

Luck, Errington recalled, was with them. UNRRA, the United Nations' relief agency, wanted radium, and was prepared to pay $20 a milligram. That kept them afloat through 1947. In 1948 there was a bulk sale to the British, which kept sales figures at the respectable level of $857,000. But of that figure only $300,000 was in normal, commercial markets. Without the windfall of the British sale things would have been very bad. Errington therefore gave thought to isotopes. He mentioned them to Bennett, who arranged a conference with Mackenzie in February 1948. It was to Errington a matter of 'self-preservation, or at any rate job preservation.'[54]

Further conferences followed, with Lewis present. Chalk River had already considered what it would do with isotopes, and established its own production branch. Shipments of isotopes to other Canadian laboratories began in 1947, and to the United States and the United Kingdom in 1948. Increased business, actual and anticipated, won the isotope production branch a new building by 1950. Eldorado was one of its customers; distribution was arranged through the Charles E. Frosst company, a pharmaceutical firm. Errington was confident he would eventually get business; to prepare, he visited an American firm in the area to investigate what they were doing, and wrote a report on how he could improve on their modus operandi.[55]

Some of the Chalk River staff saw nothing inevitable about an arrangement with Eldorado, but Mackenzie did. With Mackenzie's concurrence, Bennett authorized Errington and Lorne Gray to negotiate a deal between the two organizations. In making arrangements, the two men bore a few fundamentals in mind. Eldorado was much more a commercial enterprise than NRC. It was obvious that there would be a market, and a profitable one, for isotopes: far better to let Eldorado, which already knew the business, handle it.

Gray and Bennett reached a conclusion early in 1951. Eldorado would handle most of Chalk River's isotope distribution, with some slight exceptions. That would mean a new building to cope with the increased volume and range of products. The Eldorado board, busy financing a mine in Saskatchewan, had its doubts, but agreed, subject to NRC's promise to spend $100,000 a year for five years on research and development of isotope products. Eldorado would expand its facilities, since Errington's converted garage would no longer suit: and Errington chose a site on vacant federal land, Tunney's Pasture, just west of downtown Ottawa. (If a move to Chalk River was ever considered, it does not surface in the written record.) Eldorado would spend $300,000 on a Commercial Products building; if NRC took over the division, the mining company would be reimbursed.[56]

That same month, October 1951, Eldorado unveiled a new product, a therapy machine that directed a beam from a radioactive isotope, cobalt-60, onto a patient. The idea was not original: it went back at least to the 1930s, and had later been developed by Dr Mayneord in Britain. Errington had taken note. One day, setting off by car for Chicago with Don Green, he suggested that they stop by Ivan Smith's cancer clinic in London, Ontario, to discuss the idea. Smith's was the largest cancer clinic in Canada, part of London's Victoria Hospital. Hearing the idea, Smith approved. So did a cancer specialist in Chicago. There would be a market, the doctors predicted.

On his return to Ottawa Errington went to see Bennett. He needed $7000, he told the Eldorado president. With it, he expected to develop a therapy machine. The reason, he explained, was that a single curie of cobalt-60 was the equivalent of 1000 grams of radium. The cobalt would cost between $8 and $10; nobody had ever seen 1000 grams of radium, much less afforded them. Bennett gave Errington his $7000. With it, Errington founded a development division, hired some extra help, and worked out a design. He took the design to Canadian Vickers in Montreal and presented it to T.R. McLagan, the general manager. Could Vickers make such a unit? Probably, McLagan said. How much would it weigh and what would it be made of? With those two factors in hand, McLagan produced a price on a slide-rule. It would be just as accurate as a more elaborate calculation, he explained, and it took less time.[57]

The therapy unit took just over two years to design and manufacture. It was late, but it was nearly ready in the summer of 1951, some months before a rival unit was produced in Saskatoon by a team operating under Dr Harold Johns. Its American competition had been unable to produce on time; Errington's beam unit, called the 'Eldorado A,' was the first in the world.

It was unveiled by C.D. Howe himself in Ivan Smith's clinic on 27 October 1951. It was the first in what Errington hoped would be a long line, and it was priced at $27,000. It would operate off Chalk River's production of cobalt-60. So would its twin, the Model B, which he hoped to see the following year. To guarantee the price, it would be manufactured in Commercial Products' own shops, since Vickers had failed to come up with what Errington considered to be a reasonable price; as he commented in 1952, Vickers' 'quality of work ... was to say the least disappointing.' A 'production section,' consisting of a small machine shop, was established inside Commercial Products to do the job.[58]

The integration of Eldorado sales with Chalk River production pointed to a change of status for Errington's unit. It was still receiving radium from Port Hope, but radium was a steadily declining part of the market. 'We must spare no effort to insure that [isotope marketing] is a success because it holds for Eldorado the only presently obvious source of substantial revenue in the

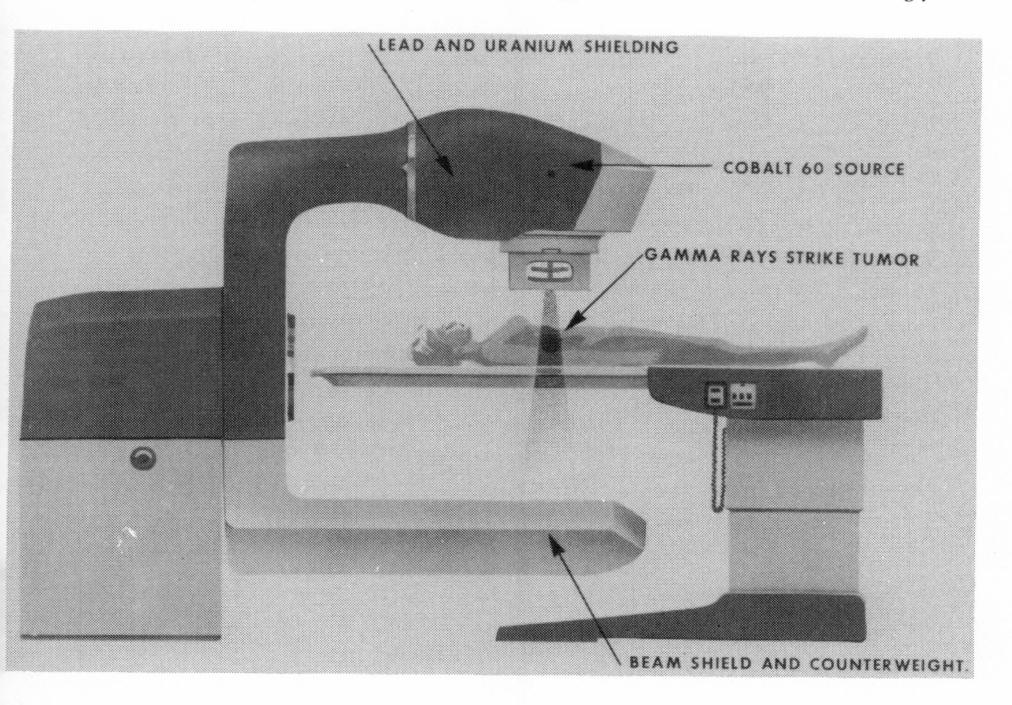

Cobalt therapy

isotopes field,' he wrote in 1950.[59] Production and profit pointed to integration. The fact that NRC was a firmly non-commercial institution, however, constituted a major obstacle. Mackenzie liked it that way; so did the NRC staff and the council's clientele, scientists across Canada. That was all very well while the council staff worked in small, pure science laboratories, or performed work at cost for industrial clients. The sheer size of its atomic energy division was outrunning the council's capacity to cope. A change, or so it seemed to Howe, was indicated.

From Council to Company

The transformation of NRC (Chalk River) into Atomic Energy of Canada Limited began casually but not accidentally. Chalk River, though operated as part of NRC, was formally the province of the Atomic Energy Control Board. The AECB, for its part, received reports from Mackenzie, its own president, in his capacity as its principal agent. It was, in the words of Bill Bennett, another AECB member, a 'highly dubious' arrangement, but it was at least realistic.[60]

In the early 1950s, however, the realism began to fray at the edges. It started with NRX, whose very success created a sense of vulnerability. If NRX broke down or shut down, there was nothing to take its place. Even if it was running, there were more experiments to design and run, and more possibilities to investigate. Foremost among them was the dream of nuclear power.

Nuclear power had always been a distant glimmer. Existing piles produced heat, but at such low temperatures and at such high cost that electricity produced from them would be small in quantity and extremely expensive. Funding the investigation of nuclear power was not a priority for governments. The government of Canada was no exception. Its priorities after 1945 were social and economic: the growth of industry through investment; social welfare programs; and, towards the end of the decade, the deterrence of Soviet expansion abroad. Although daily life, planning for the future, and civilian priorities could ignore what was happening in Europe, Europe had a habit of overshadowing even the most domestic Canadian programs.

By the beginning of 1949, the international situation looked increasingly gloomy. That at least did not allow for indefinite postponement. At a meeting between Mackenzie and Cockcroft, Michael Perrin, and Lord Portal, the titular head of the British atomic energy project, in November 1948, conversation turned to the prospects of war and what it would or should mean to Canadian and British programs. Mackenzie resisted attempts to merge Chalk River with defence research being conducted directly for the Department of National Defence, but he accepted that the NRC would once again take up the burden of war research, and possibly soon. He brought up the subject of a production pile at Chalk River with Christopher Hinton, who headed up the British production reactor team at Risley, in February 1949, and was pleased to learn that Hinton thought it might have its uses in supplementing British production.[61]

It was with these developments in mind that Mackenzie surveyed developments in the Future Systems Group that same month, February 1949. It was, he recorded, 'a most stimulating experience.' He continued: 'Laurence had prepared a very good memo and there was much lively and intelligent discussion and debate. I am beginning to see a way out of our difficulties in getting a new pile underway. The costs of course are fantastic and one must begin to think in terms of production.'[62]

The connection was simple. Chalk River needed a new research pile. NRC did not have the money for a facility that was purely research. It might, however, be able to find funds for a research reactor that also produced plutonium. The plutonium would offset Canada's obligations to its allies, principally Britain and the United States. Alternatively, it could be sold to either or both of them, and the return would help to pay for the construction of the pile in the first place. Perhaps it would pay for it all.

The idea appealed to Howe. He was not in favour of building a second pile, he told Mackenzie, unless the project could be made self-liquidating. It would be even better if a sale price could be negotiated that would pay for the townsite too, not to say the existing pile. For Howe to move, there would have to be a nibble from a plausible customer.[63]

It seemed that there was interest south of the border. Mackenzie told the Advisory Panel in December 1948 that he detected an improvement in American exchanges with Canada, perhaps because of 'the valuable information' to be gleaned from NRX. Because NRX was such an 'extraordinarily good instrument' the Americans were now 'quite interested in its operations.' Mackenzie went on: 'It provides materials and [facilities] which cannot be obtained at any of the larger U.S. plants.' Though Mackenzie warned that supplies of heavy water might become precarious if the Americans took a serious interest in using heavy water instead of graphite as a moderator, he might have pointed out the reverse: a real American interest in heavy-water piles might shore up the Canadian program and its supplies.[64]

If so, it might not be amiss to send a sample of the product – spent fuel rods – to the United States. As Norman Robertson, the cabinet secretary, observed, Canada needed American dollars more than British pounds to meet its trade deficit with the United

States: better to sell plutonium to the Americans than to the British, if there were to be a choice. Better still, the Americans might invest in Canadian production facilities, as they had in other large projects during the war. (The Americans were in fact hoping to get the British to put a gaseous diffusion plant in Canada instead of Britain.) Mackenzie was not reluctant to see the Americans invest money in Canada. He was, he told Robertson in April 1949, 'considering the feasibility of a plan.' It involved building a reactor (the term was just coming into general use) 'for the production of about ten times the amount of plutonium now being made annually at Chalk River.' He estimated the cost of such a reactor at $30 million with operating costs including amortization and capital costs at another $20 million. 'Nevertheless, if we could count on selling the United States sixty kilograms of plutonium annually at the estimated present cost to them the investment would be a very attractive one.' Howe approved the idea; now Mackenzie wanted to try it out on David Lilienthal, the chairman of the AEC. The Advisory Panel approved.[65]

Mackenzie set off on a visit to New York, Oak Ridge, and Washington, where he met Lilienthal for what he called 'very interesting' but apparently inconclusive discussions. The government considered the matter over the summer, preparing for some kind of cabinet decision. The subject was fresh in Canadian minds when the CPC did consider the matter in September. Its deliberations were enlivened by news of a Russian atomic explosion, but the minutes tell a bland story. The Canadians told their colleagues that their proposed reactor would be finished, probably, in 1955. The CPC responded by indicating that there would probably be enough uranium available to fuel the proposed Canadian reactor, among other projects – including a much expanded American production effort. Mackenzie was not entirely satisfied. 'After days of talking,' he wrote, he was not sure how well he had succeeded. Briefing St Laurent on his return was a happier experience. 'The P.M. got the meat of the situation in few minutes. He grasped the essentials very rapidly and decided what the Canadian policy should be without any fuss at all.'[66]

The policy involved what Robertson called 'a partnership for the most efficient production of atomic weapons that the U.K., U.S. and

Canada can achieve together.' He was prepared to contemplate the storage of British atomic weapons at bases in Canada (an idea that had emerged at the CPC meetings), although it would take some years to house them appropriately. This would be a signal contribution to the recently concluded NATO alliance; it would shore up Canada's status inside NATO.[67]

The fall was enlivened by another mishap at Chalk River, where some fuel rods got stuck in the reactor, just in time for a visit by the governor general, Lord Alexander. The fault was corrected, but it underlined once more the slender material base for Canada's claim to membership in the nuclear club.

When discussions resumed in Washington at the end of November, the British and the Americans were still struggling: the British inclining to collaboration in weapons production, and the Americans wondering whether they dared agree. The Canadian representatives, Hume Wrong and C.J. Mackenzie, supported the British point of view; Canada's own objectives they defined as 'principally ... civilian uses' looking to the 'eventual utilization of atomic energy.'[68]

Early in 1950 another spy scandal, this time involving Klaus Fuchs, a Harwell physicist who had spent time at Los Alamos during the war, exploded whatever hopes the Americans and British had had of re-creating the wartime atomic partnership. The possibility that Canada would secure parts of the British atomic program, never strong, vanished too.

What did not vanish was the second Canadian reactor. It was not formally approved, true. Hedging their bets, the reactor designers considered at least two alternatives, depending on whether spent fuel rods were sold directly to the United States or reprocessed in Canada. Finally, in August 1950, the AEC told Mackenzie that if the Canadians wanted to build a second reactor, the Americans would buy plutonium from it.[69]

Mackenzie immediately passed on the good news to Howe. He had talked to the Americans about a price of $170,000 or $180,000 per kilogram of plutonium, and they had not said no. Howe was pleased. He would raise the subject for cabinet approval at the next opportunity. Mackenzie could, if he wished, engage a consulting engineer, always subject to the eventual approval of cabinet.[70]

Mackenzie laid the issue before the AECB at the beginning of September. It would be advisable to engage consulting engineers. The board did not object, nor did its members differ on the man to approach: Murray Fleming, president of the C.D. Howe company of consulting engineers. The board cautiously omitted Fleming's firm's name from its edited minutes; for although C.D. Howe had severed his relations with the engineering firm he had founded when he entered politics back in 1935, some people would jump to the conclusion that patronage and influence-peddling rather than precise qualifications were at work. Howe himself had misgivings on the subject, and some of his friends warned against employing a private company with the Howe name; but he decided to allow events to take their course.[71]

Lorne Gray went to Washington to work out a final price. This took time, but Mackenzie needed the time to sell the second reactor to his peers in Ottawa. This involved taking the deputy minister of finance and one of his senior officials, R.B. Bryce, on a tour of Chalk River; their support would be necessary when the cost of the reactor, over $26 million, reached Treasury Board for approval. Howe later told Mackenzie that both men were sold on the proposal; there are strong indications that the selling involved a heavy emphasis on the military aspects of the new reactor. The finance minister observed that he understood the second reactor was 'directed primarily to the production of Plutonium for military purposes and any other purposes for which the United States would be using it, rather than for research.' With this idea in place, at the beginning of November the second reactor sailed through. It was the fifth month of the Korean war.[72]

Even if the defence aspects of the decision are put aside, more was implied than simply the ratification of a good but limited deal for a second reactor. It was not clear, even if NRU were built, what the medium-term future held for Chalk River, or for atomic science in general. The advantage of the deal with the Americans, as Mackenzie saw it, really was that it permitted 'the annual Government outlay for the project to remain at its [1950] level instead of being increased during the next 10 years of uncertainty in the development of industrial uses of atomic energy.'[73]

Finally, on 11 December, Mackenzie got word from Gordon

Dean, Lilienthal's successor as chairman of the AEC, that agreement had been reached. 'He said,' Mackenzie told his diary, 'he thought our price was perhaps a bit high but not too high for the splendid cooperation and good will that had always surrounded our relationships.' Not all the kinks had been ironed out – agreement on a price covering fuel rods from both NRX and the future reactor was not secured until late in the winter of 1951. There would be problems with heavy-water supply before 1955. Fortunately more would not be needed. Canadian engineers faced extreme obstacles in getting relevant information from Oak Ridge, prompting strong language from Mackenzie in his next round of talks with the Americans. The AEC would eventually have to square away a suspicious joint congressional committee and secure amendments to the McMahon act, to allow it to exchange information with Canada.[74]

The new reactor, which by mid-1951 had been christened NRU (National Research Universal), had had an extraordinary impact on Chalk River even before the first sod was broken on its site. It took the project through two years of intense budget calculations; it helped establish a remunerative relationship between Chalk River and the AEC; most of all, it handed Chalk River the prospect of a future beyond NRX, and guaranteed that the Canadian project could continue to qualify as a major centre of nuclear research during the '10 years of uncertainty' that Mackenzie and the scientists anticipated before any marketable design for nuclear power became available.

It did something else. Projections of the employment NRU would generate placed the laboratory's workforce at rather more than 3000 in the not very distant future. Current employment in NRC was just touching 3000, of which 1200 were at Chalk River; it would not be long before the council became an appendage of its nuclear laboratory.

NRU thus marked a turning-point. The turn had begun a year earlier, when Mackenzie's proposals for NRU passed the AECB. George Bateman, V.W. Scully, the deputy minister of national revenue, and Bill Bennett became concerned. They did not think they were abreast of the situation, and in Scully's opinion neither was Mackenzie, whose bookkeeping at Chalk River was grossly inadequate. The three men went to see Howe. They told the minister that Mackenzie was overwhelmed by his dual responsibilities as president of NRC and leader of the atomic project. He should choose one or the other. Howe accepted their advice.[75]

The minister first sounded Mackenzie on the subject over lunch on 17 November 1951. He pointed out that Chalk River's operating budget looked like being $17 million when NRU came on stream, and suggested that such a sum would be better managed by a separate agency. Although Mackenzie did not record it, Howe either at this meeting or soon afterwards made another suggestion. A single man could not handle being both president of NRC and president of the new company. Mackenzie was already sixty-three, due to retire in July 1953, and it would be appropriate if he used his remaining time to found a new institution. Howe asked Mackenzie to choose which he would like to preside over.

Mackenzie brooded over the problem and on 19 November presented Howe with his own suggestions. It was true that carrying on as at present might be incongruous. 'Normal government procedures' were inappropriate to an industrial operation. Yet combining research and development with production in a single crown company might also prove disadvantageous – presumably for the researchers. He obliquely urged a third course on Howe: dividing the Chalk River site into two enterprises, the scientific end to continue with NRC and the engineers with a new crown company, both under the overall direction of the AECB.

Howe replied the same day, dismissing Mackenzie's objections to a combined crown company. Conflict between R&D and operations was inevitable, no matter what form was adopted. The proper procedure was to have a single top executive who could decide between the two warring divisions. As for the AECB, the minister wanted its functions cut to the minimum contemplated in the act of 1946. Mackenzie would combine the presidency of the new crown company with the presidency of the board, making for what Howe called 'a unified top command of all atomic energy matters.'[76]

Acting under direction, Mackenzie and the AECB prepared for the new crown company. It would be divided into an operations and a scientific division, each with its own vice-president. Since the proposed company could not be self-sustaining, it would require, as at present under NRC, an annual vote of funds from parliament. To avoid confusion, it would get underway on 1 April, the beginning of the next fiscal year.

The company was christened Atomic Energy of Canada Limited (Mackenzie's preference was for Atomic Research and Development Limited). A slate of directors was chosen, of a form pleasing to Howe; it included Mackenzie himself, V.W. Scully, Bill Bennett, and two new faces, Geoffrey (Geoff) Gaherty, president of the Montreal Engineering Company, and Andrew (Andy) Gordon, the head of the University of Toronto's chemistry department. To these were added René Dupuis of Quebec Hydro and Huet Massue, a senior engineer with Shawinigan Water and Power. F.C. Wallace, who four years before had been an unconsummated choice of Mackenzie's to head Chalk River, disappeared from the first preliminary list to be replaced by Richard Hearn, the chief engineer for Ontario Hydro. Internally, Lewis signified his approval by becoming the scientific vice-president of the new company.[77]

Reluctantly, Mackenzie had plumped for Chalk River, as Howe desired. Steacie would succeed him at NRC; as a concession, Mackenzie kept his office in NRC's main building on Sussex Street. With Mackenzie as a distinctly reluctant captain, Atomic Energy of Canada Limited was launched. The course, however, would be set by W.B. Lewis.

Science and Commerce

The birth of AECL was less definitive than it seemed. One of Howe's first acts was to commission a study from a Toronto management consulting firm, Clarkson, Gordon; its senior partner, Walter Gordon, was placed in charge. The object of the study was to discover whether it might not be a good idea to merge AECL with its mining twin, Eldorado. Eldorado, in 1952, was expanding. It had a new mine in Saskatchewan, and it was signing new, very big, contracts for uranium with the AEC. More to the point, it made a profit, and it promised soon to make a very big profit. It did not have to go to parliament for an annual appropriation, something Howe found a nuisance. But AECL did. It made, from Howe's perspective, a 'loss,' doubly so, since it complicated the minister's life by his having to run AECL's budget past the scrutiny of his fellow-politicians, not to mention the Treasury Board bureaucrats. Eldorado was not exempt from this, but the atmosphere was somehow lighter when talk did not dwell on extracting more money from a common public, tax-based, purse.

Gordon would take a while to report, and when he did, he checked first with Bill Bennett to make sure that his recommendations were acceptable. The two companies would not merge. Bennett did not want them to and, by then, neither did Howe. Gordon found that it made no sense to combine a mining and engineering company with a scientific establishment. What they had in common was far less than what separated their interests and priorities. He did not need to comment on the dim view that AECL's scientific staff took of a bunch of rough mining engineers, far down the intellectual scale; they had already had enough to swallow in the course of 1952–3 from Eldorado. Bennett was part of the dose; Commercial Products was the other.

We shall follow Bennett's trail elsewhere. When he moved to AECL as president in 1953 he kept his presidency of Eldorado. He was directly familiar with at least part of his new company, for he was following the migration of Commercial Products.

Such a transfer had been anticipated when CP took over the marketing of NRC's isotopes; no sooner had that arrangement been made, in October 1951, than Gray and Errington sat down again to work out terms. The October 1951 agreement always had a temporary air, and Mackenzie took the view that with the birth of AECL the situation had changed. Perhaps the new company could work out its own marketing arrangements and not depend on Eldorado. So Gray and Errington started negotiations over again. Errington conceded Mackenzie's basic point. AECL, unlike NRC, could do business. The future was with isotopes, not radium, and it made sense to link CP with the company that actually manufactured its wares. After briefly considering making CP a joint subsidiary, the two men – and Bennett and Mackenzie – agreed to transfer CP entirely to AECL. In June 1952 the Eldorado board agreed.

All that remained was for Errington to work out his own arrangements with Mackenzie. He saw Mackenzie in his office, he later remembered. They had a pleasant conversation, and Errington later sent Mackenzie a three-page memorandum of what he took to be their conclusions. Mackenzie failed to respond; on the maxim that silence means consent, Errington proceeded with his own business. But with Mackenzie, he discovered, nothing was certain. After the transfer had taken place, he saw his new president again. Errington's memorandum, Mackenzie said, meant nothing. 'That,' he proclaimed, 'is just a scrap of paper.' Mackenzie's system of administration, he learned, was very different from Bennett's.[78] An unhappy year followed. But at the end of the year, Mackenzie was gone; better still, from Errington's point of view, Bennett was in. It was altogether likely, he knew, that there would be changes made.

4

A Canadian Reactor

The title 'Atomic Energy of Canada' must have seemed a hopeful bit of scientific bombast. It was, however, in tune with its optimistic times. Canadians in the early 1950s faced evidence of progress abounding: more money, more creature comforts, more jobs, more factories. They rewarded their government with huge majorities in the House of Commons, expecting, and getting, more of the same.

Yet there was a certain ambiguity about the term 'atomic energy.' It was an achievement of science, and science was greatly respected. Science had made possible advances in medicine, in communications, and in transport. The year that Atomic Energy was founded was also the first year of CBC television; and in Malton, outside Toronto, Canadian technicians were producing a Canadian-designed jet fighter plane. The fighter, the CF-100, hinted at the other side of atomic energy. Canada sheltered under a Western defensive alliance, NATO, and NATO depended on the existence of an American nuclear strike force. The government supported NATO's strategy; if asked, it would respond that Canada contributed what it could to strengthen the alliance. Economic expansion and defence, after all, went hand in hand. Though Howe did not stress it, that was particularly the case for AECL.

Atomic Fuels

The Atomic Energy Commission was at the centre of the American defence program. It was simultaneously in the business of purchasing uranium, running laboratories, and designing weapons. It had bombs for the air force. It was working on small-scale products for the army. But it was its relations with the navy that would have the most profound impact on the future of nuclear energy; those relations were bound up, almost from the start, with a most unusual individual.

Captain Hyman Rickover was an engineering officer, engaged in mothballing ships, when he was instructed to report to Oak Ridge. The US navy was interested in nuclear ship-propulsion, and Rickover was one of a number of officers assigned to discover whether, and when, a nuclear-powered ship could be built.[1]

He was an apt pupil. Like General Groves before him, he was able to combine military organization with an intensive research and development program. Unlike Groves, he was obliged to deploy political skills as well, both within the navy and in the US Congress. Like the general, Rickover became known as a man in a hurry; using his budget, he was able to overcome any obstacles in his path. Visits and phone calls, sometimes daily, sometimes hourly, helped force the pace. His subordinates learned not to question their admiral (as he became in 1952); Rickover had a vindictive streak that he used to assert single-minded control over 'his' program.

The task was to design a reactor small enough to fit in a ship, reliable enough to keep the ship moving, and safe enough not to annihilate the crew with radiation. Rickover concocted a deadline, January 1955, and then set about to achieve it. The first task was to eliminate the AEC's reactor scientists from any kind of directing role. It was, he argued, 90 per cent engineering: good engineering to be sure. Walter Zinn, the erstwhile candidate to head Chalk River, had the unpleasant task of working with Rickover; he discovered that his naval colleague was most interested in excluding his Argonne laboratory from any real power.[2]

Instead, Rickover established a connection with the Westinghouse company. To accommodate him, Westinghouse built an entire research laboratory, Bettis, outside Pittsburgh. He already had a design concept – a pressurized water reactor, where pressurized water (which could be heated to temperatures far above the normal boiling point by keeping it under pressures of 1500 atmospheres) would serve as both moderator and coolant.[3]

ZIRCONIUM

Zirconium (Zr, atomic number 40) was first identified by the ubiquitous Martin Klaproth in 1789. A metallic element, it has multifarious uses, from gems to porcelains to glass-cutting. It is of interest in nuclear reactors because of its low neutron-absorbing propensities; it is also attractive because of its resistance to corrosion in very hot water. According to John Foster, 'It is not an exaggeration to say that the zirconium alloys made the natural-uranium heavy-water power reactor possible.'

Developed under US Navy contracts, most information on zirconium and its alloys was declassified in 1955. As production increased, zirconium prices fell significantly, decreasing still further the price of CANDU-produced electricity. Canadian experimenters viewed with concern any diminution in zircaloy's credibility and played a notable role in defining and defending its performance, not only in Canadian reactors but in other varieties. Considerable attention was also devoted to improving the industrial production of zircaloy, with an eye to further cost savings.

Such a design presented problems, not the least being corrosion. Westinghouse needed a corrosion-proof substance to clad the uranium fuel elements for the naval reactor; but such a substance must not be a strong neutron absorber. Scientists at Clinton (or Oak Ridge, site of the X-10) had a suggestion. Zirconium, a comparatively rare element, seemed to be highly corrosion resistant. It was strong, too, and had what were described as 'good metallurgical characteristics.' Early tests had excluded zirconium in favour of aluminum or stainless steel, on the grounds of its high neutron cross-section; it had now become clear that this was owing to hafnium, an element always present in zirconium ore but which could be removed. Hafnium-free zirconium seemed to be close to an ideal metal for use in reactors. Rickover liked the idea, and authorized further research.[4] When that research gave good results, work could start on the rest of the fuel elements: eventually, the idea emerged that fuel need not be uranium metal, but uranium dioxide (UO_2). Progress on both fronts now depended on another phenomenon: testing the fuel prototypes in a high-flux reactor. And for that there was no better facility than NRX.

NRX's designers had conceived of 'loops,' separate coolant circuits used for testing fuel and other substances; loops fell under the operations side and allowed engineering to assert an interest in running the reactor, giving reactor use a more immediately practical cast. Loops also exacted a cost in terms of time and materials; the cost had to be paid, whether by internal accounting, by reallocation of the time a project would take, or by direct rental of time, space, and expertise.[5]

This posed a nice question to Chalk River's management. Chalk River still had a residual military purpose. Though the Canadian military took little direct interest in the site (its science was routed through the Defence Research Board, one of whose members, Omond Solandt, sat on the AECB), other military organizations did. The British continued to be actively associated with Chalk River, and relations between Lewis and Cockcroft remained close.[6] Britain's own research program paid close attention to Chalk River's activity. Conferences alternated across the Atlantic. There was a scientific benefit to both sides, but the British had in view the fabrication of atomic weapons. Since knowledge and some materials were denied them by the Americans, Chalk River became the only possible source outside the United Kingdom. British plutonium experiments were considerably delayed as a result of Chalk River's slowness in plutonium extraction.[7]

Later, in 1952, when the British were close to completing their first nuclear device, they nevertheless found a need for Chalk River plutonium, a request which, in the context of American reluctance to forward an independent British weapons program, caused some difficulty and provoked complicated negotiations among the three erstwhile nuclear partners. The British eventually got what they wanted, but for the Canadians, whose NRU project depended entirely on American willingness to supply heavy water in quantity, there were limits to accommodating the British.[8]

It helped that the Americans were willing to supply heavy water, but this supply had a practical foundation. The AEC, like the British, took a lively interest in the scientific and the operational sides of NRX's experimental program. They were intrigued by the success of the system as a whole, for it seemed to point the way to an American heavy water program, designed for plutonium production for weapons purposes.[9] They were also interested in the behaviour of materials under radioactive bombardment: this related to Rickover's reactors. Each research package was negotiated, and a Canadian price quoted. The package would be shipped up, tests run, and the result returned to the Americans, unopened. Or so Rickover thought.

That could not be done. It was unsafe to test unspecified elements in the reactor. Yet the Americans would not say what they were testing, both because US legislation specified total secrecy and because of Rickover's own obsessions with security. The problem was partially solved when the Canadian reactor operators X-rayed one of the American packages, noted the contents, and, concluding that they would not interfere with the normal functioning of the reactor, put the package in.[10] Rickover was annoyed. The Canadians had presumably learned something from their experiment, although it was unlikely that they could know all. It was a breach of security and he demanded that it never recur. The AEC, however, disagreed with the navy on this issue. It was unreasonable as well as dangerous not to specify what was going in. If security forbade disclosure to the reactor operators then obviously the experiment could not be conducted. Grudgingly, Rickover conceded.[11]

It was not the first or the last time that Rickover interfered with his own business to his own eventual cost. In the case of NRX, the main cost was delay. In the case of radioactive sources required to 'start' his ship reactors, he paid in both time and money. The sources were made in Commercial Products' Ottawa shops and they were not, according to Rickover, being made fast enough. First he phoned, to establish his presence on a daily if not hourly basis. Then, since the desired results had still not been obtained, he sent up agents called 'expediters.' Since the reason for delay was a shortage of trained machinists and workspace, and since the expediters themselves were not trained technicians, their presence was at best useless. Rather than have them stand round his already crowded shop looking over his machinists' shoulders, Errington detailed office staff to take the expediters out to Ottawa bars and expedite drinks down their gullets. The US navy was billed for the time and the drinks; it paid.[12]

Relations with the United States were handled directly through an AEC liaison officer who worked on the site. His task was to preserve and protect American property, to negotiate information exchanges, and generally to keep his eyes open; the first of the breed, Colonel Curtis Nelson, fitted right into Deep River society, where he and Lorne Gray worked together on Gray's golf course. But from Chalk River Nelson was assigned in 1950 to be manager of the AEC's proposed Savannah River site. The Korean War had broken out, and the US government decided to double its nuclear-weapons production. Savannah River had the best available land, and the longest construction season. There Nelson would supervise the building of six reactors by the du Pont company. In du Pont's opinion the best design available was heavy-water-moderated, and that, of course, made Nelson a natural choice.[13]

From the Canadian point of view, the best news was that Savannah River would require vastly more heavy water. To make it, the Americans had worked out an improved technology and built a new plant. Heavy-water technology thereby became more secure, for apart from Canada no other country had shown an interest in developing large heavy-water reactors. For the Americans, the good news was that Canadian heavy-water experience had contributed materially to the feasibility of a new weapons-manufacturing facility. It was, it seems, a happy coincidence: the Canadian reactor would have kept on turning over without any serious American interest, while the Americans had carefully kept the Canadians in the dark about their plans for weapons production.[14]

The Americans could not keep Canada in the dark about their proposed pressurized water naval reactor. The Canadians could follow, from their own direct observation, what the Americans were doing to their fuel elements. They could extrapolate from what they had towards the likely reactor size and design. It was obviously a power reactor. 'I became very conversant with the details, running the loops for them,' one of the operating staff recalled. 'We were making it all work under actual conditions in our thimble.'[15]

What was of the greatest interest was American progress in metallurgy. It was not Chalk River's greatest strength. Aluminum was enough for NRX and it would be sufficient for NRU; but a power reactor worked in a different temperature range, where corrosion of aluminum was a serious problem. If Chalk River went for a power reactor, Rickover was unconsciously pointing the way.[16]

If it worked once, at a temperature of 500° Centigrade, why not again? It was a question the Americans were already asking themselves, in Westinghouse and General Electric, the two electrical conglomerates, and in the AEC. It was a question that was being asked at Chalk River as well.

The NRX Accident

NRX's performance seemed generally reliable. Gib James, the man in charge of NRX's operations, praised the flexibility of the pile's design, which allowed for modification when the need arose. There had been need, however, and there were even problems of a serious kind. One problem was the purity required of the heavy water, where even a small amount of light-water contamination could shut the reactor down. It was also necessary to avoid the accumulation of deuterium (heavy hydrogen) in the helium used as a cover gas. With some ingenuity these problems were surmounted by 1950 or 1951; but there were still others. It was difficult to measure reactor output, and difficult to get a fix on leaking rods because suitable instrumentation was only beginning to be developed at that time. And there were problems with the shut-down system. NRX could be shut down by the insertion of eighteen (later twelve) boron-carbide rods into the reactor core. The rods were raised and lowered by air pressure, and held in place by an electromagnet; they were operated from the control room where six switches raised and lowered six banks of between one and four rods. The rods fitted into tubes through which they travelled into or out of the reactor, and the space between rod and tube was not great. Very small particles in the tube could slow or even jam the rod; this happened with monotonous regularity, on the average about once a month. Between 1947 and 1952 one or more rods jammed on sixty-two occasions.[17]

There were also control rods, originally four in number. These rods were used to adjust the balance between settings of the heavy water level; but one, it was discovered, would do, and the other three were given over to the production of isotopes.[18]

Because of its experimental and industrial load, NRX remained in constant demand. This record reinforced the prevalent belief that a new reactor was needed as soon as possible, before the existing pile reached the term of its useful life. It did not need to be emphasized that the NRX complex, including the separation and isotope laboratories attached to it, was a finely balanced instrument. A breakdown in any part of the system posed a threat to every other part and challenged the viability of the Canadian project as a whole. The first serious accident, in 1947, we have already seen.

It was in the newer chemical separation unit rather than in the reactor that Chalk River had its next serious accident, on 13 December 1950. In May 1950 a new effluent evaporator system started up, with the object of treating wastes from the extraction plants so that they could be sent on to the river. It was not easy, although the process was, in principle, straightforward. There were glitches, sometimes leaky valves, and a back-up of waste. The system seemed to be functioning normally on the 12th and 13th of December, but after lunch on the 13th, while the effluent was being tested for an excessively high beta count, there was an explosion. 'This was not so surprising,' Les Cook wrote, 'since residues of Trigly were undoubtedly in the waste. And as every schoolboy knows, carbon and ammonium nitrate made gunpowder long ago.' One man was killed outright, while the force of the explosion ripped the clothes off two more. Four men were sent to hospital.[19]

After the accident, serious questions were raised about emergency procedures. There were concerns as well about the design of parts of the chemical unit, which one observer called 'a potential death-trap' because of the limited number of exits available. The explosion involved some radioactivity, but an investigating board concluded that in its origin it was 'the type of accident which might have happened in an ordinary chemical plant which had no connection with atomic energy.' One problem, it was suggested, was that the operating crew had been paying too much attention to radioactivity while ignoring the ordinary but very real chemical hazards. It was, nevertheless, worrisome and in the short term it was certainly troublesome. The laboratory was closed and the substance that had caused the eruption, ammonium nitrate, was replaced after considerable experimentation.[20]

That the reactor centred around a highly radioactive core was obvious from the first. The danger of allowing NRX or any other reactor to run out of control was so great that the Chalk River project featured, from the beginning, a rigidly hierarchical system of authority and responsibility. For this reason, if for no other, the operations section's control over the reactor was absolute.

In overall charge of the reactor was a superintendent; under him came six 'senior supervisors,' each in charge of a section: rod work, control and safety systems, process and service systems, reactor structure and research, loops, and reactor physics.[21] Around them was a series of precautions to guard against radiation hazards; the organization itself looked to the prevention of dangerous accidents.

The hazard meant in the first instance exposure to radiation. It was of two kinds: *internal*, discovered in the 1920s when workers painting luminous dials licked the tips of their paintbrushes to give them a better point, resulting in osteogenic sarcoma or bone cancer; and *external*, more commonly found in workers in radium-producing plants, with leukaemia or other cancers a possible result. Thus, while radium was becoming a valuable tool in the fight against cancer, it was becoming recognized as a hazardous substance.[22]

Radioactive substances by definition emit radiation, more specifically *ionizing* radiation, and what is desirable inside a reactor – diffusing neutrons and alpha, beta, and gamma rays – is unwelcome outside, because of radiation's ability to travel through, and to alter, the structure of materials (whether liquids, gases, or solids) it encounters. The result, in humans, can be biological changes; how much change, what kind, and at what odds are central questions that ultimately help to determine the feasibility of large-scale nuclear projects. The outward sign lay in the guards, fences, and radiation meters surrounding Chalk River and, particularly, the active area. Less prominent, but as important in the long run, was the study of what was essentially an unknown subject: radiation medicine and health physics. Radiation medicine, as the term implies, is the study and treatment of and by radiation; it includes questions of dosage and effect. Its practitioners are usually MDs. The most important activity in health physics is the development and use of instruments to measure and detect radiation hazards. Its personnel are usually physicists.

The agenda before radiation doctors – or health physicists – was to ascertain what exposure to radiation did; to determine from a data base that was at best incomplete and at worst idiosyncratic what the probabilities of damage to a human system from radiation were; and, above all, to estimate how much radiation could be described as probably (not certainly, as we now know) safe. The answers to these questions directly determined how reactors were to be designed, and the conditions under which they could be operated. When NRX was designed, an exposure of 15 rads per year was considered to be acceptable. A decade later, when NRU was designed and NRX was being operated, the permissible limit had been reduced to 5 rads per year. It remains at this level. ('Rad' stands for radiation absorbed dose and it measures exposure.) In the longer term, such limits would govern the spread, if any, of reactor technology into the larger field of electric power.

Enough was known of the dangers of radiation to cause the Montreal project to be highly selective in its locale in 1942 and 1943; the knowledge that the project members had to hand, largely from prewar sources, was supplemented in 1944 by the authorized exchange with the Manhattan Project of 'information necessary for the guarding of the health of the operators at the Montreal Plant.' This information Cockcroft channelled through a health section, originally located in the Technical Physics division of the Montreal laboratory.[23] The location was not as implausible as it may seem, for one of the first duties of the health service was to secure measuring instruments for their work, and for that purpose technical physics was a logical venue. The health section also had to set about compiling records on personnel in the lab, setting forth their exposure to radiation; such records would at some later date assist in compiling other, broader standards for radiation safety. By the time NRX came into operation the medical-biological area of Chalk River had come under the sway of André Cipriani.

The first major application of health physics in Canada occurred in 1945, when the Montreal project was required to investigate conditions at the Eldorado mine on Great Bear Lake, and at the uranium refinery at Port Hope, Ontario. The news from both sites was depressing, and from another site in Scarborough Township, outside Toronto, it was downright alarming. At Port Hope the problem was inadequate shielding and ventilation; at the mine it was inadequate ventilation again, with an increased hazard because of the accumulation of *radon* gas, a by-product of exposed uranium; to breathe radon-bearing air is to inhale intensely radioactive particles, with a much increased chance of lung cancer.[24]

Experience during the war, according to J.S. Mitchell, who had been seconded to work on health physics at the lab, showed that 'the effects of radio-active radiations on health are more, rather than less, important than was previously appreciated.' The prewar standards now seemed to be excessive; to Mitchell in October 1945 the safe level for daily exposure was no more than 0.05 roentgen per eight-hour day, that is about 12 rads per year.[25]

Precautions against radioactive exposure at Chalk River were therefore rather more than an afterthought. The feasibility of the whole project depended on the protection of the staff from radiation produced during normal operations, and on measures to prevent any extraordinary release of radiation into the atmosphere. The first category prescribed shielding, containment, and decontamination measures; the second contemplated ways and means of shutting down the pile in case of malfunction.

The question of safety in the NRX pile was handed to George Laurence in the summer of 1944. In an emergency the operation of the pile would be governed by shut-down rods designed to bring the reaction to a complete halt even without the moderator dumped. Laurence proposed two sets of rods, one to descend when the flow of cooling water decreased at its outlet, or when gamma ray emissions reached a designated level. The other set would be released under more 'abnormal' operating conditions, such as a cut in electric power or a failure in the compressed air system that was to be used to drive the control rods. Precise, constant monitoring of the power level of the reactor and of a number of other important operating parameters was necessary. The rods would be released and the reactor shut down if any important abnormality was detected. Interlocking controls ensured that the heavy water level would never reach criticality unless a sufficient flow of cooling water was also being passed through, one of Ginns's ingenious additions to NRX's design.[26]

How much shielding, and what kind, was pondered by two physicists, Donald Hurst and Bruno Pontecorvo. The shielding was especially tricky at the top and on the bottom, where channels for cooling had to be included in the design. To protect reactor operators, NRX was encased in concentric shields: first the graphite reflector, designed to bounce back neutrons into the core, thereby increasing the efficiency of the machine; then rings of water-cooled cast iron and concrete, the three totalling eleven feet. It was necessary that the reactor core be entirely surrounded by shielding. Neutrons and gamma rays in particular had to be absorbed and stopped, such other radiations as beta and alpha being easier to deal with.[27]

There was some initial scepticism among the engineering staff as to the desirability of elaborate secondary and tertiary safety systems. The comparable pile at Oak Ridge was described as over-designed by one of the DIL engineers, designed to respond to conditions that would never occur. But caution if not experience demanded that safety measures be elaborate even at the risk of complicating the pile mechanism.[28]

How the safety devices functioned was a matter of considerable interest to those standing around NRX in July 1947. The shielding was, as predicted, adequate, though when NRX's power was raised to 1 megawatt in 1948 it was thought advisable to increase the shielding in a few places because of a rise in background radiation. In Lewis's opinion the problem was not so much with radiation from within the reactor, but rather with radioactive products leaving the reactor. Airborne activity was discharged from a 200-foot smokestack, creating what the director called 'an invisible cloud' of argon-41, which was neither sufficiently diluted nor adequately dispersed under certain unfavourable atmospheric conditions; to meet the problem he recommended increasing the height of the stack.

For the moment NRX remained, as Les Cook called it, 'a success story.' The reactor was regularly operating, not at one megawatt but at twenty. To Lewis, however, the real interest of the reactor lay in exploring the effect of radiation on its fuel bundles, with all that that implied for the design of a future power reactor. The reactor was periodically shut down to allow measurements to be made relating to the condition of irradiated fuel; one such shut-down was in progress on Friday, 12 December 1952.[29]

BURN-UP
The degree to which the atoms in nuclear fuel have been consumed by the fission process in a nuclear reactor.

The experiment involved the comparison of an unirradiated rod with some highly irradiated rods. The level of heavy water in the calandria was almost at its full power level, a necessary condition for the experiment. However, in order to compensate for the absence of xenon poison, six of the eighteen shut-down rods had to be inserted, so that they were not available for their protective function. Normal cooling to some of the fuel rods had been turned off and replaced by hoses; the unirradiated rod was air cooled. These arrangements had been tried before during similar experiments; they were considered to be quite safe with the reactor at low power.

So it was, but only if the reactor did not return to high power. Unfortunately, an operator made a crucial error. He had been instructed to lower a seventh shut-off rod to permit the heavy-water level to be raised further. This was an unusual action, and it necessitated the manipulation of valves in the basement, identified only by code numbers. He chose the wrong valves. The result was that several of the lowered rods rose, causing the reactivity to go above critical. The supervisor in the control room was warned by indicating lights. He phoned the operator to stop, and rushed down to make sure that he did.

In the basement he partly corrected the valve operation, but all the shut-off rods were now left without 'accelerating air,' which normally aided gravity and greatly speeded up the insertion of the rods. The rods which had risen should have gone back fully, but some of them stuck because, as noted earlier, they had become 'sticky.' Indicators in the control room showed the shut-off rods descending. Unfortunately, they did not show that the rods had not gone very far down – not far enough to arrest the reactivity. That need not have mattered greatly, except for the next mistake. The supervisor phoned up to the control room and told his assistant to press two numbered buttons, four and one. The assistant put down the phone and pressed these buttons. However, the supervisor had meant to say four and three; he screamed a correction into the phone, but too late. Button number one caused more shut-off rods to rise; the reactor went far over-critical. The power and heat output doubled every two seconds.[30]

CONTROL RODS
Rods of neutron-absorbing material inserted into the core of a reactor: the degree of insertion determines whether the chain reaction is quenched, sustained, or divergent.

In the control room it was obvious that something had gone very wrong, and the assistant supervisor tripped the pile. The trip signal should have caused all the raised rods to drop fully back under gravity, but most of them did not. This was the decisive hardware failure; the 'stickiness' of the shut-off rods should have been remedied long before, but the 'accelerating air,' considered to be an added safety feature because it speeded up the rate of shut-down and enabled the rods to overcome more resistance or back-pressure, had in fact made the stickiness seem to be of minor importance and had thus contributed to an actual reduction of safety. A later appraisal of the incident noted that the assumption that the rods would free-fall without accelerating air had 'never [been] thoroughly tested and, in fact, was not true.'

The reactor was left in a runaway state. In the basement the supervisor now heard 'a quick noise [as] of air-activated pistons, followed seconds later by a dull thud.' He saw water pouring down, 'seized a beaker, took a sample, and ran.' The sample, when analysed later, was indeed heavy water.

Up in the control room one of the physicists concerned with the experiment sized up the situation and knew what could and should be done. He quietly urged the assistant supervisor to dump the moderator. He could not do this himself, since there were (and always will be) strict rules about 'who runs the ship.' This action consisted of throwing a switch on the control panel which opened valves in the basement, allowing the heavy-water moderator to drain into a storage tank provided for the purpose. In the original design, dumping of the moderator had not been rehearsed as an operating procedure. The pipes and valves were on the small side, and the reactivity was reduced too slowly to affect this particular mishap very much. But it did decisively prevent any continuation or recurrence of the excursion.

As the power rose above the normal cooling capacity of the reactor, the cooling water in the fuel channels boiled. The steam absorbed fewer neutrons than water, so that a further increase in reactivity took place. Many of the aluminum tubes which formed the fuel channels melted, allowing the cooling water, under pressure from both ends, to squirt out into the moderator space and into the basement. The power rose to 90 megawatts before the combination of downgrading and dumping of the heavy water turned the situation around. Hundreds of gallons per minute of water, highly contaminated with fragments and particles of irradiated fuel, poured down into the basement and continued to do so until remedial work started several days later, by which time four million litres had accumulated there.

Outside the reactor building, staff learned for the first time that something was wrong. Instruments in the chemical separation building went off the scale. 'This was startling news,' Cook wrote. 'I got on our public address system into all our chemistry labs, told everyone that, whatever it was, it would be best to close up shop, put everything safely away, and prepare for an evacuation (the first one ever). I was hardly off the mike,' Cook continued, 'when fire sirens sounded. The fire trucks soon roared past us up the road, followed by Andy Cipriani running on foot. Soon the evacuation siren began to go, and we joined a stream of people heading for the bus terminal.'

Word travelled fast. One story current in Chalk River had

Mackenzie in Ottawa learning of the accident from a reporter. Whether that was true or not he soon got the word from Keys, who was holding the fort in the administration building while everyone else, barring the operators and physicists, was evacuated. (Lewis was at a meeting in Ottawa.) Keys told Mackenzie that there had been an explosion in the reactor, and that radiation levels around it were high, especially in the basement.[31]

Public reaction was confused. There was no danger of a nuclear explosion, Canadians were told. Matters were under control. NRX was out of commission, true, but it would soon be put back into service. Clyde Kennedy, just hired as AECL's first public relations officer, found his talents put to the test. It was well that the news he brought was, basically, good. While there was publicity, the public had no reason not to accept official reassurances. What stuck in the public mind was the fact that the reaction had been brought under control, rather than the fact of the accident itself.

Chalk River rebounded. The first step was to limit the consequences. The contaminated water was piped along a new pipeline to a sandy valley, away from the Ottawa River. New hoses were connected to the damaged fuel elements, to keep them cool. Then the task was to discover what, exactly, had gone wrong – a combination of design flaws and errors in judgement. Next, Lorne Gray mobilized for the clean-up. Unlike a regular industrial accident, it would be necessary to scrub away an invisible residue, the radiation, especially in the basement of the reactor building, and the only way to do that, in the age of atomic power and computers, was by using a mop and pail. Exposure in the basement had to be strictly limited, and entry there was limited to one time. Gray stood at the door with a stop-watch, sending in relays of accountants, machinists, and soldiers, all suited up in protective gear, each to a prescribed patch of floor carefully mapped out and identified beforehand during a whirlwind training session.[32]

The news of Chalk River's accident caused some resonance in the world outside Canada. The Americans were vastly interested; Rickover sent a crew of navy personnel including a very junior lieutenant Jimmy Carter to lend a hand and to report back on what was happening. They also helped with the clean-up, since the supply of useful Canadians at Chalk River was running low. The publicity made a point that it was indeed possible to bounce back even from a reactor accident; equally, it was made clear that the Canadians had handled a tricky situation with dispatch and then aplomb.[33]

RADIATION STANDARDS
Radiation standards existed before the Manhattan Project was dreamt of. The amount of energy released in an x-ray dose was quantified as early as 1908, the key being the measurement of 'ionization units'; this was done with reference to human tissue through a cleverly designed 'ion chamber' constructed by Leo Szilard in 1915. The 'unit of x-ray intensity' was formally labelled the roentgen in 1928, after the discoverer of x-rays, Wilhelm Roentgen. By then x-ray operators sheltered behind a protective shield while doctors and physicists debated how much an operator could 'tolerate.' Tolerate, and tolerances, now entered the language of health physics. *Tolerance* described how much radiation an individual could safely and regularly receive, and it was applied as a work standard to those engaged in radiation work. Put another way, it established a 'threshold' below which it was presumed a worker could safely operate. In 1937 an international congress in Chicago established the safe level of exposure at one roentgen a week or less. Before and during the Second World War, therefore, it was assumed that each individual had a threshold below which he or she could tolerate exposure to radiation; this assumption held sway until roughly 1950.

Once the initial clean-up was complete, the dismantling could begin. The reactor had to be taken apart, piece by piece. The wiring and piping went first. Then the undamaged rods were extracted. That left the calandria and the damaged rods, and whatever else lay inside. Some of the shielding was removed, and a new, rotatable, lead shield placed atop the vessel. Special periscopes were inserted through the shield to inspect the calandria's interior, the better to extract the damaged rods. Two were removed safely, but the third broke, pouring sludge into the basement. The remainder had to be 'pulled bodily from the joints below,' in Wilfrid Eggleston's description.

The calandria itself went next. It was too hot to handle or even approach; an hour's exposure to it was judged to be lethal. After a night's dress rehearsal, seventy men using long ropes moved it onto a skid. This sledge was towed in what Lorne Gray called 'a funeral-like cortège,' supervisors ahead in cars, a grader towing the sledge, and health inspectors behind measuring for contamination. The calandria was towed to a distant hole, still on the company property, and buried in sand.[34]

The rest of the reactor, miraculously, did not have to be replaced, apart from one of the shields. A new calandria was manufactured and installed. The thermal shields were modified to permit a higher power rating – 40 megawatts. Donald Hurst and Arthur Ward, two of the physicists on the project, told a United Nations conference in Geneva in 1955 that the accident proved that 'the worst reactor runaway so far recorded could be handled in safety.' That was encouraging, especially when it was considered that 'problems which would have been considered almost insuperable before this accident are now considered [merely] difficult and annoying.'[35]

When it was announced that NRX would return to service and, more important, when it was announced, fourteen months after the accident, that the reactor had started up again, it seemed that Chalk River had plucked a resounding victory from the jaws of disaster. But the victory had an ironic twist for AECL. As Lorne Gray told a group of engineers in Hamilton, Ontario, in September 1953 – before the reactor was back in service – 'the NRX reactor, which was originally given a life of perhaps five years, may now be maintained for many years of useful service.'[36]

The NRX accident was not the only reactor 'incident' to occur in Canada, but it was the worst. It was an important formative experience for the world's nuclear technology; it led to a great deal of soul-searching and the adoption of a policy which amounted to the treatment of technological safety as a major subject in its own right. In the aftermath of the 1986 Chernobyl accident, its implications bear consideration. As in Chernobyl, Three Mile Island, the 1960 SL1 accident at Idaho Falls, and the 1957 Windscale accident in Britain, the reactor was shut down or at low power when the decisive events occurred. In all cases, a chain of errors was involved. It could be seen with hindsight how the accident could easily have been prevented at almost no cost; the new approach had to be concerned with translating this hindsight into a very broad foresight.

For some people, this 1952 mishap confirmed that nuclear reactors could never be made safe; for others, it showed that nuclear accidents were not the end of the world, and could be managed like all other kinds of accident. Thirty-five years later, this division of opinion remains.

In the case of NRX, short-term remedies, mainly involving the shut-off rod system, were developed and applied before the reactor restarted. A more complete redesign was implemented a few years later. This involved redesigned shut-off rods and an augmented moderator dump system, but above all it was based on a new philosophy which aimed at the integration of design, testing, maintenance, and operation. This was the first practical application of systematic safety engineering, and the principles developed have been routinely applied ever since in the Canadian program.[37]

Gray's masterful reaction to the disaster was his finest hour. Mobilizing support from above and below, he concentrated on the reconstruction of the reactor. The almost universal reaction of operators, designers, and scientists to the accident was to clean up the mess and carry on. That the reactor should continue to function was psychologically important. It demonstrated that Chalk River could cope with adversity, and it also deflated the importance of the set-back. No permanent damage had been done to the physical fabric of the site; more important, morale was sustained. Gray, not for the last time, put his money on morale.

NRU

The NRU reactor was planned to serve first as a back-up to NRX, and then to be its ultimate replacement. It would, at the same time, be a more powerful reactor, it would apply the lessons learned in the operation of NRX, and it would be a more versatile scientific instrument. It would pay for itself. On that understanding the minister, the cabinet, and the NRC all approved the project. 'The new Plant,' according to Mackenzie in September 1951, 'would be operated as a financial venture. If this Plant became a profitable venture, then further advances could be made.'[38]

As C.E. Beynon, an engineer working on plant design, summarized the situation, 'early in the design of the reactor it was felt that, if the NRU ... was really to make a name for itself in the research field ... it would be desirable to provide a special high flux facility not likely to be found elsewhere.'[39] It would also be bigger, operating at 200 megawatts (thermal), and would incorporate the lessons learned in the operation of NRX. It would produce 60 kilograms of plutonium a year, valued at $12 million, and it would consume 100 tons of uranium a year, costing about $3.4 million. It would be used for about seven times as much isotope production, with a proportionately greater return. For scientific purposes, the unit would include horizontal 'through tubes.' It was hoped to start up the reactor in mid-1954. In the summer of 1950 Chalk River staff estimated that NRU would cost $30 million (miraculously reduced to $26 million by the time approval was granted in the fall), and would cost rather more than $9 million a year to operate.[40]

There was no publicity. NRU was, because of its plutonium production, tied into national security. Publicity on reactor design and function might give a saboteur an unwelcome opportunity. Should the Russians take an interest, it was hoped they would be forced to make the same mistakes as the Canadians had, if possible at the same or higher cost.[41]

The Howe Company appointed a staff of engineers to supervise the task: Winnett Boyd was the principal contact with AECL. As general contractors, they chose the Foundation Company of Canada. The engineers approached their task with enthusiasm and optimism. They considered Lorne Gray's $26.6 million price tag 'very high': it would be 'impossible to spend this money on this job.'[42]

The course of development did not run smoothly. The direction of flow of heavy water, the design and manufacture of heat exchangers, and the problem of refuelling provoked lively debates among Howe Company designers and Chalk River staff, with Mackenzie uncomfortably trying to hold the ring. The question of whether to procure from an American or Canadian firm also exercised the engineers; it was complicated by American security regulations as well as by the consideration that a Canadian firm, even without experience, should be encouraged to get into the business with an eye to the future. In the case of heat exchangers the American firm got the job.[43]

Time passed. This may have resulted in better design, but time was also an enemy. There was a tendency for design people to drift away, and their successors had to be trained all over again. Occasionally the Howe Company was blamed for taking up time exploring alternative designs; at other points its staff took the offensive, demanding that the scientists come up with an agreed design for given components and then stick to it. Lewis indignantly replied that freezing the design at any time before the last possible moment would reduce both immediate creativity and eventual efficiency. It was almost a ritualistic battle; the scientists wanted to design the best possible equipment for a job, while the engineers were more concerned with doing the best possible job on an existing design. When the necessary contracts were let, the ability of the manufacturers to deliver within a realistic time also became an issue. More time was consumed in excavating foundations for the new reactor, where it was learned that only by doubling the workforce over previous estimates could the job be done on schedule. By August 1952 work was about twelve months behind, and counting.[44]

The main item that was counting was the cost. Between September 1952 and October 1953 the estimated cost rose from $26.6 million to $31.6 million, *exclusive* of heavy water and uranium. Ian MacKay, of the AECL engineering staff, explained in January 1954 that changes in design had led to a substantial increase in the size of the reactor building, while other items such as thermal shields had also evolved in design. Nevertheless the costs attributable to the C.D. Howe Company were '40% to 50% higher than expected.' And if the cost of uranium and heavy water were added, the estimate rose to $41 million.[45]

It did not stay there. It is pointless to reproduce the slowly ascending expense curve. As the curve rose, the date of completion receded, through 1955, into 1956, and beyond. Even when the calandria was delivered – surely a good sign – there was further delay so that a new thermal column window could be installed. In October 1956 the estimate was $51 million.[46]

It was only in September 1957 that Lorne Gray, by now

vice-president for administration and operations, reported to AECL's executive committee that the job of filling the reactor with heavy water had begun, and that the Foundation Company had turned NRU over to his division.[47] At 6:10 AM on 3 November NRU went into operation.

The reactor was imposing, located in the tallest building on the site, with a steel frame 90 feet (28 metres) high. Inside, the reactor was roughly in the centre, flanked by maintenance and experimental physics areas. Above moved a 75 ton crane, and below that a tall flask, on top of the reactor. NRU was designed for operation at 200 megawatts (thermal), with a high neutron flux in the moderator. Above and below the core were stainless steel 'header blocks,' and surrounding it were steel thermal shields, encased in thick (2.9 m) concrete; inside them was a light-water reflector. Through the concrete and the shields extended twenty-seven horizontal experimental holes, accessible from the side of the reactor.

NRU was moderated and cooled by heavy water, pumped in at the bottom and out at the top. The amount of heavy water available had been one of the major determinants of pile design. The designers had assumed 60 tons, and had then designed the pile around them. The fuel was loaded and unloaded at the top, through a giant 'rod-transfer flask,' which moved above the reactor and necessitated the height of the reactor building. The reactor, like ZEEP and NRX, and like the concurrent generation of British reactors, was heterogeneous. The provision made for replacement of fuel elements, clad in thin aluminum, was one of the most interesting and significant features of the reactor. NRU's could be replaced one at a time without interrupting the operation of the reactor – on-power refuelling. The transfer flask, because of shielding, weighed 240 tons, rather more, as the company pointed out, than a diesel locomotive. The rods were then lowered into a water-filled trench (the water acts as a shield against radiation), and moved along the trench to a temporary storage basin.

Profiting from the NRX experience, as many features as possible of NRU were designed for easy replacement, from the shell of the calandria to the aluminum reflector surface. A basement room contained a carriage onto which the bottom header could be loaded, along with the lower shielding. Sensitive ion chambers, developed by H. Carmichael, measured the power of the reactor from complete shut-down to full power.[48]

The reactor was controlled by sixteen absorbing rods containing cadmium or cobalt. Certain circumstances would evoke what were called 'absolute trips,' the immediate shutting down of the reactor; these included excessive neutron flux or excessive power, which might indicate too much reactivity. Lesser alarms triggered 'conditional trips,' where repairs might be effected while the reactor was kept at low power.

NRU was and remains an experimental reactor, although, as we have seen, Mackenzie expected it to have an economic function. Experiments could be conducted in the horizontal holes, or in three permanent vertical 'loop' arrangements, which permitted access (or extraction) from both above and below. Ordinary fuel positions – also vertical – could be used for experimental purposes, and a sophisticated x-ray system permitted the experiments to be studied while in the reactor. There were also two 'tray rods,' which extended vertically through the reactor, designed to hold forty capsules, usually cobalt, each; for smaller experiments any capsule could be removed and replaced through a compressed air system, which would deliver the capsule to the chemistry laboratory or to either of two outside buildings.[49]

The reactor did not, on the whole, disappoint. Its operating record eventually pleased its proprietors, despite difficulties with fuel assemblies. In May 1958 a fuel rod broke apart as it was being unloaded. Although the accident was quickly dealt with, the resultant contamination shut down the reactor for six months. The problem was finding enough people to spend two minutes each – the maximum allowable – in the reactor building. By drafting in soldiers from Petawawa and air force personnel, the clean-up was eventually completed; there were, subsequently, charges that some of the military were contaminated in the course of the clean-up with serious damage to their health.

The reactor returned to full power in November and thereafter gave good service. The fuel rod problems stimulated a search, ultimately successful, for a fuel element that included a bond between the uranium core and the aluminum cladding.[50]

Those who worked the reactor had no doubts that its cost was

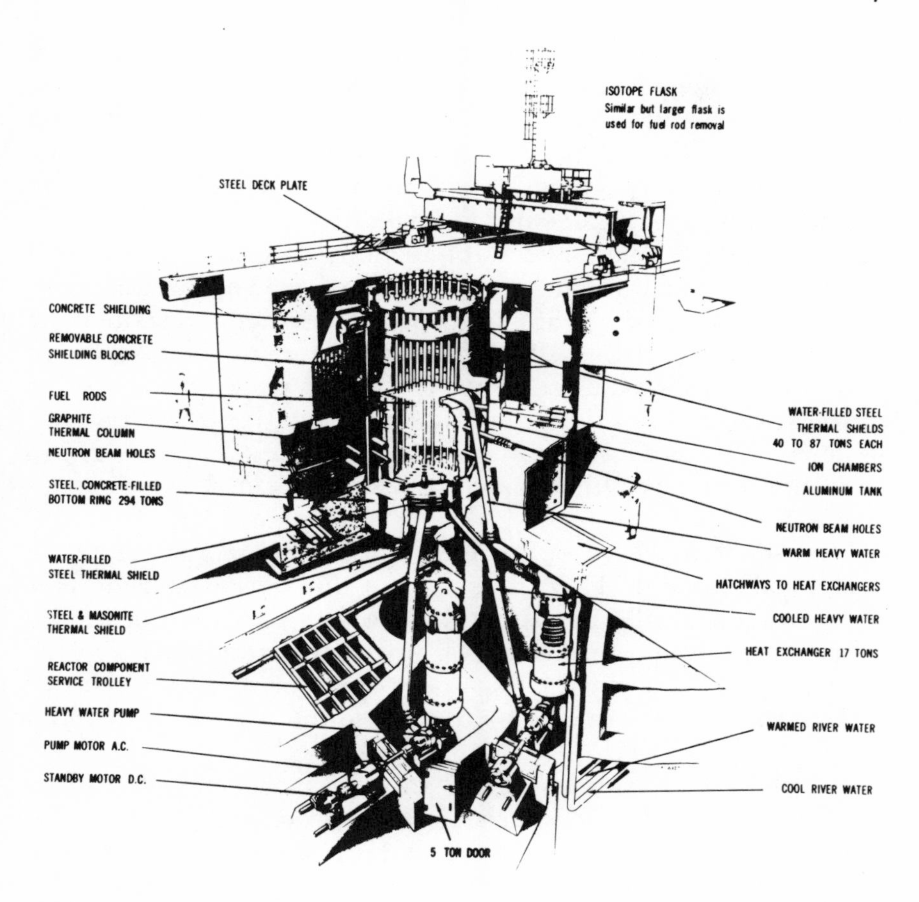

NRU reactor

well spent. For Chalk River it was a shot in the arm, as well as an insurance policy that nuclear research would not suddenly be abandoned. It kept Canada in the nuclear game, a fully paid-up participant in the small international club of atomic powers. It allowed independence and reciprocity with the Americans and the British; if Canada were not in the 'big leagues' that some dreamed of, then surely it occupied a very solid middle position without the expense and complications that a weapons program entailed.

One other feature of NRU deserves to be stressed. It was the first Canadian-designed reactor. NRX had been a British production with Canadian input. It owed its success to the ingenuity and perseverance of a group of ICI engineers, operating with the support and resources that Canada could afford to give them. NRU was different. It originated inside Chalk River, and it responded to the specifically Canadian need for an alternative reactor to NRX. It was built primarily for the purpose of preserving the Canadian program intact. In the process, it brought together and refined the skills of a group of Canadian scientists and engineers. Although NRU was not designed to be the prototype of a power reactor, some of the features of its design inevitably had an influence on later designs. This was not surprising, for many of the later power group were the same. Having created a reactor team, Lewis did not disband it.

It must not, however, be assumed that the NRU team operated in a national vacuum. There was during this period considerable interchange with the British, and some with the Americans; and there was an echo of the wartime tripartite arrangements in the establishment of a reactor that was to pay for itself, or so it was hoped, through the sale of plutonium to the American weapons program.

NRU therefore assumes significance as a transitional project; there is more significance, perhaps, in the process of designing and building it than in the final product. Because construction of NRU was prolonged, it became tempting to consider the reactor as something of a sideshow, although not, as one critic has called it, 'an abortion.' Had NRU been the end of the nuclear game, there might have been some justice in a negative verdict from the standpoint of money-conscious public policy; but it was not the end.

The Gleam in the Eye

In the fall of 1949 W.B. Lewis received a slightly out of the ordinary group of visitors. Members of the Canadian House of Commons, they were the first serving Canadian politicians to visit Chalk River. It was a tame enough affair. The members were shepherded around the plant, allowed to view NRX, and then seated for a presentation by the director.[51]

Lewis is not remembered as a good speaker. Nervous and sometimes hasty, his style often jangled his audience's nerves; but when occasion demanded he could rise to it. For the MPS, he concocted an oration called 'The Gleam in the Eye of the Atomic Scientist.' The gleam in his eye, he explained, was not atomic weaponry. The presence of weaponry might account for the fact that the United States was spending $250 million a year, employing 60,000 people, while the British were committing $80 million with 20,000; Canada, in contrast, was spending $20 million a year, and Lewis's estimate of 5000 atomic-related jobs was probably an exaggeration.

His gleam illuminated the future, Lewis told the members. It was his job to look at where atomic science might be in fifty years – or more. His audience should not doubt that the effort was worth it. Heat from uranium and thorium already helped heat the Earth, from inside the planet's crust. The director foresaw, and freely predicted, that uranium would come to equal water, oil, or coal in importance as an energy source, but the road to usable, economical atomic reactors was long and difficult. It might involve uranium, or it might use thorium; it might be a breeder, or it might be a more orthodox kind of reactor. But he could foresee a time when a reactor could produce vastly more energy than a hydro plant of equivalent cost. And unlike hydro, reactors could be sited anywhere.[52]

Energy is the capacity for doing work. A horse ambling along a path is using energy; so does a workman wielding a hammer; so do a barrel of gunpowder and, incidentally, an atomic bomb. A bomb features the uncontrolled release of energy; controlled and confined, energy – even atomic energy – may be put to useful work. A reactor generates a very considerable amount of heat. Controlling the heat occupied a very large part of reactor design; cooling circuits and heat exchangers were central to the functioning of NRX. But the heat they handled was dissipated, a waste product of the reactor function.

Canada had, it was believed, plenty of energy. Most of it was of the non-renewable variety: coal, oil, and gas. These fuels were mined in some quantities. Recent history, however, underlined their disadvantages. During the Second World War, as a government survey showed, 'output [was] relatively small in comparison with domestic requirements.' It followed that a large amount of Canada's coal was imported from the United States – 61 per cent of total consumption in 1945. Canada's own coal was situated in Alberta (most of it) and Nova Scotia, while oil was just beginning its boom, after the discovery of a new oil field at Leduç, Alberta, in 1947. The memory of oil dependence lingered, and it would find its reflection in public policy.[53]

Public policy in the energy field was something of a patchwork. When Canada's constitution was drawn up in the 1860s, energy was firmly lodged in the private sector. It was something one cut down, or that one dug or pumped out of the ground. It was distributed according to the laws of supply and demand. Except in time of war, jurisdiction over mines and resources rested with the provinces. When electric power became feasible in the 1870s and 1880s it fell into the hands of the provinces. Abridged during the Second World War, when the federal government assumed direction over energy supplies, this control now returned to its nine constitutional homes across the country. Coal, then, was provincial, unless it crossed provincial boundaries, and unless it required a subsidy from Ottawa – as it frequently did. Imports and exports of coal and oil belonged in Ottawa's field, even when a provincial energy agency was doing the importing or exporting; so did the balance of payments, over which the central government brooded. The balance, in 1946 and 1947, was not at all healthy.

Electricity seemed to be an exception. Electrical power in 1946 and 1947 was overwhelmingly generated in hydroelectric stations; only 2.5 per cent of Canada's electrical production came from thermal (coal or oil burning) stations in the latter year; only in Prince Edward Island did thermal stations bear more than half the burden of electricity generation.[54]

Electricity, whose importance for energy had not been dreamt of when Canada was founded in 1867, had been acquired by the provinces, which farmed out their jurisdiction, in some cases to provincial utilities, and in others to large private power companies. 'Large' was in almost all cases the operative word, for electricity demanded huge power sites, miles of transmission lines, and millions of dollars of investment, a truism repeated in every government statement on the subject. Like oil and gas, electricity moved across the frontier, some northwards, but most to the United States. Exports fell into two categories, 'firm,' which flowed in respect of strict contractual requirements, and 'surplus.' Surplus does not mean bankable: 'interruptible' also defines the term. In the late 1940s, with heavy demand placed on industry to fulfil investment and consumption, there were too many interruptions for comfort.

The short-term perspective is that branch of policy that politicians can hope to influence; the longer term tends to belong to prophets of various stripes who contend manfully with Lord Keynes's famous dictum, 'In the long run we are all dead.' To which might be added the strong tendency for prophetic visions to be costly, something that does not add to the popularity or influence of their begetters. Such prophets exist at the margin of policy, hoping to exert pressure at opportune moments on politicians; as the current phrase has it, to slip through the window of opportunity before the politicians exercise their natural tendency to slam it lest any more money escape through it.

The Second World War and its aftermath were such a window of opportunity. The war was important because it created a new technology – indeed many new technologies – and because it gave governments the ability to pay for it through an augmented economy and ingenious new methods of taxation. What had been counted in the tens of thousands before the war could be counted in millions after it. The war also expanded governments' ability to plan ahead, and on a larger scale, than they had done in the cash-starved thirties, while upheavals in the international system gave both politicians and civil servants unusual opportunities in devising new ways of dealing with the unfamiliar. In part because they were unfamiliar or even untried, many governmental initiatives partook of the miraculous, or so it must have seemed to a public on whom new names, practices, and ideas were raining down. Unfamiliar acronyms like UNO, UNRRA, UNESCO, concepts like jet aircraft, rockets, or plastics, pulsed with hope and chromium-plated modernity. Old institutions, even old countries, were vanishing and that seemed hopeful too, at least to the liberal temperament. The British Empire was among the casualties, though few outside government suspected it as yet. Some Canadians naturally remained attached to the old and familiar, but had few avenues along which to make their sentiments count. And while the new institutions attracted support, they had yet to find their own groove in public opinion. Special interests, in other words, were operating at a temporary disadvantage; those who had custody of the national interest were not.

To Lewis, a stranger to Canada still, the intricacies of the nation's balance of powers, and their fluctuations, were not entirely relevant. But they had created his project; they were sustaining it; and they, like he, were looking to its future. The future would be arriving sooner than either Lewis or his federal patrons believed.

5

Nuclear Power Decisons

Canada was, as the phrase went, 'a businessman's country' in the 1950s. It was a country where conditions were good for business: where there was no capital gains tax, for example, where necessities were relatively cheap. Above all, it was where affairs were managed moderately but progressively. It was not surprising to outsiders and even to Canadian observers that it was a country of unusual political stability. The symbol of that stability was the prime minister himself, Louis St Laurent, and, after St Laurent, his minister of trade and commerce.

The minister who concerns us we have already met. C.D. Howe was continuously in office, in the cabinet, from October 1935 until June 1957. He and St Laurent were on the most cordial terms, each man respecting the other's abilities and prerogatives. For St Laurent and the Liberals, Howe was their principal (though not their only) link to the business world; as minister of trade and commerce after 1948 Howe shared his overview of the Canadian economy with the minister of finance. It was well that he did, for Howe's network of contacts in corporate offices was second to none; after all, many of their occupants had worked for him during the war: Howe's 'boys.'[1]

Howe's experience during the Second World War made him sensitive to issues of energy; he had an uneasy feeling that in times of crisis Canada might find itself at the wrong end of a very long supply pipe, with little trickling through. As trade minister, he worried over oil and gas imports; but as a professional engineer he had kept his interest in science and technology. We have encountered him as NRC's minister and C.J. Mackenzie's patron, and in that capacity Howe funnelled the money necessary to keep Chalk River abreast of the international nuclear game.

Howe's longevity in office made him an unusual minister. His memory of events dwarfed the collective recollections of his own civil servants, while his breezy self-confidence overrode caution and obstruction. That by itself would have made him a significant factor in Ottawa's system of decision making, and some observers argued that 'dictator Howe' must surely think himself immortal.

Interestingly, Howe did not. He had outlived many governments, in Canada and abroad, and he guessed that his own was no more immortal than the rest. He turned over the idea of leaving government in his mind several times in 1953 and 1954. St Laurent's importunities dampened his resolve, but sooner or later time would take its toll. If he was to have a legacy, it should be proof against the vagaries of fortune; if he was to leave anything behind he preferred that it be embodied in a form that would withstand political and economic tempests.

The creation of AECL in 1952 was an example of Howe's prudent management. Giving the atomic laboratory corporate form might have seemed a bold assertion of confidence in the future of nuclear power; but in the more mundane world in which these matters are decided it might equally have been a desire to avoid complications by divorcing Chalk River's destiny from a cumbersome and increasingly unwieldy Ottawa bureaucracy. Even more prudent was his desire that his scientific stepchild find a way to live within its means; we shall deal with that below.

Howe was not unaffected by the attractions of a bright new technology; the merely novel held no terrors for him. At the same time, he was not inclined to put his faith and money in an infinite and indefinite proposition. As an engineer, he enjoyed finishing a project more than he liked beginning one, and he was attentive to budgets and schedules. Howe preferred, if expense was involved, to get others to share and even to shoulder the risk and the burden. The more interests involved, of course, the more secure a project both from a financial and a political standpoint. If they could be given a common administrative direction, so much the better. Howe's perception of the short term therefore surpassed that of the average politician, and by dint of special circumstances he was able to do more than even the above-average. Those special circumstances meant that in the case of atomic energy the politician could turn an occasional ear to the song of the prophet.

But the prophet was not the only one who counted. Decisions on atomic energy in Canada between 1953 and 1957 were made by essentially three men. Howe was one, and the most important one. In the triumvirate the minister stood first. The second was C.J. Mackenzie. Mackenzie was on his way out, awaiting his sixty-fifth birthday in 1953, and 1953 was too soon for any new developments to be brought to the sticking point. It was Mackenzie's successor, W.J. Bennett, who would have to recommend what direction AECL's future should take.

W.J. Bennett

Bennett was not a scientist nor even an engineer. This did not make him unique in his generation, or among the directors of atomic projects. Neither the UK director, Edwin Plowden, nor the American, Lewis Strauss, had any special scientific or technical background.

Bennett was a railwayman's son who hailed from Schreiber, in northern Ontario. After high school in the Lakehead, he attended the University of Toronto, taking a degree in philosophy, English, and history. He liked to say that philosophy had taught him how to think, history had taught him how to remember, and English had taught him how to express his thoughts and memories in lucid fashion. He was planning to enter law school when his finances permitted, and was working to that end, when he came to C.D. Howe's attention during the 1935 election. When Howe was made a minister, he summoned the young man to Ottawa. Aged twenty-three, Bennett became Howe's private secretary.

Private secretaries had many duties; Bennett's included managing Howe's political backyard. When Howe became minister of munitions and supply there was less time for that. Administration and liaison bulked larger, fending off mad inventors and dealing with as many problems as he could spare the minister, whose time was an increasingly scarce commodity. Bennett therefore got to know many of Howe's 'boys,' old and young. The war was an education in the personalities and practices of Canadian business. It was a supplement to his knowledge of how the Canadian government was run, not merely at the political level but lower down, by civil servants mostly Bennett's age and with a similar background to his own. At the end of the war Bennett decided to strike out for the private sector; Howe dissuaded him with one temporary emergency after another. The greatest emergency was Eldorado, whose president had, in Howe's eyes, proven his incompetence and whose administration was horrendous.

Bennett accepted a temporary assignment to Eldorado, as vice-president and managing director. The next year, 1947, he was president; the year after that, the company was back on a sound financial footing and, with the Cold War in full swing, there was a market for as much uranium as Canada could produce, at whatever price Canada could justify. Between 1948 and 1953 a series of discoveries gave Canada a vast reserve of uranium ore, while a series of contracts with the uranium-starved AEC guaranteed Canadian uranium producers a healthy and risk-free return on anything they could produce.

Bennett had responsibility for the conception and realization of a uranium policy that proved to be very much to Howe's taste. It did not work without help, which proved to be available from a hard-working expert Board of Directors drawn largely from the private mining sector. This too was a Howe characteristic, and it pointed up the minister's belief that a government project should not exist in isolation from the interests and expertise that Canada had to offer. By extension, one might suppose, if those interests proved insufficient, so too might be the minister's interest in sustaining a given project.

Bennett understood Howe's approach. Better, he shared it, a fact that the minister appreciated. It made Bennett a useful trouble-shooter, someone capable of turning incoherence into policy, and sometimes failure into success. Where success was not possible, Howe knew that Bennett would cut his, and the government's, losses, undeterred by misplaced sentiment. Bennett was, in that sense, closer to his minister than was the man he replaced at AECL, for he was not closely identified with 'pure' science as Mackenzie had been. That might (and did) seem a disadvantage in the eyes of Chalk River's scientists and visionaries, but to the extent that Bennett's scepticism reassured his minister, it would redound to Chalk River's advantage.

The first question before Bennett was whether AECL should continue to exist in its present form. It had two defects, as far as Howe's policy was concerned. It was not self-sustaining and it was not independent. It was not self-sustaining because it did not make enough money to pay its own way. Even the sale of isotopes and the rental of reactor space to the Americans did not turn a profit, while the research operation was dependent on a parliamentary vote of funds – two votes, in fact, each susceptible to debate in the House of Commons.[2]

AECL's dependence on the government for its daily bread irritated Howe. He contrasted the research company's situation with the happy circumstances of his other atomic operation, Eldorado, which was netting an increasingly healthy annual profit, with dividends to the shareholder, the crown. Howe had no illusions about AECL's profitability, but he was inclined to wonder, not urgently but persistently, whether merging Eldorado with Chalk River might not simplify matters by creating a single, healthy balance sheet. But although the idea intrigued Howe, it did not interest Bennett. His first job as AECL's president would be to frustrate one of Howe's less strongly held preferences.

By the time Bennett came to AECL he had served on the AECB for five years; he was therefore familiar, in a general way, with atomic energy and with the most important features of Chalk River's atomic program. Because of his contacts with the Americans he also had a sense of the size and urgency of the Americans' atomic program, not to mention contacts inside the AEC. Bennett was also a founding member of AECL's Board of Directors, which gave him an opportunity to cast his eye over Chalk River and to meet its principal personalities, including its scientific director W.B. Lewis.

If Bennett was the conduit, the man who knew how to put things together and make sure they connected at the Ottawa end, Lewis was the catalyst. If anyone could communicate a vision of the future it must be Lewis, but Lewis's talents did not include an ability to make the future happen. As Mackenzie's vice-president for scientific affairs, he gave the concept of atomic energy in Canada a content and a direction that concentrated its appeal and added a missionary resonance to it. By himself, or left to the resources that Chalk River could muster, Lewis would probably have failed, though possibly only after some hundreds of millions of dollars had been spent.

It was necessary for Bennett to understand what Lewis was about. The two men met frequently, in Ottawa and in Chalk River. After hours they relaxed with Lewis's high-fidelity sound system, if they happened to be in Deep River; Bennett moved up from Ottawa every year to enjoy the woods and water with his large family and, not coincidentally, to keep tabs on what was happening at Chalk River. Bennett shared Lewis's taste for music; but better than the scientist he understood his Ottawa audience. As luck would have it, Lewis had been given an intermediary, who could interpret his scientific sounds and filter them for distant ears above; like any good sound system his interpreter took out sounds he judged discordant, and arranged the rest in a coherent symphony that could be played at the Treasury Board and the Department of Finance; it helped that the impresario was C.D. Howe.

An Atomic Power Proposal

There were two tasks facing AECL's administrators in 1952–3. The first, and the easier, was coping with the company's form. The second was to discover and formulate the company's purpose.

The first task we have already encountered: whether AECL should be merged into an atomic conglomerate with Eldorado, a possibility that had occurred to C.D. Howe. Howe even mused on the idea in public; Bennett successfully derailed it. What Bennett wanted was an amendment to the Atomic Energy Control Act, and in June 1954 he got it. Forgotten was the 'holding company'; instead, AECL was brought out from under the AECB, now headed by Mackenzie, and authorized to report directly to its minister rather than to the board. This emancipation of the company bore

Bennett's personal stamp; Howe's unflattering remarks about the board – it might disappear entirely in a few years' time and in the meantime would revert exclusively to 'control' functions – reflected the minister's sceptical view of committees and regulatory agencies even when, as in the case of the AECB, they were headed by someone regarded as one of his 'boys.'[3]

This development was barely noticed at Chalk River, where most of the staff regarded politics and the distant head office with incomprehension if not disdain. For Chalk River the important change had come two years earlier, when NRC departed and AECL arrived. Now it was time to make another decision, and for once the scientists' timetable matched the company's.

The second item was, essentially, what the company would do to justify its existence. It was now sailing atop the waves of political prejudice, and it needed something more than the temporary, artificial harbour of pure research to shelter in. Although Canadians liked the idea of the disinterested scientist working selflessly in a laboratory, they did not really see why they should pay for it. They might not object for the moment, for the memory of the scientific triumphs of the Second World War was still relatively fresh and the possibility of a new war none too distant. But as these two factors receded they would need some new reason for supporting science, and especially the country's single most expensive scientific establishment, the Chalk River laboratory.

That establishment had admittedly not been doing badly. It had a famous reactor, and it would soon have another. It would learn from that one, and was already doing so, in terms of design and construction. It might learn more when NRU was finished, but that would take time. Events outside Canada were moving on. There were rumours of power reactors: more than rumours, as they knew from the experiments the laboratory was running for the Americans. The rumours were useful, for they helped to provide a necessary justification for a further effort. Chalk River's morale might require some incremental change, especially in an effort to keep abreast of the next stage of reactor development. Perhaps a small power reactor should be the next stage. It need not be a full reactor; a pilot plant would do to serve as a training ground for a new generation of nuclear scientists and engineers. It was clear that

Chalk River would not be allowed to grow simply because that was good for the laboratory; but if power reactors were the object, then opportunity became possible. For such a reactor more money, and more personnel, might be available.[4] To get them, however, Chalk River needed a further mandate. To use a later bureaucratic term, Chalk River needed to be 'tasked.'

Coincidences do occur. Although as we have seen the establishment of AECL as a separate company had much to do with its budgets and its size and relatively little to do with the state of its objectives, those objectives had been quietly maturing in the years since 1946. By 1952 they had reached a stage where they could be actively considered, and AECL's new embodiment offered a forum for such consideration.

The objectives were not purely Canadian. Scientists were well aware that nuclear technology could, optimally, be used for power. The trouble was defining, and creating, such conditions when they were, and must be, dependent on variables that they could not hope to control. To achieve them they must move the familiar terrain of science into the bog of economic prediction, a task for which they were not well suited. But it had to be tried, and when Lewis came to Canada he brought with him one of the first examples. An official in the British Ministry of Supply had set down his thoughts on 'heat and power from nuclear energy.' His conclusions formed the basis of discussion at Chalk River in the winter of 1946–7 and were reflected in Lewis's first Director's Report the following March.[5]

Lewis presented his readers with a blunt comparison between nuclear power and alternative forms of energy. Coal he described as 'easy', with 'fair ... conversion to mechanical [energy]'; oil was 'very easy'; and hydrostatic (what is commonly called simply hydro) was susceptible to 'efficient conversion to electrical [energy] by turbo-generators but only near the place of storage, and involving some capital outlay.' Nuclear fissile matter, on the contrary, was 'very difficult' to convert into energy. 'Involves health precautions and protective structures measuring several feet and weighing several tons,' Lewis minuted. 'Conversion from heat to mechanical [energy] is further complicated by chemical effects.'

That said, nuclear power had one great advantage. It had about a thousand million kilowatt-days of stored energy per ton of fissile material, far more than a ton of coal or oil.[6] It had other advantages. Unlike hydro power, it was not necessarily aquatic; it could be 'made available anywhere.' Admittedly the difficulties associated with generating nuclear power, principally size and expense, made it impractical to service transport units such as trains and trucks (but not large ships), or to site nuclear piles in isolated communities with small demand. But, Lewis concluded, 'nuclear power may be expected to displace coal and oil primarily in large fixed installations particularly electric generating stations in areas where water power is not available.'

The reactor Lewis envisioned was a breeder type, probably using thorium to produce uranium-233. Given that assumption, he asserted that with coal at $10 a ton it would still be profitable to spend $30 *million* a ton on thorium; 'even if 98% of the thorium were wasted it would still be profitable to operate with a total expenditure up to $600,000 per ton or $300 per lb. of thorium. Most of the expenditure would be in the repeated chemical processing necessary for removing the fission products and recovering the Uranium 233.'[7]

Breeders were, however, a remote possibility. A senior scientific committee estimated in December 1946 that feasible nuclear power would not be available for a generation and that it would not replace conventional power until the late 1970s, ten years more than the first British prediction. Chalk River had other things on its agenda and, while the breeder remained a subject of ultimate fascination, it did not become an object of immediate desire.[8]

By the same token, Chalk River did not settle on any single pile design. That, Lewis reported in September 1947, was a matter of 'uncertainty.' The uncertainty was sufficient to justify several lines of applied research having to do with fissile materials, moderators, coolants, and fission products from the pile. Not all the research could be immediately undertaken, and its direction veered from time to time as the Future Systems Group – whose title accurately describes its function – determined.[9]

The concentration of effort on NRU during 1949 and 1950 absorbed most of the practical energies of the future systems members; but it is interesting that the early proposals for NRU included the ideas of 'energy as usable heat,' and 'energy as electric power.' Neither idea could be financially justified, and NRU emerged without applied trimmings.[10] The breeder, Lewis told the parliamentarians in November 1949, still looked like the best bet, as it had in 1946. It still remained a distant prospect, however, Lewis suggested to another audience, because its prospective parents, Britain and the United States (he included Canada for balance and, possibly, local relevance), had spent too much time worrying about the production of plutonium and too little on research.[11]

That looked like the re-emergence of the old pattern of reliance on Britain and the United States for guidance, although Lewis stressed that there was no substitute for independent and 'venturesome' research. What direction would such research take? Chalk River had considered a variety of reactors, but its experience was all with heavy water. It might prove to be technically too difficult to tackle another design, especially once NRU was well underway. George Laurence, whose relations with Lewis made up in persistence for what they lacked in warmth, suggested in March 1950 that it was time to abandon the breeder and turn to designing a different, non-breeder power plant, 'suitable for Canadian requirements,' and not those of Britain and the United States. The true line of development was the one proposed for NRU, natural uranium and heavy water, a line Laurence would consistently uphold. At the time he wrote, the decision to finance NRU was still hanging in the balance, and the suggestion that it might be both a research reactor and a prototype power reactor therefore came conveniently; but Laurence's argument is too close to what later developed to be ignored.[12]

We cannot know whether Laurence's argument had any great impact on Lewis, but by the summer of 1951 the laboratory director had indeed altered his focus. At the end of August Lewis produced DR-18, 'An Atomic Power Proposal.' What he proposed was 'a heavy-water natural uranium reactor pressurized to over 1500 lb. per square inch and with the coolant heavy water heated to 550°F (288°C). The total amount of heavy water would be about 60 tons and the uranium charge about 15 tons.' The reactor would produce 400 megawatts (thermal).

Lewis bolstered his preferred design with some speculative comparisons with the cost of coal-fired plants, although he admitted that his figures for the capital cost of a uranium reactor were infirm and would therefore require to be 'examined further.' Those figures, $14 million for the reactor proper plus $6 million for the heavy water, must be supplemented with the annual cost of uranium supplies, which might vary between $10 and $25 per pound, given Lewis's expectation that the price of uranium would rise (as, in the short term, it did).

This did not compare well with the capital cost of a coal-fired plant. The only thing that made the comparison worthwhile, Lewis argued, was the discovery that large amounts of heat could be extracted from uranium metal in NRX *without any reprocessing*. Reprocessing had been an ominous and constant cost in previous estimates; now it appeared that reprocessing could take a more limited, and less costly, form. The effect, in Lewis's mind, was to make uranium about a quarter as expensive as coal. Moreover, the NRU design showed that constant operation was possible (he presumably meant on-line refuelling), with consequent significant savings, while new developments in metallurgy suggested that temperatures as high as 550°F could be achieved by substituting zirconium sheathing for aluminum. Given aluminum's propensity to form an alloy with uranium at high temperatures, this was an important development. Equally important, though he did not need to stress the fact to an audience of scientists and engineers, the temperature could be kept high by keeping the heavy water under pressure, turning the calandria into a giant stainless steel pressure cooker or, as the term was, 'pressure vessel.'[13]

Lewis expanded his thoughts in a supplementary paper, DR-20. These two papers attracted the sympathy and qualified support of Christopher Hinton, who from the United Kingdom followed the design and development of reactors with acute interest. Hinton approved the idea of a pressure vessel, and using heavy water as both moderator and coolant: 'the best possible thermal reactor is the one you are proposing,' he wrote to Lewis in February 1952. If all went well, it should be possible, using the Canadian scheme, 'to be generating power from atomic energy in five years.' But Hinton's faith in the desirability of the breeder reactor was not easily shaken. It was still 'the real long term solution to the problem of producing power from atomic energy,' and the British were even then starting design studies in the area. In the meantime, he asked to be kept informed of Canadian progress in what he termed 'the most promising line in thermal reactors.'[14]

Work continued at Chalk River. Lewis seems to have become increasingly convinced that 'the Canadian line' – heavy-water moderated reactors – was the appropriate one. The line was not yet more closely defined. To assist in the definition, 'future systems' were confided to two groups, Nuclear Physics and Reactor Engineering, and Metallurgy and Chemical Processing. In July 1952 Lewis convened the Nuclear Physics group, which included Laurence, Donald Hurst, H. Carmichael, and Ian MacKay, among others. It prepared a shopping list for presentation to the newly constituted Board of Directors. At the top of the list was a 500 megawatt (thermal) reactor, heavy-water-moderated and using natural uranium; but such a reactor should not mean abandoning the possibility of enriching the uranium or of using liquid metal as a coolant, should that prove more promising than heavy water. Only afterwards, as second priority, did they list a fast fission breeder reactor of the kind that continued to preoccupy the British.[15]

The chemists had a rather different agenda, reflecting the priority Lewis ascribed to producing remunerative plutonium from NRU. Power reactors were not, however, forgotten. They had before them a British plan for a natural-uranium, graphite-moderated reactor, as well as the Chalk River plan for a heavy-water system, in which there would be two products, power and plutonium. Interest in plutonium as a by-product died hard; a less efficient reactor could always be paid for through plutonium sales along the lines of the NRU contract with the Americans. The chemists continued to be intrigued with the possibilities of what was called a 'recycling power reactor,' a high-temperature reactor with a core of plutonium or uranium-233, surrounded by thorium-232 or uranium-238, which because of its high temperatures would have to be cooled by liquid metal.[16]

Some of these propositions were regarded as very long-term. Lewis warned the group studying fast fission reactors that there was no prospect of such a reactor for a long time and that they must design their experiments accordingly.[17] But some, and especially the heavy-water power reactor, were ready for the next stage. At the beginning of August 1952 Lewis got ready to present his shopping list to C.J. Mackenzie and the Board of Directors.

Future Power Piles

Dr Mackenzie came to Chalk River on 12 August 1952 to attend a panel meeting on 'future power piles.' It was, he wrote in his diary, 'very interesting. Lots of questions still unanswered and problems to be solved, but we have an excellent group working on these systems.'[18]

What happened next is clear enough; but how it happened, and why, remains something of a mystery. At the beginning of September Mackenzie saw Howe. He told the minister that he had invited 'Hydro engineers to form a team to work immediately on power reactors.' Howe's reaction, the AECL president recorded, was favourable, but there our information stops.[19] It seems reasonable to conclude that Mackenzie did not need very much convincing. The year before he had addressed the Canadian Club of Montreal on 'atomic energy as a potential source of power,' and although he refrained from saying that it was imminent, he left no doubt that he was in favour. The speech attracted some notice, especially in Toronto, in the offices of the provincial electrical utility.

Mackenzie did not formally consult his Board of Directors as to what should be done. That probably did not matter very much, for some of the people he needed to help out with his proposed power reactor team were themselves already on the board, and if he did not seek them out in one capacity he presumably consulted them in the other. When the board was selected, Howe chose four representatives from Canada's electrical utilities, Dupuis, Gaherty, Massue, and Hearn, three from Montreal-based utilities, and the fourth, Hearn, from Ontario Hydro. They in turn had little doubt that they had been chosen because of the likely connection between nuclear energy and electric power. Gaherty was with the Montreal Engineering Company; but he was also president of Calgary Power and associated, through the Killam interests, with electricity in the maritime provinces. Hearn was chief engineer at the Ontario

Hydro Electric Power Commission (the formal designation: we shall use Ontario Hydro), an influential post in the provincially owned utility whose chairman, Robert Saunders, was better known as a politician than as an administrator.

Both Hearn and Saunders were attracted by the idea of nuclear-generated electricity. Ontario was running out of hydro capacity, and in the late 1940s was having to turn to coal-fired thermal generating stations. Coal had to be imported from the United States – not far, but expensive, foreign, and possibly, in a crisis, unavailable. Coal stations were not expensive to build, unlike hydro and, as it turned out, nuclear plants, but they cost a lot to fuel, also unlike hydro and nuclear stations. It was worth considering other forms of thermal generation if they promised any reduction in costs. There was not, they found, much information to be had from Chalk River, but there were occasional reports from the United States that suggested that atomic power generation might be an eventual alternative.[20]

"We study everything,' the Hydro chairman told the *Globe and Mail* in March 1950, but for the moment the commission would continue building its new thermal generating plants to burn coal or oil. Ontario Hydro's research director, W.P. (Percy) Dobson, was told to be ready for a new assignment, only to be frustrated by the secrecy surrounding atomic research. But there were others outside the charmed circle who shared his interest, including representatives from both Canadian General Electric and Canadian Westinghouse. The best they could discover, Dobson reported, was that the Americans were building something, but refused to say where or how big. Dobson could merely tell Ontario Hydro's executive committee that a trend had been established, and that he was waiting to see what it was. It did not seem too urgent in any case; technical problems abounded, nobody seemed to know what to do with nuclear wastes, and cost estimates were guesswork at best. But the Americans were beginning to stir; task forces from the Atomic Energy Commission and private industry were being set up south of the border, and there was a danger that Canada would be left behind.[21]

Under the circumstances, Dobson reported in September 1951, it might be time to make an approach to Keys and Mackenzie in order to get closer to the real sources of information. Perhaps Canada could follow the American model and bring provincial utilities, private manufacturers, and federal science closer together. And so, in January 1952, Dobson and Hearn headed for Ottawa to converse with Mackenzie. It was a propitious time, and either during the meeting or shortly thereafter Mackenzie invited Hearn to join the board of his new company.[22]

It was an invitation that pricked Hearn's curiosity, and an opportunity that he would prefer to take; but he had been long enough in the game to keep his administrative fences mended. He needed permission from his superior, chairman Robert Saunders, but he needed a higher sanction too. Premier Leslie Frost, a Conservative, was fortunately on good terms with C.D. Howe, AECL's Liberal minister. He was happy to meet with Howe in January 1952; also present, Hearn later recollected, were Mackenzie and Hearn himself. The subject was co-operation between the yet unborn AECL and Ontario Hydro, and the object was the creation of a nuclear powered electrical system for Ontario. Frost and Howe agreed that it was a good idea, and if it proved feasible something would be done; in the meantime, Hearn could join the AECL board with the approval and blessing of both the minister and the premier.[23]

If Ontario Hydro was interested in AECL, the reverse was also true. As Mackenzie began his conversations with Hearn and, on a different plane, Saunders, Laurence was writing from Chalk River to Hearn to ask if he could send down some of the project's engineers to consult the Hydro staff about turbines, steam generators, and the like. Hearn responded positively.[24]

The next development was the decision by a meeting under Lewis's chairmanship of the Nuclear Physicists and Reactor Engineering Group at Chalk River that it might well be time to start discussions with Ontario Hydro. If a power reactor were to be ready by the end of the decade, the engineering should start soon. The best way to do that, the meeting decided, was to ask Hydro to send up a small group of engineers to Chalk River. There, inside the wire and cleared for secret information, they could learn about atomic energy and join in the early design work for the projected power reactor.[25]

Both sides were agreeable, but it would take time, an interlude that was used by the Chalk River staff to firm up what they wanted. It should be a 'no frills' reactor, Ian MacKay wrote on 19 August, with a specific power output. Perhaps it should be considered as a peak power station, which would permit it to be shut down on weekends for refuelling and maintenance. It should use natural uranium fuel, at least at first, until it was demonstrated whether this was the best kind; it should be used for the longest possible time, even at the risk of producing a high, and possibly unsaleable, plutonium isotope. Lengthy irradiation of plutonium resulted in a less than ideal product from the point of view of the potential customer, the AEC.[26]

Although Lewis did not accept MacKay's conclusions about plutonium – they were admittedly speculative – he was nevertheless thinking along roughly the same lines. And so he told the Board of Directors when they convened at Chalk River on 30 September. He had a list of seven projects he wanted considered, ranging from 'a natural uranium–heavy water power and plutonium producing reactor of 400 MW' to 'exploring more actively the possible methods of producing heavy water.' His program would be costly, Lewis admitted, and would mean more people on the payroll, the expansion of Deep River, and 'new links with engineering industrial firms.'[27]

The board took the news well. Subject to the money becoming available, something that depended on the government, Lewis should proceed. The immediate requirements, such as the hiring of graduates of the class of 1953, could be met. The following month, Lewis and Mackenzie set out for England, where Mackenzie was to visit various atomic sites and make inquiries. The subjects included the availability and suitability of British industrial firms for Canadian needs. The result of those inquiries, the AECL president told his board in December, was to confirm his view, and Lewis's, 'as to the soundness of the plans being made for research and development on the Canadian project.'[28]

What plans? As of December 1952 there was an idea, cautiously held by Lewis, that a natural-uranium, heavy-water-moderated reactor was the best possible choice. The best way to proceed was not to expand Chalk River but to involve industry, by which was meant the most interested consumer – Ontario's publicly owned electric utility.

The first start with Ontario Hydro was a false one. A team was sent to Chalk River from Toronto. They busied themselves around the laboratory, listened to lectures, and read what was available. They concluded, to Lewis's dismay, that a homogeneous reactor (the slurry type that dropped in and out of reactor designs between NRX and NRU) was the ticket. They reported their conclusion to Toronto, and received a bemused reaction.[29]

The documentation of what happened next is surprisingly sparse. Early in 1953 W.P. Dobson was appointed Ontario Hydro liaison officer with AECL. A Hydro engineer was given a security clearance and sent up to Chalk River to learn what he could of the ways of the atomic laboratory; a small team would follow. But nothing concrete occurred until the summer was far advanced. AECL had other, more immediate priorities: salvaging NRX after the December 1952 accident, and supervising the agonizingly slow evolution of NRU.[30]

Lewis did not count the time wasted, and scheduled a Chalk River symposium for early September, with the purpose of bringing his 1951 proposals up to date. The discussions ranged widely, from the properties of zirconium – promising – through the latest applications of neutron physics to reactor design to the economics of fuel and power. Not least, they should discuss 'power systems planning for Ontario,' which would be followed by comments from invited electrical power representatives. To them Lewis would say that while it cost 30 cents to derive a million BTU (British Thermal Units) from coal, he calculated that it would cost only 20 cents, or thereabouts, to do the same from uranium metal. With world energy consumption steadily increasing, he foresaw a day when nuclear energy would be the best if not the only variety available.

What Lewis wanted, he said, was 'to construct a small heavy water power reactor for which the capital cost may be relatively high for its power output but the operating costs reasonable.' The high cost could be written off to development, and the cost of the operation of the reactor written off to experience. If the reactor were sited so as to compete with existing inadequate or costly power, it might even make some money. But until something was done to realize his dream, all this remained speculation.[31]

The great event of the summer, therefore, was an anniversary: C.J. Mackenzie's sixty-fifth birthday, an event which effectively removed him from the presidency of the company as soon as the board could get around to ratifying Howe's choice of his successor. That choice had long since been determined: W.J. Bennett. It is not known whether Mackenzie approved the choice, although it was known that relations between the two men were not close. But they had known each other since Mackenzie's arrival in Ottawa in 1939, and they had always been able to work together. They would be together still, since Mackenzie retained his job as chairman of the AECB, and signed on as a consultant to AECL at $10,000 a year (he resigned from the Board of Directors).

Bennett's inheritance was, as far as policy was concerned, somewhat inconclusive. What should happen had been determined some time ago, but the way of making it happen had not. Public pressure was building up as reports reached the press from the United States and Great Britain about progress being made towards nuclear power in those two countries; opinion was becoming receptive to a properly Canadian power program. But, apart from some tentative approaches from Mackenzie to Ontario Hydro and to British industry, nothing had been done. Any formal study would cost money; some of the money would have to come from Hydro. Delay was inevitable while Hydro contemplated its options.

Mackenzie's approach to the British indicates the direction his mind was taking. When he visited Britain in the fall of 1953, his discussions at Harwell and elsewhere in the United Kingdom were more than academic; interestingly, he may also have been speaking on behalf of Ontario Hydro, which had strong links to British engineering firms and which was thinking of involving companies already producing for that country's own atomic project. The British derived the impression that Mackenzie had in mind turning design over to a British firm. The strongly affirmative British reaction suggests that Mackenzie's approach, which was made during the autumn of 1953, was more than a straw in the wind.[32]

Bennett was not committed to Mackenzie's viewpoint; he knew

little of his British gambit. He was committed to changing things. Mackenzie, he thought, was slapdash, if not actively disorganized. So Bennett began by overhauling AECL's administration, establishing, as in Eldorado, an executive committee of the board. At first the committee consisted of Gaherty and Hearn; later Andy Gordon, the head of the University of Toronto's chemistry department and a director, sat in on some meetings. Donald Campbell, the treasurer and later corporate secretary, Lewis, and Lorne Gray often attended. The executive committee did not greatly diminish the board's role, already negligible under Mackenzie; but it did create a cabinet system on which Bennett (who remained in charge of Eldorado, and ran AECL's operations from his Eldorado office) could rely. For Bennett it afforded instant expertise on a senior level.[33]

Simultaneously Bennett moved to reorganize Chalk River. Keys, at sixty-three, was made 'scientific adviser to the president.' He could move into a gentle and dignified semi-retirement, interpreting Chalk River to the outside world. His vice-presidential job was formally divided between Lewis, vice-president for research, a job he was already doing, and Lorne Gray, who in 1954 was made vice-president for administration and operations. Lewis had no interest in that end of Chalk River, so the new arrangement, as Bennett recalled, worked well for all concerned.[34]

The first priority was arranging matters so that the engineering study could go ahead. In November Hearn provided $200,000 for a two-year feasibility study, and in December the arrangement was ratified by AECL's Board of Directors. Lewis was asked to draw up the necessary qualifications for the staff, and Hearn undertook to do the rest. His principal engineer, he understood, would be joint head. This may seem trifling, but it was of great importance to Hearn and to Hydro. Hearn was known to complain of 'general weakness at Chalk River on the engineering side.' Moving in one of his own meant that he would have greater confidence in what the feasibility study produced. The man chosen, Harold Smith, Hearn regarded as his 'best and brightest.'[35]

A group was constituted to conduct feasibility studies for the design of a pilot nuclear power plant. It consisted principally of staff from Chalk River, Ontario Hydro, and Montreal Engineering. It should report in about eight months. Provided its report was satisfactory, construction of the pilot plant could begin in the spring of 1955.[36]

The decision to constitute the feasibility study would prove to be extremely important, but it was not the only one, and it had ramifications. Because of the involvement with Ontario Hydro, that company's priorities and budgets became a necessary part of AECL's schedules and preferences. The study did not involve the C.D. Howe Company, whose future in the nuclear business was to prove extremely limited. Finally, the study did not involve any other country, either through its atomic energy authority or through its industry, in a major way. The umbilical cord to Britain was thereby severed.

That was certainly not the original intention. Under Mackenzie, relations with the British had continued close. In the fall of 1953 he even suggested a link between British reactor design and manufacture and AECL. When Lewis toured Britain in May 1954 he confirmed Mackenzie's general line, telling the newly constituted Atomic Energy Authority that AECL 'were considering placing an order with one of the big British heavy electrical firms,' and inquiring 'whether the Atomic Energy Authority ... would be willing to do the necessary supervising of the work of manufacture in this country.'[37]

Hearn confirmed Lewis's suggestion in a meeting at Risley on 12 May. 'He is strongly of the opinion,' the British minutes record, 'that the design work should be a collaborative effort between Canada and the U.K. and that it would be deplorable if Ontario Hydro were forced into the position of having to introduce an American firm to collaborate with them.' The British were understandably excited. This was big business at a time when the British government was strongly encouraging precisely this kind of high technology, lucrative export. Even if they had to divert their research effort, they told Lewis and Hearn, they would be happy to co-operate. In a third meeting, on 20 May, they cautiously asked the two Canadians to refer their proposal back to Ontario Hydro to ensure that it had their support; presuming that the response was favourable, 'the U.K. Atomic Energy Authority would consider the best way of redeploying their resources so as to meet the needs of the Ontario Hydro project.'[38]

When Plowden toured Canada in June he took up the subject with Bennett. Bennett was markedly less enthusiastic. He told the AEA chairman that he was 'interested' in collaboration, but that the British must understand that, 'for obvious national reasons, [he would] need to build up a Canadian engineering team who would eventually be able to design reactors in Canada.' Plowden nevertheless continued to hope; as late as mid-September he wrote that Risley and Harwell would be 'deeply involved in this project.' Final arrangements awaited his meeting with Bennett at the end of the month.[39]

The Canadian visit, when it occurred, had a sobering effect. 'It was clear,' Hinton stated on 23 September, 'that [the Canadians] had completely changed their plan.' They now intended to do the design themselves, using their own team. What had happened? It was possible that the minister, Howe, had interposed. His opinion of the British was always a variable commodity and in 1954 he was annoyed with them over the negotiation of the International Wheat Agreement and over uranium purchases. Howe favoured Canadian industry and Canadian technology, when he could; and it is not unreasonable to suggest that he decided to apply a Canada-first principle to the building of Canada's first power reactor.[40]

The proposed collaboration with the British limped on, but the life was slowly draining out of it. A study team would be sent to Canada the following spring (its arrival was postponed because the Canadians were not ready for it) but its composition was reduced and its purpose most unclear. Even the possibility of a major industrial advantage was lost when, in the spring of 1955, the Canadians announced that the contract to build a prototype reactor had gone to a Canadian firm. Under the circumstances, Hinton lamented, he saw little point in devoting scarce British resources to an academic exercise with the Canadians.[41]

But that is to anticipate. For the moment, Bennett and his directors had taken only the first step. But it was a first step in a constant direction. It is therefore important to see how that direction was defined, and why.

The trend towards a heavy-water, natural-uranium design was well established by 1953. Bennett, as president of a uranium company, knew that Canadian resources were great and growing;

in October 1953 it was estimated that Canada's uranium production would quadruple within two years, and rise by another 450 per cent by 1958. Much of the production would come from Ontario, where large deposits had just been staked.[42]

Despite the obvious coincidence, it is difficult to establish a direct connection between the uranium boom and the natural-uranium, heavy-water reactor. George Bateman, who was a member of the AECB and a confidant of both Bennett and Howe, argued late in 1953 that Ontario's power needs dictated the speedy development and adoption of atomic power in that province. To Bateman this suggested a need for a clear, co-ordinated policy on uranium and other forms of nuclear activity; but although he intended to have 'several of us ... get out heads together and develop specific proposals to lay before the Minister,' there is no indication that this was ever done. Certainly no overall policy was adopted. Howe might not have appreciated the effort to urge him to develop one, while Bennett did not welcome interference in his company's affairs, even from Bateman.[43]

The next question was whether to involve a utility. This proved to be controversial. René Dupuis, who sat on both AECL's and Quebec Hydro's boards, resisted any such idea; his was a watching brief, not a licence to participate. It was very simple, he told Bennett. Hydroelectricity was a provincial matter. Quebec Hydro, and by extension the government of Quebec, had no objection to the government of Canada dabbling in atomic energy. If AECL wanted to go ahead and build its own demonstration power reactor, he and Quebec Hydro would approve. But an alliance with Ontario Hydro would not do. He would have to resign.

Bennett tried to placate Dupuis: Ontario's involvement could be twinned with Quebec's. Dupuis disagreed. The federal government had no business in a provincial field, and a line had to be drawn. Scientific experiments were one thing, but generating power in co-operation with a province quite another. When, on 17 December 1953, the board approved the joint feasibility study of power reactors with Ontario Hydro – a meeting Dupuis did not attend – the Quebec Hydro director resigned. Although Howe wrote to Quebec's premier Maurice Duplessis to praise Dupuis's services and to ask for a replacement, no reply was ever received.[44]

With participation from private firms, Ontario Hydro and AECL proceeded with their study. Hydro wanted to establish a single, simple design and stick to it. As Lewis explained in May 1954, 'Ontario Hydro were anxious not to have to deal with many different kinds of reactor in the course of their atomic power developments.' They were persuaded, as Lewis himself was, that the heavy-water model would be 'the most economical,' and that therefore it made most sense to follow that line. Now it was up to the study team, dubbed the Power Reactor Group, to prove that a reactor could be designed that was competitive with coal at $8 a ton. When that happened, the reactor 'would be accepted.'[45]

To the study team the contributors named their most promising engineers. Montreal Engineering, Gaherty's company, sent John Foster, their brightest young man. Hearn's choice, Dr Harold Smith, we have already met; he had managed the conversion of southern Ontario from 25-cycle power to the standard 60-cycle. AECL contributed Ian MacKay, and others besides. They would try to come up with a power reactor.

Ontario Hydro was not at first committed to a demonstration reactor; it was the 'feasibility of competitive nuclear power' that was at issue. Lewis urged speed, and speed, to his mind, meant a small prototype. In July 1954 Hearn conceded the point: a very small reactor first, a large one later. 'My impression,' Smith later wrote, 'was that this agreement was reached largely on the basis that haste was of prime importance in the work and that delay involved in training the study group to perform the requirement [the large reactor] was not tolerable.' To the British, Hearn and Lewis explained that 'limited financial resources,' and the lack of a military program off which the civilian power program could feed, meant that 'the establishing of public confidence in the practicability of nuclear power' was 'an essential first stage,' making it 'possible for industry to find the money for a full-scale nuclear reactor programme.'[46]

The demonstration reactor was still waiting on the conclusions of the Power Reactor Group (PRG). By May, it had about twenty members, but there was no doubt that Lewis, who presided, was the key to its operations. He had to bridge the gap between engineers and scientists without causing the engineers to feel that they were regarded as failed scientists, and without making them feel that they had fallen into the hands of unworldly savants. He had to be inspirational as well as intellectual. The engineers got four or five lectures a week, each an hour and a half in length, plus two or three lectures a week on reactor theory. It was, in Percy Dobson's opinion, 'an essential foundation' for the feasibility study that was to follow. As far as most of the outside engineers were concerned, the effort succeeded. 'Excellent,' Smith said at the time; Lewis, John Foster remembered, was 'brilliant': sometimes acerbic, but capable of feats of diplomacy as well as prodigies of explanation. Apart from Lewis and Smith, Ian MacKay was the dominant figure; the group, recalled Foster, worked 'under [his] wing;' but it worked under the direction of Smith.[47]

The PRG turned into the Nuclear Power Group (NPG). NPG's purpose was at first purely exploratory; Lewis, however, treated it as a missionary enterprise. As we have seen, there was no thought that it would evolve into something larger, and that it would produce its own designs. But when Bill Bennett decided, at the end of June, that conversations with the British over design were premature, the NPG began to take on new significance. Instead of sending Canadians to Risley, AECL would rely on home-grown talent. But first the talent had to show that it was willing.[48]

Lewis did not start with any clear design in mind, although heavy water seems always to have been the moderator of choice. 'Lewis,' Hinton wrote in September 1954, 'is attached to the heavy water type with an almost bigotted [sic] devotion,' but whether that was true at an earlier stage, and when it became true, are open questions. He had already excluded the homogeneous reactor that had attracted the first Ontario Hydro mission in 1953. Lewis aimed at completing the study by 1 May 1954, while Smith seems to have had no such deadline in mind.[49]

At the first meeting Lewis outlined the issues as he saw them. The fuel, Lewis said, might be natural, unenriched uranium, or it might be the enriched variety, using plutonium for the purpose. The coolant might be deuterium (Lewis thought heavy water would work better in large reactors), but ordinary water was still possible. As for the reactor core, it might be a pressure vessel – effectively a large pressure cooker – or it might feature pressure tubes, or it might not be pressurized at all. As for access to the core, Lewis thought it should be bi-directional, from both top and bottom, for reasons of economy and for ease of maintenance. Finally, the break-even figure should be a comparable coal-fired plant using coal costing $8.50 per ton – 50 cents more than the figure he gave the British two months later.[50]

1 May came and went, but fundamental questions remained. What was the best sheathing for the fuel elements, aluminum or zirconium? Should the reactor be an extrapolation of NRX? Was a reactor that was merely another version of NRX worth doing? Would the power produced be economical, Smith's principal concern? How best to make heavy water? And what about a new deadline of 1 July?[51]

For the deadline they produced NPG 1, 'A Preliminary Proposal for Heavy Water Concentration in Steam Power Plants,' by John Foster and George Haddeland of Shawinigan Chemicals. It considered how to run a distillation process to produce heavy water in a steam plant. It was the first of a series that they hoped would eventually lead to a formal proposal. But not, it was clear, by 1 July.[52]

The delay sharpened Lewis's proposals. A small reactor made more sense, he announced in mid-July, and was more likely to be funded than a pilot plant. That should have pleased the engineers, who were looking for a firm design with which they could work and on the basis of which they could prepare estimates of time and funding. Lewis now added a larger objective: two big power reactors for which preliminary design studies should begin almost immediately, even as work on a 'small reactor' (labelled SR) proceeded. The SR might be finished by 1958, and its designers and builders might then graduate to the large reactor (LR) project. Yet there was still a nagging doubt, and not only among Lewis's audience. What if, as Lewis put it, 'the type of reactor [changed] with new developments?'

The PRG almost agreed on natural uranium, not to mention a heavy-water moderator. They were not agreed on a coolant, though opinion was trending to heavy water. Just at this point the Americans were consolidating the design of a boiling light-water reactor (BWR); a few months earlier their design might have been enticing. (Lewis did not ignore the idea; in 1955 he tried to interest the British in a boiling heavy-water reactor.) The delay suited Lewis very well; he wanted agreement on a timetable and on his basic concepts, and he got it. 'In *general*,' the PRG's minutes recorded for 14 July 1954, 'there was *reasonable* agreement that the programme was sound and that it should be adopted.'[53]

What program? Not the design program, where uncertainties lingered on rod sheathing, coolant, calandria, enrichment, or the materials to be used in making the pressure vessel. What seems to have been agreed on was a small reactor to overlap with two large reactors (each 100 megawatts [electric]), the former priced at $15 million, the latter at $60 million, the whole to be completed by the end of 1961. It was a large-scale repetition of the building of ZEEP and NRX. As one way of saving money and adding personnel, the United Kingdom might be involved in a second feasibility study on the large reactors. What there was was further developed at a meeting on 19 July; again, agreement was on outlines and magnitudes rather than on details. But agreement there was, and it was agreement on a heavy-water system – Lewis's system, in fact if not in so many words.

Lewis gave the news to the executive committee in a meeting at Chalk River on 26 July. (Interestingly, it was not *called* a meeting of the executive committee and was not recorded in the latter's minutes.) Besides Bennett, Gaherty, and Hearn, Mackenzie, Gray, Laurence, MacKay, and Smith attended. Lewis told his audience he was selling his whole program and not bits and pieces of it. If any major part were omitted, then the consensus it represented would dissolve. Lewis presented the small reactor/large reactor outline. It might, he observed, seem strange to proceed with a small reactor while developing two large ones, but he had a good reason. NRX and NRU would not test necessary control methods or the proper balance between cheapness and reliability in fuel fabrication. At the same time, the true economies of nuclear power using natural uranium could not be demonstrated in a small reactor; only a large one could do that.

Lewis did not argue for natural uranium as the most desirable fuel. A recycling reactor with some enriched fuel was attractive at least to demonstrate economies of production, always assuming that such economies were not distorted by an unexpected rise in the price of plutonium to be sold to the Americans. Answering worries about the well-known shortage of engineers and designers, Lewis argued that the existing team of NRU-trained personnel could cascade from one project to another – to the small reactor first and then on to the large ones.

It would be costly. How costly nobody really knew. The program must be paid for by government. Government must make up its mind quickly, for if it did not and a promising reactor design was not ready by, say, 1959, there was a risk that industrial concerns in the United States would be ready with a better design and a better offer. It was necessary to face up to a reactor development program that would cost $75 million by 1961.[54]

The executive committee reacted cautiously. They would study the matter, and they did. As Bennett wrote in November 1954, 'This problem has engaged the continuous attention of the Management and the Executive Committee of the Board since the month of August.'

First they scheduled meetings with the British and Americans to discover what they were up to, and what, if anything, they thought of the Canadian proposals. With Lewis in tow, Bennett, Gaherty, and Hearn visited Britain; Bennett himself handled meetings with the Americans. There were no surprises. The British and the Americans, like the Canadians, placed their long-term hopes in breeder reactors. Like the Canadians, they did not propose to let these hopes interfere with present prospects, and were ready to sponsor 'short-term' reactor development, the Americans along three parallel lines (pressurized water reactors, or PWRS, boiling-water reactors, or BWRS, and sodium graphite reactors) and the British with what came to be their Magnox type, a military-civilian design under development at Calder Hall in northern England. (The term 'magnox' derives from the magnesium oxide alloy used to sheathe fuel elements. The reactor was otherwise a graphite-moderated, gas-cooled pile.)

A few months later, in February 1955, the British announced a twelve-station nuclear power program at an estimated cost of £300 million ($840 million Canadian) with three types of reactors (some liquid rather than gas cooled, and one 'advanced'). In comparison, Lewis's $75 million was exceedingly modest, an advantage in Bennett's and, ultimately, Howe's eyes. 'The limitation of resources of money and men does not make it possible for Canada to undertake the design and construction of more than one prototype reactor,' Bennett wrote. Diffusion of effort would make it impossible to do any part of the job well. The same applied to the speedy construction of larger reactors. There, the controlling factor may well have been Ontario Hydro which, according to the British (our sole source at present on this matter), had decided 'that they cannot at present find the money for a full-scale reactor.' But if the data accumulated from the operation of NRX and the construction of NRU could be put to work, Canada's fund of knowledge was at least comparable to what the British and the Americans had on any one model. That indicated the direction in which Canada should proceed.

That Canada should proceed was almost axiomatic. 'It is recognized,' the board was told, 'that the Company must maintain its research and development effort, if Canada is to hold her position in this new field.' Besides, it would not cost very much. The British and the Americans would make available information on the lines of research they were undertaking, a very large saving. The British had made a modest commitment to a joint preliminary design group, and would detach staff for the purpose. The small reactor, to which the board committed itself, would cost about $8 million if it were 10 megawatts (electric), or $10.25 million if it were 20 megawatts, the more likely alternative. It would take another $6.2 million to fund the preliminary design study for the large reactor, making a grand total of $16.45 million over the next five years. This money the government was prepared to commit.

Bennett and his colleagues had also examined British and American management practices. The British had concentrated research and development in the hands of the Atomic Energy Authority but were now reaching out to private industry. The Americans, in contrast, had contracted out most of their program, military and civilian, to private industry. Even Oak Ridge was run by a private corporation rather than by the government. Howe and his officials preferred an approach that spread the risk of failure and allowed industry to invest the benefits of success. The executive committee indicated its approval of the American practice. They also preferred to make haste slowly, one project at a time.

So did the board, which authorized a small reactor and stipulated that seven private Canadian companies be asked to submit proposals for its design and construction. AECL would supply them with all necessary data for developing their bids, take responsibility for supplying nuclear fuel and heavy water, and supply 'some key personnel' to the contracting firm. The firms bidding on the contract must also understand that AECL would not underwrite the development of such items as heat exchangers, turbines, and generators; if those did not now exist there would be no time to create them. The spring of 1955 was set as the deadline for starting detailed design work. The prime contractor was to rely on Canadian sources of supply as far as possible. Since developing this new technology would afford a competitive advantage to the winning firm, it should be prepared to contribute to the cost of design and development in light of its anticipated profits. Finally, when the new station was operating, AECL would handle the nuclear side only; anything mechanical would remain the contractor's responsibility.[55]

It had been just four months since Lewis broached his atomic propositions to the board. The board had moved speedily in response. Money was available, a new demonstration power reactor had been authorized, and a long-term program was in prospect. The long-term program, however, looked to *one* large reactor and not, as Lewis would evidently have preferred, to two.[56] The program would be more broadly based than had previously been the case, by involving a private company for both design and construction. It would have greater authority than the Howe Company in the building of NRU; that, Bennett later explained, was because of the endless divisions of opinion between the C.D. Howe Company and AECL, with the subsequent refusal of the Howe Company to accept responsibility for the design of NRU. As Bennett observed to Gaherty, he wanted to avoid any repetition of the problems over NRU. Chalk River's ability to interfere would be circumscribed.[57]

What was not clear, or so it seemed, were the details of the reactor the contractor would design and build, the location of the reactor, and the identity of the contractor. The first item was becoming clearer. The PRG had agreed by the end of January on specifications for the pressure vessel, the deuterium coolant, zirconium or zircaloy sheathing for the fuel rods, and fuel temperatures (1100°F). Agreement was close on the power rating, with Lewis supporting an increase to 20 megawatts (electric) from 10. Better still, there was agreement on what Smith called design principles. First among them was a pledge not to alter the design if it involved extra cost. Other principles were no refuelling under power, and access from the top only.[58]

As design was proceeding, so was the search for a prime contractor. Bennett sent out invitations on 3 December 1954. By the beginning of March the replies were in. Two considerations applied. First, AECL had to evaluate whether a company had the technical wherewithal to live up to its commitments. This criterion eliminated Canadian Vickers, Dominion Bridge, and John Inglis in a preliminary round, and Orenda and Canadair in a subsequent round. Second, Bennett had to assess the total cost of each proposal, including the contribution that each company was willing to make. There, one company stood out because of the size of its proposed contribution: Canadian General Electric, which had offered $2 million towards the cost of research and development. At a meeting on 17 February 1955 the board approved the CGE bid, and on 11 March Bennett recommended it to Howe.[59]

The 17 February meeting of the board was held at the Ritz Carlton in Montreal. There were two principal items on its agenda: the contract with CGE, and the question of which utility to choose as partner. Given Hearn's early interest in the idea and his presence on the board, this might have seemed obvious. But it was not, as Bennett learned. In December 1954 the AECL president got a call from Hearn in Toronto. There was a small problem. Robert Saunders, Hearn's boss, was showing signs of cold feet. Perhaps Ontario's destiny was not nuclear after all. Perhaps Saunders's vanity had been wounded the previous fall when the British had snubbed his attempt to tour British atomic facilities. An interview with Saunders proved unsettling and, as far as Bennett was concerned, very unsatisfactory. For whatever reason, the alliance with Canada's premier utility looked like it was coming unstuck just as the Board of Directors was being asked to ratify it.

Bennett decided to invoke the third member of the executive committee, Geoff Gaherty. Gaherty concocted an alternative proposal from Nova Scotia Light and Power, one of the Killam stable of utilities. A proposal from the maritime utility was waiting for the board as it assembled. Obviously, to Gaherty as well as to Bennett, the Ontario choice would be better, but AECL needed a partner, and if the partner was to be found in Nova Scotia they would live with it.

Because the board was meeting in the morning, Bennett spent the night in Montreal. Hearn, coming from Toronto, took the overnight train. Before starting, he placed a call to Bennett. They arranged for breakfast. At breakfast the news was good. Saunders would back the proposal. Hearn had no time to put it in proper form, but he would do so.

Bennett therefore postponed discussion on the utility question and adjourned the meeting to 10 March at the Royal York in Toronto. When the directors reassembled there were two proposals on the table. One was Gaherty's, which everyone understood to be pro forma; and there was Ontario Hydro's. Hydro's offer had three points. It would pay for the conventional parts of the small reactor. It would buy steam from the boiler at an equivalent price to steam from an emergency plant in Scarborough. It would use its own labour force as far as possible in civil and other conventional work at the plant. The board accepted these conditions, while disclaiming any liability in case the plant failed to work and with the hope that the plant would be built close to Chalk River. It also noted the advantage to be expected should it prove possible to scale up to a 100 megawatt (electric) plant; since Ontario was almost the only conceivable customer for such a plant its sponsorship of the small demonstration reactor was a happy coincidence.[60]

It all happened remarkably fast. This was, in part, because the government, of which AECL was a creature, was strong and stable, and because the company's president stood well with his minister. The Howe system specialized in quick decisions; to spend $16 million (or rather more when the cost of a site, of heavy water, and of various sundry items is added in) was not difficult when it seemed that Canada was in a race for the development of a usable nuclear reactor.

At the same time, the decisions that were taken were cautious. The British and the Americans would get their reactors into service first. The Canadians reflected that if their own design was unsuccessful they could still share the Americans' (or the British) success. They hoped that if the heavy-water demonstration reactor was successful, the British and Americans would share Canada's, in the most practical way: by buying a Canadian reactor system.

First, however, it would be necessary for the St Laurent government to share their enthusiasm.

A Canadian Reactor

The federal cabinet met to consider Canada's proposed reactor program on 23 March 1955. Bennett had circulated a paper explaining what AECL proposed to do, and ministers had read it, or not, as they chose. Eighteen of them, headed by the prime minister, Louis St Laurent, attended for what promised to be the fullest opportunity to date to discuss atomic energy. For atomic energy, while not exactly a stranger to cabinet agenda, had not usually impinged very much on the responsibilities of other departments.

Howe's proposals were never lightly received and, given his standing and seniority, they were deferentially treated. But they were not always approved, especially when the minister of finance had views. For an open-ended commitment like atomic energy, caution was indicated.

Bennett believed in 'putting you in the picture.' The cabinet, via Howe, had the opportunity to read Bennett's script. Atomic energy had been part of the war effort. Though it had obvious military implications, it was always 'expected that atomic energy would eventually be used for peaceful purposes.' AECL now had at its disposal enough data to design a demonstration power reactor. This reactor, fuelled by natural uranium and moderated by heavy water, and producing 20 megawatts (electric), would be ready sometime in 1958. AECL wished to proceed to design and build this reactor, as well as conduct studies leading to a 100 megawatt plant further down the road. (Only one such plant was mentioned at this stage.)

Why fund such a program and why now? That, Howe urged, was because there must be a further effort 'if the position of Canada was to be held.' The time was opportune. Ontario Hydro would share some of the burden, and he intended to share out responsibility for actually building reactors between the appropriate utilities and such manufacturers as could be interested. Seven were, he noted, although the final nod had fallen on CGE. For all this the price tag was $13 million, shrunk from the $16 million the AECL directors had approved; of this, $8.25 million would be footed by the federal treasury. He had included some of the federal costs in the government's expenditure estimates for 1955–6. These were estimates only; they could rise, and if they did it would be AECL that picked up the tab.

The ministers were nervous about the fairness of the deals that had been made. Was Ontario Hydro being favoured over other utilities? Would the technology now being contemplated be made available to all comers? There was concern over the scale of the proposed reactor. What about smaller ones, perhaps to be used on the Distant Early Warning (DEW) radar line then being built in the Canadian Arctic?

Howe replied that if Ontario needed nuclear power and had to turn to foreign suppliers for the necessary technology, 'there would be good reason for criticism,' especially in view of the money already spent on Canada's nuclear program. At the same time, it was necessary to plan for plants on the 100 megawatt (electric) scale; both Britain and the United States were 'going ahead along these lines,' because only at the 100 megawatt level could capital costs for a nuclear station possibly be recovered.

Howe pointed out that this new program could assist in international economic development. Scientists and engineers from 'underdeveloped countries' – what would later come to be called the Third World – could be invited to participate. 'This would be concrete evidence that Canada was willing to share her resources and experience with less fortunate people in the world.' It was, perhaps, no accident that this item showed up again a week later, when the cabinet was asked to approve a 'most important gesture,' namely, the gift of an atomic reactor to India under the existing Colombo Plan for aid to underdeveloped Commonwealth countries. This development will be considered at greater length in chapter 10.

The cabinet approved the program without modification. It asked that any announcement stress that Ontario was not the sole beneficiary of this new federal program, and that all comers – at any rate in Canada – could share in the anticipated benefits.[61] That there would be some was the assumption that divided Howe's new program from the scientific experiments that had gone before. With the cabinet decisions of March 1955, AECL was clearly entering a new age, with larger budgets, higher expectations, and different standards of performance.

Practical considerations militated in favour of accepting a nuclear power program. Ontario was fast running out of large hydro power sites and Canada's wealthiest province was increasingly forced to rely on thermal generating plants dependent on American coal from across the lakes. The St Laurent cabinet, and especially Howe, were sensitive to considerations of energy deficiency. Just over a year later, in June 1956, they sacrificed their immediate political future to ram a natural gas pipeline connecting the Alberta gas fields with central Canada through parliament. Atomic energy was at least less controversial than that.

Atomic energy was a much longer-term bet. Howe admitted to his colleagues that nobody knew whether an economic level of electricity generation could be achieved, either by the system that AECL was proposing to use or by any other. One could only learn from experience, and the demonstration reactor – and even more its 100 megawatt (electric) successor – would provide that. As an engineer, Howe was sceptical of exact predictions; for scanning the future, he preferred to rely on instinct and, as he told the cabinet, 'hope.' Hope as Howe understood it was a more exact commodity than the pseudo-scientific calculations of wizards, whether academic, bureaucratic, or political.

It was perhaps fortunate for AECL that Howe, aged sixty-nine, was their minister, and fortunate that the question of government support arose at this precise moment. He had an unusual propensity for risk-taking, and he accepted that risk meant mistakes. Nor did it hinder AECL that Bill Bennett was close to the minister or that Howe had a very high regard for Bennett's judgement. That AECL was 'Howe's' company made its passage through the bureaucracy easier; under Bennett, AECL was a company that dealt with ministers rather than civil servants, however high.

There were other, equally favourable, factors at work. Atomic energy was at or near the peak of its popularity. Widely viewed as Canada's and the world's best hope for scientific advance, it was easy to sustain as a point of national pride. While the heavy-water, natural-uranium system was only gradually becoming identified with Canada, it was obvious that Canada had an expertise in the area that permitted it to hold its own both in terms of international prestige and concrete diplomatic advantage. In a world where a small but wealthy country had to work hard for notice, atomic energy held strong attractions.

6

A Competitive World

Geneva Summer

In the summer of 1955 the old League of Nations buildings up the hill from Lake Leman in Geneva played host to a conference whose like had not been seen for some time. From all over the world scientists and bureaucrats and politicians had descended on the old Swiss city of Geneva to exchange ideas about the scientific miracle of the age. Better still, they were meeting under the auspices of the United Nations. Their purpose was to explore the possibilities of atomic energy, and, while exploring them, to consider just how much better they might co-operate with one another.

The question was hardly novel. Prior to 1939 it would not have arisen but, as we have seen, during the war the older traditions of scientific interchange were abolished in the interest of allied strategic superiority. After the war, secrecy and security, enforced by periodic spy scandals, kept the lid on. By 1952, however, the restrictions imposed by existing American policy had become irksome to the Americans themselves. Increasing consideration was therefore given to expanding the information and range of co-operation that might be made available to allies, including Canada.[1]

There was, as it happened, a model to hand. Since 1947 the United States, the United Kingdom, and Canada had been co-operating in the release of commonly held information. It was, in part, a matter of enlightened self-defence. It was far preferable to regulate the disclosure of information by all three countries than to risk some kind of competitive, and selective, declassification. Declassification conferences were held on an annual basis; the most recent had occurred at Chalk River in April 1953. There was, of course, another model: the still-surviving Combined Policy Committee whose frequent deadlocks paralysed it for all but housekeeping.[2]

The year 1953 seemed a time for new beginnings in American policy. The Truman administration had been swept out in the elections of November 1952, and in January 1953 the Republicans under General Dwight D. Eisenhower took command of the government. Eisenhower had commanded allied armies in Italy and France during the war and later briefly served as NATO's supreme commander. He was anxious to emphasize that his administration could be more imaginative than his predecessor's, and he saw atomic energy as a focus for his efforts. The coincidental death of the Soviet dictator Stalin in March 1953 suggested that the time was ripe for a new approach, and the next month Eisenhower proclaimed his desire to seek 'international control of atomic energy to promote its use for peaceful purposes only.'[3]

He then made his message more precise. After conferring with a small circle of advisers, including the new chairman at the AEC, Lewis Strauss, and with the partial concurrence of the British prime minister, Winston Churchill (back in office since October 1951), he appeared before the General Assembly of the United Nations on 8 December 1953. 'Ike' wanted to inspire his audience, and he did. He rehearsed the well-known history of nuclear weapons and the failure of international control. Nevertheless, peaceful nuclear power could still become 'a great boon for the benefit of all mankind.' To utilize the power, Eisenhower proposed the creation of an international atomic energy agency under the United Nations, to serve not only as a bank for accumulated atomic knowledge but as a storehouse for such fissionable material as the three existing nuclear weapons states – the United States, the United Kingdom, and the USSR – might wish to donate. The new agency would 'serve the peaceful pursuits of mankind,' Eisenhower proposed. 'Experts would be mobilized to apply atomic energy to the needs of agriculture, medicine and other peaceful activities. A special purpose would be to provide abundant electrical energy in the power-starved areas of the world. Thus,' the president hoped, 'the contributing powers would be dedicating some of their strength to serve the needs rather than the fears of mankind.'[4]

Eisenhower's catch-phrase, 'atoms for peace,' caught on immediately. It took considerably more time to turn it into a coherent plan. Eventually, the Americans decided they wanted an international treaty, involving the Russians, and that would take time. They also knew that they had somehow to keep the momentum of 'atoms for peace.' To make a real exchange possible, the Atomic Energy Act of 1946 was amended so as to permit freer exchange of 'restricted data' on the industrial applications of atomic energy, as well as greater co-operation with American allies.[5]

Strauss had another idea. He wanted a big conference, to which the world's atomic powers would bring evidence of their peaceful achievements and, perhaps not accidentally, of their nuclear prowess. He did not expect any serious consequences from such a conference: quite the contrary. The demonstration of American nuclear technology could be impressive, and the export of limited quantities of such knowledge could be very useful to the United States. At first the AEC contemplated exporting research reactors only, useful for training, research, and possibly medical purposes. By October, opinion had veered towards dedicating what was called a 'power reactor experiment' in the United States to the future international agency, with the aim of orienting 'foreign reactor programs toward the United States – at a time when the trend seems to be somewhat away from United States leadership in this field.'[6]

Strauss secured Bennett's and Plowden's co-operation, while postponing the question of sponsorship to a later date. The United States, the United Kingdom, and Canada established a steering committee for the conference, to which were appointed Isidor Rabi of Columbia University, Sir John Cockcroft of the United Kingdom, and W.B. Lewis of Canada. They were told to plan for a conference the following January. It would, for a variety of reasons, take rather longer.[7]

The steering committee met in England in August 1954. It developed what it considered a satisfactory conference program, practical rather than theoretical, and reported this recommendation to its principals. These in turn agreed that a conference should be summoned under UN auspices to Geneva the following spring, 1955. The project was then unveiled by the American secretary of state, John Foster Dulles, in a speech to the UN General Assembly on 23 September. Dulles preferred to keep the conference separate from the atomic agency negotiations. If the United Nations could be fobbed off with the conference, it might refrain from interfering in the agency, a subject Dulles believed more nearly related to American interests. The Canadian government, which was more concerned about the success of the agency than it was about the conference, concentrated on keeping the agency proposal in recognizable form while fighting off assaults from the Soviet Union and from the American delegate to the United Nations, who was showing a tendency to formulate his very own policy on the issue.[8]

The immediate effect of bringing in the United Nations was to widen the sponsorship of the Geneva conference. The steering committee was expanded to include Brazil, India, France, and the Soviet Union. India, a neutral in the cold war between the Soviet Union and the United States, got the task of presiding; the Indian president was Homi Bhabha, a distinguished physicist who, like Lewis, hailed from Cambridge, and who had returned to his native country to head up its atomic research agency.

Despite the fears of some American political leaders, the AEC expected an easy triumph at Geneva. They sent over a small working reactor, with Lewis Strauss in charge. Britain also sent a large delegation, as did Canada. Canada's was headed by Bennett acting in his joint capacity as head of both Eldorado and AECL. Bennett brought along a delegation of thirty-six, in which Lewis inevitably figured very prominently.

Conférenciers were expected to do more than enjoy the older sights and creature comforts of Geneva while gawking at the marvels of the conference's exhibition hall. (This was nonetheless educational: the Belgians were convinced that a particularly pure specimen of uranium imported before the war from the Belgian

Congo, and subsequently stolen by the Germans, had made a reappearance at Geneva as a Soviet product.) It was a conference, and at conferences scientists and engineers read papers. Lewis had marshalled thirteen papers that ranged from uranium geology through reactor control; they were sound and respectable, though perhaps not wildly exciting. Neither were the Americans'. As for the Russians, nothing much was expected, because of the secretiveness of the Soviet system.

It was therefore a considerable surprise when, just before the deadline for submissions, the Russians brought forth a set of detailed papers that threatened to outclass American output for the conference. According to the responsible French delegate, Bertrand Goldschmidt, the Americans had been holding back awaiting the proper moment, and the moment had come. They then sent in their 'main' papers. Some of the Canadians preferred to believe that the Americans had been caught flat-footed and had had to recast and rewrite their submissions.[9]

There was a fair amount of useful information exchanged at Geneva, but for some, like Rabi, the great achievement was to demonstrate that exchange on any terms was possible. For others, like Bhabha, it was closer to an evangelical meeting than a scientific conference. As he told the final session, 'The feasibility of generating electricity by atomic energy had been demonstrated beyond doubt.' Goldschmidt, in retrospect, called it 'a remarkable success,' both from that point of view and in opening up discussions and the flow of information.[10]

The direction that success would take was not yet clear. 'At my low level,' a Chalk River engineer recalled, 'I was impressed at the lifting of the oppressive weight of secrecy.' Now the plans and capabilities of the world's nuclear states could be judged more accurately. The powers were divided between those that favoured enriched uranium, the United States and the Soviet Union, both of which had built separation plants at vast cost to procure U-235 for their military programs, and those that preferred natural uranium, in part because they felt they had to. The state or size of their economies left them little choice. Britain and France were the members of this club; 'later,' Goldschmidt wrote, they were joined by Canada and Sweden. It was another reminder that the final direction of Canada's program had yet to be chosen.[11]

There was no question that the cause of atomic energy was greatly advanced by the Geneva conference; so enthusiastic were participants that they held another one three years later, where the flow of information was even greater. All nations, including the Soviet Union, got useful information and, sometimes, good publicity. The world learned that the USSR already had a functioning atomic power station, at Obninsk, with an output of 5 megawatts (electric). It could hardly be called economic, but at least it did work. The Americans were building a new, highly publicized, 60 megawatt (electric) reactor at Shippingport, Pennsylvania. And Shippingport was lagging behind the British reactor at Calder Hall, scheduled for completion just over a year away, in 1956.

If optimism was the mood, 'demobilization' was the reality. Eisenhower had intended to demonstrate that nuclear power was beneficial, that it could be used in civil societies as a pathway to the future rather than an epitaph for the present. Its military connotations, which had fed nuclear programs and helped to justify their existence, were discreetly muted or swept under the rug, a hangover from a bygone and somewhat disreputable age. What had been concealed could now be boasted about, and the result was a quantum leap in publicity, mostly favourable, about atomic energy.

AECL's *Annual Report* for 1955–6 caught the mood. 'Access to information is no longer a problem,' it burbled. 'Much of [it] is now available in the published literature and particularly in the published proceedings of the Geneva Conference.' This was an exaggeration, but a pardonable one, for the opening up of information did present a notable contrast to the immediate past. To handle the new bonanza the company established an Office of Industrial Assistance and stepped up visits to and conferences at Chalk River.[12]

There was one other lasting, and in this case concrete, consequence from the enthusiasm for 'atoms for peace' in 1953–6. A twelve-nation conference finally convened in New York under American sponsorship in February 1956. It sketched a plan for a comprehensive and powerful international supervisory organization for nuclear exchanges (to be called the International Atomic Energy Agency or IAEA), one that would act, in Goldschmidt's phrase, 'as a broker rather than as a banker in matters related to the distribution of nuclear materials.' Such distribution would follow the new agency's own firm rules.

The reasoning behind the US propositions, which had been extensively canvassed with some allies, including Canada, was simple. Nuclear energy was a great opportunity, and correspondingly attractive to energy-starved developing countries. The provision of nuclear energy was in effect a seller's market. At least part of the attraction was the military potential that possession and manufacture of fissile material conferred. Thus the potential for power and economic development must negate the potential for destruction. The means was safeguards, legislative or contractual provisions intended to secure good behaviour from states using the agency's facilities.

It might have been just another case of great powers imposing their conception of order on lesser states. The smaller states might not like it, but were essentially powerless to do anything about it, at least for the short and even medium term, because of the retardation of their own science and research facilities. Unluckily for American intentions, the great powers were neither all-powerful nor united. Channels already existed between nuclear states and non-nuclear states; the link between Canada and India, which we will discuss in chapter 10, is such a case. The great powers could not have policed the whole world, and as became evident, they could not agree on how to do it.

The IAEA had not even been born when fierce quarrels broke out between the Soviet Union, allied for this purpose with India, and the United States and its allies. Even American allies such as France were not firm, preferring to follow their own nuclear destiny towards atomic weapons. To such countries it seemed that a coalition of retired burglars was trying to close the thieves' market to those still active in the profession. Or, to put it another way, the IAEA and its safeguards were a neocolonialist, imperialist plot, designed to replace old-fashioned imperialist fetters with invisible, nuclear ones. The emerging nations would thereby become the dependents of their former (or soon to be former) mother countries.[13]

Cursed at birth, the history of the IAEA was at best rather mixed. Like other international agencies, it became prey to bureaucratic

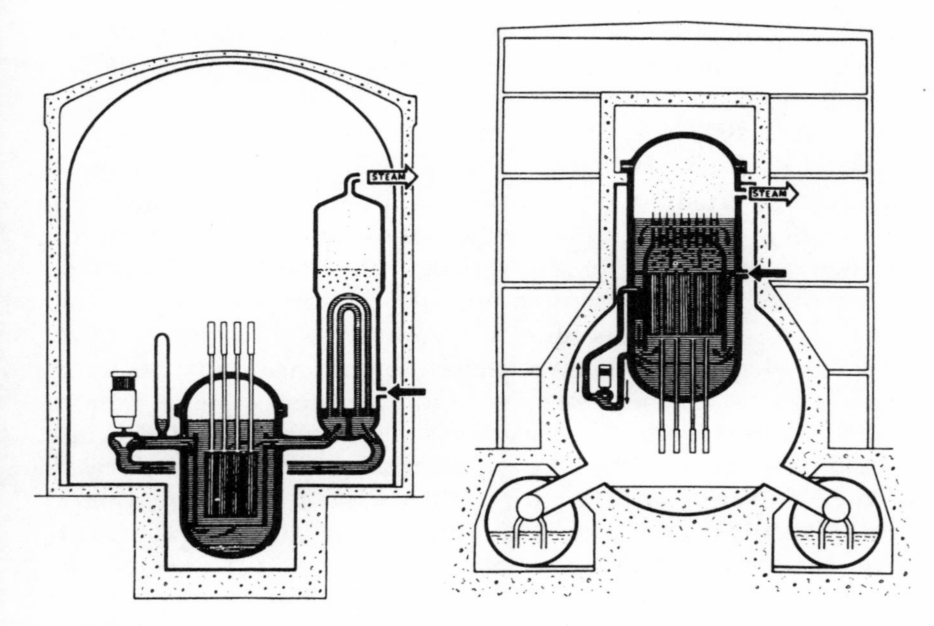

High-temperature gas-cooled reactor. *Nuclear Engineering International*

Boiling-water reactor. *Nuclear Engineering International*

gamesmanship practised on behalf of special national interests. This would have been frustrating had its responsibilities, like those of UNESCO, been principally confined to the preservation of waterlogged cities and the enhancement of internationally designated (on a pro rata basis) places of beauty. The IAEA was given functions as well as a mandate, a remnant of its idealistic origins, the bringing together of the peoples of the world in a direct, businesslike way to talk about something that was important, indeed vital, to them all.

On small issues it had some success, but on the larger questions that had called it into existence, very little. The fact that the IAEA did not work very well therefore became a kind of standing reproach to those of its servants whose aims in life were not confined to enjoying the cosmopolitan delights of the city that harbours it, Vienna. Its aspirations were not laid to rest, nor could they be. We shall see why they did not prosper as Eisenhower had hoped when we consider the international aspects of Canada's nuclear program.

In considering why the idealism of 1953 became the sordid reality of 1956 it is well to bear in mind that the reopening of atomic energy to discussion meant opening it, for the first time, to competition. Competition, as well as co-operation, was the name of the new game. That Geneva signalled the start of competition rather than a continuation of an older pattern of co-operation among the English-speaking atomic powers was not at first evident.

It was Bennett's purpose, shared by his scientists, to keep relations between the erstwhile atomic partners close as well as cordial. The exchange of information was mutually beneficial, and should proceed as before, through annual conferences at Harwell or at Chalk River, or through regular exchanges with the Americans. The British and Americans stationed liaison officers at Chalk River, as before. The Canadians sent their own scientists and engineers to Savannah River or to Harwell. Yet while co-operation endured in some spheres, it masked the fact that after 1955 the three countries were increasingly moving in a contrary and even competitive direction. As Hinton's office wrote to AECL in 1956, 'we are not prepared to disclose any commercially valuable information,' and accordingly Anglo-Canadian discussions on a wide range of topics 'will be basic.'[14]

Bennett hoped at first that British, Canadian, and American reactor design would be parallel. He hoped that the most practical as well as the cheapest line would be the Canadian one. But the effect of the cancellation of the tentative joint program with the British in 1954–5 was to encourage the AEA to consult its own narrowly defined interests, which were competitive and not complementary. 'We were resigned to enriched uranium,' Fenning later commented. The British had already decided against the heavy water reactor, and the results of the joint study were of little concern. As Hinton explained to Lewis in the spring of 1955, they simply had too much on their plate because of military demands. After his annual meeting that September with Plowden and Cockcroft, Bennett decided that there was 'no way in which we can force the issue even if we do not consider the [British] decision a sound one.'[15]

The British were not through yet. They had it in mind to improve their capacity for enriched uranium, a fantastically expensive project, which, because it was heavily consumptive of electricity, was difficult to site in the United Kingdom. Instead, it should be located close to one of Canada's great energy sources. Hamilton Falls in Labrador was a possibility; so were the headwaters of the Columbia River, or a large, low-grade coal deposit in Alberta. The Canadians should make the initial investment and recover their costs through sales of their product to the British.[16]

The Canadians, according to Bennett, took it seriously. Consulting engineers were engaged, and duly recommended the Columbia site. Location was not, however, the major problem. Money, of the order of a few billion dollars, was. The British, it seemed to Howe, were taking no risk while Canada was assuming quite a large one. If the British had offered to pay part of the capital cost, he would probably have agreed. Since they offered no such thing, he declined.[17]

The enrichment plant was not directly tied to AECL and its concerns. Its construction would, however, have had an impact on Canada and probably on Canadian atomic policy. Natural uranium was preferred as a fuel for largely political reasons. If an enrichment plant had been built it would most likely have been used, not only for the production of weapons-grade materials but for fuel too. And had that occurred, the history of the whole Canadian reactor program would have been very different, 'a second string to our bow,' as Bennett put it. But it did not happen, and the British and the Canadians diverged. When the two programs met again, it would be as competitors.[18]

That left the Americans. Walter Zinn of the Argonne National Laboratory was believed to be an enthusiast for heavy water, which raised the possibility of a joint study between Argonne and Chalk River. The Americans did not say no, but neither did they say yes. That, wrote General Kenneth Fields, the AEC general manager, would depend on Zinn and Lewis; and he omitted to give Zinn any directions in the matter.[19]

Bennett proceeded with what he could get, which was considerable. The Americans by amending their Atomic Energy Act had opened the door to co-operative agreements with suitable countries. Canada was suitable and got an agreement, negotiated personally by Bennett, to the surprise and irritation of the Canadian embassy in Washington, which had to insist that it at least be allowed to witness the signature. Information flowed north from the various US atomic projects: information on heavy water, on suitable alloys for fuel cladding, and even (and unwittingly) on fuel, as we have seen. That did not inhibit the development of a keen spirit of competition. To Lewis, writing to the Board of Directors in November 1955, it was essential to keep Chalk River, 'the main resource' of the company, in a competitive condition. Canadian public opinion, well aware of the country's 'present advanced technological position,' would expect no less and would surely accept no less. 'We have to maintain our position alongside the nuclear technology of the U.K. and the U.S.'

Canada should not only proceed with the small demonstration reactor, which had come to be called the Nuclear Power Demonstration or NPD, but also prepare for the construction of one or two large reactors, of 200 or even 250 megawatts (electric) size, by 1962 or 1963 when, it was thought, Ontario Hydro would have to look for a new source of power. To do otherwise risked dispersing the engineering staff who would be trained on building NPD, for they would be lured to new projects in the United States. In effect, Lewis was arguing that having started to build reactors, AECL did not dare stop as long as the Americans were going ahead; and the Americans were doing just that.[20]

The board took some convincing. The directors chewed over Lewis's program of piling one reactor on another in meetings in November 1955 and again in their next meeting in March 1956. The expense of just the first new reactor, NRU, seemed both infinite and endless; the second, NPD, might go the same route, and now Lewis wanted two more, much larger than anything yet built anywhere. NRU alone was forcing the company to pare its expenses wherever it could, and in January the executive committee learned via CGE that its contribution to NPD might be $1.5 million more than anticipated. Should the extra cost delay signing the final contract with CGE and Ontario Hydro?[21]

Bennett passed the news to Howe, and, given the sums involved, asked for a new authority to proceed. It was granted, and the contract was signed. With that complication out of the way, the board and Bennett still faced Lewis's ambitious power program with very limited finances. These funds were commensurate with the modest power program approved by the cabinet the year before. There was no doubt that the minister would ensure that the company's treasury was topped up to match NRU's escalating costs as well as those of the NPD project, but the large reactor development was another matter.[22]

Bennett therefore sought to define the company's objectives, in the context of Lewis's five-year program. Its basic purpose was to develop reactors suited to the needs of those utilities that wished to purchase them, in co-operation with those utilities. Bennett had constituted an advisory committee on atomic power, consisting of utility representatives from across the country; this group received reports and figures from the NPD project. It had a double function, for besides spreading the attractions of nuclear energy before a wider audience, it emphasized that AECL's mandate was truly national and helped defuse resentment of Ontario and Ontario Hydro. (For investment-shy utilities, it was also a brake on rash spending on something not yet proven.)

After reviewing the history of the project and the reasons for the heavy-water Canadian design, Bennett argued that power plants everywhere were still at an experimental stage. Yet because the demonstration reactors would not by themselves furnish adequate (as Bennett put it, 'final') information for reactor design, operation, or cost, it was necessary to go a step beyond and proceed with research and development for the next, larger generation. He had employed the same reasoning in his 1955 submission to cabinet, but the R&D program was now clearer and firmer.

The second objective was also familiar, but more emphatic and explicit: 'To carry out the power reactor development programme in such a manner as to ensure that the Canadian manufacturer will be in a position to design, fabricate and construct power reactors and their components for the domestic and the foreign market.' To this end the manufacturer, CGE, would be intimately involved in all aspects of 'design, development, fabrication and construction.' No distinction was made between the manufacturer's participation in the NPD project and participation in designing and fabricating the later large reactor project.

In the event, Bennett suggested passing both a minimum budget for the five-year program up to the minister, and an implied warning that another serious decision would be required, probably during 1958, as to the manner and timing of proceeding with a large power reactor.[23] But when he testified before a House of Commons committee in June 1956, Bennett simply assumed that 100 or 200 megawatts (electric) of nuclear power would be installed and operating by 1961.[24]

To the politicians, Bennett and his staff explained Canada's natural-uranium and heavy-water program. The program was risky. They could not be sure, yet, that natural uranium and heavy water were the best route to go, but they admitted a strong preference in that direction. They had no doubt that some form of nuclear power would soon be needed, probably in the next decade, and although its growth might be slow for the fifteen years after 1956, they were certain that, during the 1970s, demand, spurred by need, would justify the expenditures of the present.[25]

Their testimony represents an aspect of AECL's work to which we have only briefly referred, and which should now be examined in greater detail. It was an area for which the company had no specialist to hand: economics. Physics determined whether atomic power was conceivable, engineering dealt with whether it was possible, but only economics could show whether it was justifiable. Scientists, those on the administrative end, were sensitive to the issue, and not at all reluctant to try their hand at interpreting their physical data in ways that would impress and stimulate the most jaundiced accountant. Engineers on the project, as elsewhere, prided themselves on being more practical than the scientists and better able to generate costing information. We have already seen the two groups clash over the scientists' desire to build a better atomic widget – NRU – in the teeth of the engineers' worries over exploding costs. When the costs did explode, the engineers made their point, even if the scientists liked to refer to the experimental nature of the design and construction of NRU.

Scientists were prone to plan for the long term. Most of the Earth's usable energy was being depleted, at an increasing rate. Oil, gas, and coal would eventually be used up and nuclear energy's desirability would be universally admitted. Early calculations exaggerated how soon this might be, but even where a sober examination was possible fervour crept in, for uranium seemed to hold out a promise of abundance, no further away than the closest reactor. Reactors could overcome the handicap of long-distance transmission and its associated costs, for they could be built almost anywhere. The 1940s and early 1950s concentrated on the eventual availability of breeder reactors and thorium conversion in order to compensate for the apparent scarcity of uranium; but while nuclear scientists continued to laud the prospects of a breeder system they were duty bound to acknowledge, as Lewis did to the 1956 parliamentary committee, that such engines were twenty to forty years off.[26]

This was not as depressing by the 1950s as it might once have been; large deposits of retrievable uranium had recently been discovered and new mines opened up. That some of these were in Ontario does not seem to have affected the earliest calculations of Hydro's planners in Toronto; that some of them were in Canada was very much to the point. Security of supply was best guaranteed by an energy delivery system entirely within Canada's borders; and Ontario Hydro was sensitive to complaints that its thermal power stations used coal that was 100 per cent American.[27]

The key economic consideration in Canada was not so much availability as comparability – that is, nuclear power must be at least as cheap as the non-nuclear alternative, which in the case of Ontario meant coal-burning thermal stations such as the Richard L. Hearn station on the Toronto waterfront or the J. Clark Keith station in Windsor. Nuclear stations were, however, very expensive to build; any saving from a nuclear power system must come from the fuel. There uranium seemed to have a strong advantage, but was it enough to balance the heavy initial cost, including interest charges, for the reactor to burn it?

There were three questions. Was nuclear power economically defensible? Lewis thought it was, but how many agreed with him? If it was, how much would a nuclear power station actually cost? Finally, would such a station be safe? For the answer to that, Bennett depended on Cipriani, and on the staff at Chalk River, especially Laurence, who was specializing in this matter. The Nuclear Power Group had gone as far as it could to answer the second question. Their preliminary study, NPG-5, dated 31 January 1955, reduced the reactor to its basic characteristics and then tried to correlate them to the necessary costs that must follow: cost of pile material, cost of heat-transfer equipment, cost of turbine plant, and cost of electrical equipment. But even though they could estimate, in a crude way, what a demonstration reactor would cost, this answer helped very little in telling Bennett and the board whether the longer-term enterprise was worthwhile.[28]

For that Bennett turned to an economist with the additional and unusual qualification of professional engineer. He worked in the same building, Number Four Temporary, for the same minister, as an economist in the Department of Trade and Commerce. Jack Davis was thirty-eight in 1954, a graduate of the universities of British Columbia, Oxford, and McGill, where he had taken his PhD in economics. He had served in a number of war industries, landing in Ottawa in 1948. In 1954 he had just become head of Howe's economics branch. Howe released some of Davis's time to Bennett, and later handed him on to the Royal Commission on Canada's Economic Prospects, headed by the minister's detested ideological rival Walter Gordon, to survey Canada's energy future in a volume published in January 1957.

By then Davis had been studying the problem for over two years. He published an essay in the area just at the time of NPG-5. Entitled *The Economics of Atomic Power Reactors*, it suggested that the costs of nuclear reactors were too high to justify – unless they could be subsidized from the military budget either directly or indirectly through the sale of plutonium. Only under the most favourable conditions (implicitly artificial) could nuclear power be cheaper than its thermal, coal-fired, alternative. Under no conceivable circumstances would it be cheaper than hydro power. Naturally it was conceivable and obviously desirable that reactor design improve to the point where nuclear power might be commercially competitive, but that was unlikely before 1970 at the earliest.[29]

That was enough for Shawinigan Power, which promptly concluded that there was no need for nuclear power in southern Quebec. It would keep its oar in the development of the Canadian reactor program, and perhaps contribute something, but purely as a distant precautionary measure.[30]

Davis's advice to the Gordon commissioners was essentially the same as the earlier paper, although it was slightly more optimistic in tone. First, as had been apparent, there was only one Canadian market sufficiently large, prosperous, and desperate for energy in a few years' time to justify investment in nuclear plant, and that was southern Ontario.

Second, Davis accepted that reactor design was incremental, and that experience in construction as well as operation should lead to a significant decline in cost. That would be even more the case as between NPD and the still unconfirmed large reactor; if that reactor was of the 200 megawatt (electric) variety its costs should be of the order of $400 per installed kilowatt as compared to the anticipated $1000 per installed kilowatt at NPD. Since it was true that the larger the station the greater the economy to be anticipated, experience alone was not reflected in the lower per-kilowatt cost of the large reactor. Even if experience, in five-year tranches, each with improved reactor designs, allowed a saving of 10–20 per cent for each reactor generation, it would be some time before reactors could be built for $200 per installed kilowatt, 'a level of investment,' Davis suggested, 'at which electricity produced as a result of nuclear fission might prove to be really interesting.'

Third, only one component of fuel costs, the burn-up, which represents the fraction of the energy in the uranium which can in fact be released, promised to reduce those costs significantly. If the number could be moved up from 3000 megawatt-days per tonne to 5000, as some thought possible by 1965, then the result might be very attractive. (The eventual CANDU figure would be 8000.) If this and various other projections proved true, then Davis could envision nuclear power costing less than conventional steam-generated power by 1969, and possibly less than hydro power by 1976. But while this suggested an optimistic objective, it would take at least thirteen years for a pay-off.[31]

With Davis's answers in hand, the board looked again and again at the longer-term implications of the Lewis five-year program. They could manage with the certainty that the NPD would not be economical; that was all part of proving nuclear power would work.

But before they arrived at a definitive answer, they heard further news, and it was depressing. The estimated cost of NPD was rising; it was up from $13 million to $20 million, and that was a pattern that looked familiar to a management still shell-shocked from the interminable delays and cost overruns connected with NRU. AECL might outlast one NRU; could it hope to survive two?

Building NPD

The superficial history of NPD, the Nuclear Power Demonstration reactor, is briefly told. The engineers of the Nuclear Power Group were divided into two. One contingent with Ian MacKay left for Peterborough, northeast of Toronto, where CGE had decided to place its Civilian Atomic Power Division (CAPD); this party took employment with CGE. The remainder under Harold Smith stayed at Chalk River to work on the design of a 200 megawatt (electric) reactor.

NPD was to be a co-production of Ontario Hydro, AECL, and Canadian General Electric. It was located at Rolphton, in Rolph Township, on the Ottawa River, just downstream from the Des Joachims power plant and thereby close to a power source of its own. It was ultimately designed to produce 20 megawatts (electric). Construction started in September 1956 and the plant started up in 1962.

The least controversial aspect of NPD's history was its site. Convenience dictated a location close to Chalk River, with its technical resources; it was more important that it be easily reached from the laboratory than from large cities, or from abroad. To some that was the point; far from large cities, far in fact from any city, its location was actually a safeguard in case of an accident. Hydro had doubts about locating its nuclear experiment so far from head office and direct scrutiny; but AECL's strong preference for Chalk River as the proper place for the Nuclear Power Group proved crucial.

Des Joachims, the damsite twenty miles further up the Ottawa, was among the most modern of the Ontario hydro system, with 480,000 horsepower. It was also among the last to be built, a point that must have occurred to the designers who chose it as the most convenient source of land and power for NPD. Though it was close to Chalk River, it was a mile from the village of Des Joachims, and even farther from anywhere else. The location did bring its disadvantages; it had to be moved during 1956 because of uncertainty as to whether it was actually in Ontario or Quebec.[32]

But it was there that the politicians and their aides converged in September 1956 to turn the sod for the new plant. Leslie Frost, premier of Ontario, came, as did C.D. Howe. Dick Hearn, Ontario Hydro chairman since Saunders's death in a plane crash, and Bill Bennett, still president of AECL, were also on hand. The politicians picked up the ceremonial shovel and advanced on a patch of sod that covered the rocky ground. They bent over, turned it, and officially launched Canada's first nuclear power reactor. Cameras flashed and whirred and reporters scribbled down the happy news. Then, after it was over, the sod was scooped up; like the politicians, it had to be trucked in, for Rolph Township's rocky soil was better adapted to jackhammers than to spades.[33]

The project they were inaugurating had changed in several respects since its approval in principle in the spring of 1955. It had grown, as Lewis wished, from a 10 megawatt (electric) station to a 20. And its fuel had been transformed from uranium metal to uranium oxide. It would soon change even more, so much so that it was turned, literally, on end. But first let us examine how its first alterations came about, and what impact they had on the direction of the NPD project.

The process by which NPD was changing is startlingly reminiscent of the continuing, indeed continuous, design and development of NRU. The difference was that there were more actors, and the action was spread out over a larger geographical stage, from Chalk River to Peterborough, with occasional scenes in boardrooms in Ottawa and Toronto. It might be thought that with the inspiration stage over, the engineers and scientists, applied and pure, of Chalk River would return to their laboratories and drawing-boards to concentrate on research and design of the next project. This did not occur. NPD was characterized, from beginning to end, by cross-fertilization from one workshop to another; and therein lay most of the explanation for the steady rise in its estimated cost that so shocked AECL's board.

The reactor was to be enclosed in concrete shielding, in a pit excavated on a flat near the Ottawa River. Inside the concrete shell were three compartments, the central one housing the reactor proper, with steam generators on either side. The reactor had to pump heat to the generators and this was done by means of a heavy-water cooling circuit (the heavy water conserved neutrons better than the light-water alternative). To prevent boiling, the whole of the cooling circuit was kept under high pressure. The reactor was placed inside a 'pressure vessel,' a kind of pressure cooker, made of a steel alloy, 13 cm (5 1/8 inches) thick, containing both the heavy-water moderator and the fuel rods, which were encased in the latest American discovery, 'zircaloy,' an alloy of the metal zirconium, chosen for its low neutron absorption cross-section. Zircaloy had to do more than permit neutrons to pass through it; it had also to resist corrosion and other damage from heat and radiation, and it had to be durable in very small thicknesses.[34]

Zircaloy was the fruit of co-operation with the Americans. So was another innovation, the abandonment of uranium metal in favour of uranium oxide as fuel. This, like zircaloy, sprang from the American Bettis laboratories which were developing fuel for the US nuclear submarine program. They needed, by definition, a long-lasting, reliable fuel; they found it, during 1954, in uranium dioxide. In April 1955 Hyman Rickover, the head of the US navy program, plumped for innovation: uranium dioxide it would be.[35]

In Canada, however, Lewis had his doubts, and as late as April 1955 was firmly in favour of uranium metal for the large reactor program (and presumably for NPD too), even though some of his chemists were intrigued and attracted by the American discovery. Nor had Lewis given up on retrieving plutonium from the demonstration reactor at the cost of more frequent refuelling – no small consideration since the reactor was not designed for on-line fuelling and faced a refuelling shut-down every few months as it was. The plutonium had value, of course, as far as the Americans were concerned. It might also be used for a fuel recycling system, a concept to which Lewis kept returning throughout 1955.[36]

The issue was resolved by Rickover's renting space to test his fuel elements in NRX. The Canadians as usual discovered what was going in, and as usual they discussed their findings. Whatever the new experiment consisted of, it was only half as dense as metal, and, as John Foster observed, 'the inference was clear.' The oxide showed 'exceptional performance,' and Lewis was convinced at last. 'In October 1955,' Foster remembered, 'he came to Peterborough and told us that the fuel would be uranium dioxide.'[37]

For the rest, the reactor was designed with familiarity in mind. It would not have on-line refuelling. Like reactors abroad, it would have a pressure vessel made of special steel. Because of the unfamiliarity of Canadian suppliers with such an object, it would be made by Babcock and Wilcox in Scotland. The reactor would be refuelled from above, into a fuel rod flask moving across the floor of the reactor hall above. Inside the pressure vessel the fuel was contained in tubes, through which the fuel-rods proper and the cooling heavy-water circuits were to be run.[38]

The original timetable had NPD active by early or mid 1959. This deadline was, in the nature of things, difficult to meet, especially when a final decision on the reactor's site was not made until the summer of 1956. The decision to double the reactor's rating from 10 to 20 megawatts did not speed things either, since that necessitated rearranging the fuel into a larger number of pencils of smaller diameter within the same containing tube. The thickness of the fuel cladding was another matter for dispute, and an important one, since the neutron density of the reactor could be increased by a thinner, less absorptive, sheath. The Peterborough designers believed that they had gone one better than the Americans by recommending a 0.02 inch thickness, down from the 0.025 standard in the United States. But Lewis had decided that it would have to be thinner still – 0.015 inch (0.038 cm), and at a review meeting in Chalk River he made his decision stick.[39]

Not all the decisions for NPD were made with such panache, or finality. Even the decision to opt for a pressure vessel was by no means easy. Pressure tubes had an attraction in 1955. However, there was no important cost advantage in using them, and the only material so far proven was stainless steel, which was too greedy of neutrons. At the same time, it would not be easy to extrapolate the 20 megawatt (electric) heavy-water pressure-vessel design to 200, since the pressure vessel would be almost impossibly large and thick walled. (Reactors using enriched fuel have much smaller cores for the same output, so pressure vessels are used for much larger outputs.)[40] It was therefore inevitable that pressure tubes remained a subject of investigation at Chalk River, where Lewis urged more work to test their resistance to stress and irradiation effects over an adequate service life. It was inevitable, too, that if Harold Smith's team found a way for pressure tubes to work, NPD would be the guinea pig of choice.[41]

> **PRESSURE TUBES**
> Pressure tubes are small pressure vessels used in heavy-water reactors to contain the coolant under pressure as it flows over the fuel elements. Several hundred pressure tubes passing through a calandria fuelled with heavy-water moderator make up the core of a typical CANDU reactor.

While Smith and his team were worrying about the reactor's intestines, others were concerned about its armour. Was there enough shielding around the reactor in case the pressure vessel cracked? There was enough concern to cause the NPD Coordinating Committee in December 1956 to order a second shell, in addition to the protection provided by the reactor pit with its concrete top; the committee made its decision at short notice and after less than a day's discussion.

The cost of the new shell promised to be of the order of $250,000, a consideration that shocked some of the engineers on the team. E. Siddall, a project engineer, complained that 'It should be a matter of deep concern that the cost of NPD already appears to be far above the original estimates, and is getting higher all the time. The normal rules of competitive engineering must eventually apply in the nuclear field,' he continued. 'Such rules hardly permit the expending of an extra $250,000 with so little consideration as in the case of this shell.' And, Siddall maintained, an extra shell would do very little to enhance safety, either against an explosion or against escaping radiation.[42] John Foster took up the case for the defence. He made four points: 1) A relatively simple condition can create a risk, the magnitude of which cannot be established. 2) The risk can be protected against at an estimated cost of $250,000. 3) The protection cannot be added to the plant as an afterthought. 4) Opinion regarding the balance between risk and the cost of protection is not available until the design is completed.[43]

The various supervisory committees ('technical' and 'co-ordinating') agreed with Siddall rather than Foster. There was reasonable protection already, both inside the plant and in terms of NPD's remote location. That saved a potential $250,000, but that was where the good news ended. CGE had been doing a close re-estimate of what would be required for NPD, and the result was beginning to worry AECL staff. The margin of financial safety was diminishing, and a request from AECL to CGE in December 1956 to scrutinize 'contingencies' resulted in the complete disappearance of that item from the estimates.

The co-ordinating committee's next meeting, on 1 February 1957, was 'somewhat disturbed at this information,' as its minutes chastely described it. CGE told its AECL partners that 'a complete re-estimate was required,' and then described and defended in detail what had occurred. Contingent items had been confirmed as necessary; new equipment, not foreseen in the first estimate, was required; and some of the Hydro share was redistributed to AECL's account. The construction camp at Rolphton had cost $400,000 more than originally projected. The result was $2 million up on the original estimate, for construction alone; research and development had risen by another $400,000.

The co-ordinating committee restrained its enthusiasm for this news. Admittedly some of the increases were perfectly logical and entirely justifiable, but it would be well to look at the matter in more detail before going forward. Its decision acquired more force when Lorne Gray and W.B. Lewis wrote to Ian McRae, the CGE executive in charge of the Civilian Atomic Power Department in Peterborough, to tell him to make no new commitments for the next two months, pending a complete review of NPD's costs. A wide range of technical decisions, already taken, was to be reviewed and, possibly, changed. Even that, Lewis and Gray recognized, might still add up to $20 million, as CGE predicted, but clearly such a large cost increase must be justified. Writing personally to McRae, Gray was more explicit: the most recent estimates from CGE commanded 'no confidence.' It was now up to CGE to prove that they were valid.[44]

There was a price for the review of the costs, as Gray and Lewis recognized. Two months' delay in making new commitments could well result in as much as a year's delay in completing NPD. McRae agreed; to try and make it up new deadlines were established.

The two months' hoist, from February to April 1957, was caused in the first place by cost overruns, but at a deeper level it signified what Lewis called 'the further development of technical thinking.' It followed that the review then taking place was the last opportunity to change design so as to avoid any further increases in costs; it might even be an opportunity to save money, as Lewis wrote on 13 March.[45]

Two weeks later, Lewis drafted a reconsideration of NPD. It was a litany of achievement, favourable portents, and drastic conclusions. First, Lewis noted, the price of heavy water was down, to $28. Next, recent experience with deuterium as a moderator was encouraging. Third, the uranium oxide fuel was another happy sign, and every report showed that it worked. Fourth, zircaloy was a notable advance. Fifth, it looked as if 8000 megawatt-days per tonne could be achieved, without reprocessing, thereby disposing of one of Lewis's earlier and perennial preoccupations. Sixth, moderator control alone would suffice, eliminating control rods. And seventh and most important, 'there is [a] prospect of making satisfactory pressure tubes from zircaloy.'[46]

There was a full complement of officials present for the next meeting of AECL's executive committee in Ottawa the next day, 27 March. The executive committee first digested the latest estimated cost, including fuel, of $24 million. That was a blow, but not, as it turned out, the main business of the meeting. More important were three alternatives presented by Lewis, Gray, and Harold Smith. They could go ahead on the basis of the present design. They could cancel the project and chalk it up to experience. Or they could change NPD's design to incorporate what the minutes called 'the new design concepts which have been developed in the preliminary design study for the large reactor,' and which Lewis had summarized for his own benefit the previous day. These new concepts, the executive committee was told, would result in a probable and substantial saving over the existing design of NPD.

That being so, Bennett and his directors, Hearn, Gaherty, and Gordon, plumped for the third alternative. Agreeing that the high cost of the existing design of NPD did not justify the data they could hope to get from it, they conceded that some kind of demonstration reactor was still necessary, so that the project ought not to be cancelled out of hand. That left the third choice, and they took it.

NPD as it had been was effectively dead. It took its place in the files as NPD-1; NPD-2 replaced it on the drawing-boards.[47]

NPD Leads to NPPD

The decision to redesign NPD was daring. It was also costly: a year of detailed design work had to be redone. Orders had to be cancelled, principally the one for a pressure vessel from Babcock and Wilcox in Scotland. Pressure tubes had to be designed, ordered, and manufactured. The new design involved more than pressure vessels. It would now be possible to incorporate on-line refuelling in NPD. It was true that this and other features would require intensive development; at least there was now time to do that.

These improvements were attractive; more important was the attempt to bring the large reactor design and NPD back into harmony. What then could be learned from the small reactor that could usefully be applied either to the first large reactor or to future reactor generations? NPD after all was a 'demonstration' reactor; if what it demonstrated was not worth the knowing, then what was the point of having it?[48]

The news was therefore not all black. AECL's finances would have been distinctly shaky had work gone ahead as scheduled on NPD-1. Now the money that had to be spent on the project in the fiscal year 1957–8 dropped from $3.5 million to $0.5 million. Because NRU was going to cost an extra $1.4 million, AECL would finally save just over $1 million on its existing budget, without embarrassing itself and its minister by returning to Treasury Board to ask for supplementary funds.[49]

The new design was to be ready by October. The re-examination was not limited to points of design. To Smith, Gray, and Bennett, it was obvious that the company had barely escaped falling into the same predicament that it had stumbled into with NRU, committed to costs that it had not anticipated and could not control. It was time to consider whether the company was organized in the best possible way and, if it was not, what could be done to improve its controls over both NPD and the large reactor still to come.

To Lorne Gray and Harold Smith, a large part of the problem lay in the hiving off of detailed reactor design to Peterborough and to CGE. In a letter in March 1957 Smith described to Bennett the bulge in the estimates to $24 million, a subject he knew was bound to get the president thinking furiously, and then went on to explain what he (and by extension other members of the Chalk River staff, including Gray) thought had happened.

The most obvious defect in the NPD program was that the estimates sent to cabinet in 1955 were preliminary and necessarily incomplete. At that time, to take a most glaring example, nobody had known what zircaloy would cost. Costs for development, for the site, and for housing construction workers were excluded. That was inevitable given the pressure the company faced to come to some kind of a decision, fast. Since CGE had no previous experience in the business, much of the time and at least some of the expense went for training its Civilian Power Department in how to make reactors. Some at least of AECL's staff deplored the cost-plus contract given CGE; supervision from the AECL side had not kept costs under control. Lastly, the time limit set for completing the project was grossly unrealistic, although admittedly less so than Lewis's initial three-year deadline. If one were trying to quantify what had gone wrong, these factors must be given weight.

Smith was not critical of CGE as such, although it did not take much perception to discern that in his view assigning the contract there (and perhaps assigning a contract at all) had been a mistake. CGE's estimates, when developed, were more accurate than those in NPG-8 even though, in an effort to appease the climate of opinion at Chalk River, they had promised completion within four years instead of the more probable five. Smith did not blame Lewis, but he condemned the speed with which the five-year program had gone forward in the summer and autumn of 1954. Haste had, in this case, made classic waste.

Both haste and waste should be brought under control. It was clear that Chalk River as presently constituted could not do this. Bennett must therefore 'establish a central control organization for the *entire programme*,' meaning both NPD and the large reactor. This was no more than common sense, particularly if NPD's design were adapted to the large reactor's, because there would be less duplication, and consequently smaller costs. It might inject helpful competition into the exercise by emphasizing that NPD would be more than a one-shot effort, encouraging thoughts in industry of a large construction program to come.[50]

Smith had another idea in mind. He wanted to move the control group away from Chalk River and down to Toronto. It was more than coincidental that the move would extract his team (for it would be his to direct) from Lewis's direct influence. Canada's reactor program was moving from research into industry. And in industry the customer was, eventually, always right. The customer, Smith did not need to say, was Ontario Hydro.

Bennett accepted Smith's conclusion. The bare NPD estimates, $13 million in 1955, $14.5 million in February 1956, showed that there had not been enough time to absorb the full costs of the program. The upward creep of CGE's figures in the fall of 1956 had not triggered an alarm until December, when the disappearance of contingencies suggested that the margin for error had vanished. The estimate had since risen again, to $20 million without fuel, housing, and transportation, or $24 million including these items.

In April 1957 Bennett informed Howe that he and his board, with Ontario Hydro's concurrence, had decided to scrap NPD's original design and to merge NPD with the large reactor program. This would firm up the tentative commitments that AECL had already made to the large reactor. Taking up Smith's suggestion of a new control organization, Bennett proposed, and Hydro agreed, that it 'be set up in Toronto under Hydro's auspices'; its function, he wrote, would be 'to co-ordinate design and engineering development.' It would be more efficient and less costly than continuing along separate developmental paths, just as Smith had suggested.

Bennett did not wish to suggest that he was displeased with CGE's design work to date, but he was sadly disappointed in their inability to give him an accurate cost while there was still time to recover from it. If CGE were faced with the threat of cancelling its present contract in the event of its failure to comply with AECL's wishes, its attitude would 'not be too difficult.' At the same time, while CGE's stock was descending, Ontario Hydro's was rising. That 'long-term relationship,' Bennett concluded, was 'essential' to AECL, and everything must now be done to see it 'firmly established.'[51]

Howe approved Bennett's proposals on 20 April. He had other things on his mind. On 12 April parliament was dissolved and an election called for 10 June. Meanwhile on 13 April AECL told the media that redesign was going forward, as CGE had agreed the previous day.[52]

In Ottawa, Bennett had a couple of months' leeway to develop the policies he was already contemplating. The Liberals expected to win the election, as with every contest since 1935. The polls told them so, and they had plenty of money to spend to make assurance doubly sure. Bennett decided that it would be useful to take the new chairman of Ontario Hydro, J.S. Duncan (Hearn had retired), to England in May. He wanted to get Cockcroft's opinion of the Canadian reactor designs, and he hoped to persuade the British scientist to do a detailed evaluation of the Canadian program. Meanwhile he and Duncan, a wartime veteran of the Ottawa scene and, like Bennett, a friend of Howe's, could relax and chat. What about? Why, about 'the long-term relationship which should be established.' Such a relationship, Bennett hardly needed to explain, would not be 'of the pattern we have already established in connection with the NPD project.'[53]

We do not know, except by inference, what happened on the trip to England. We do know what happened in the election of 10 June. The St Laurent government was defeated; not least among its casualties was C.D. Howe, beaten in his own riding of Port Arthur. No party had a majority in the new parliament, but the Progressive Conservative party under John Diefenbaker had the largest number of seats. Diefenbaker was asked to form a government and on 22 June he was sworn into office. To replace C.D. Howe he asked Gordon Churchill to become minister of trade and commerce, and Churchill accepted.

Canadian atomic energy had never had a minister other than Howe. Churchill had not been especially interested in the subject while in opposition (he had been an MP since 1951) and nothing in his background – law and teaching – indicated an interest in either science or industry. Churchill was conscious that he was replacing

Howe, who loomed as large to Conservatives as he did to members of his own party; some said he was too conscious, for he suspected the worst of some of Howe's left-over appointees, including his own deputy minister, Mitchell Sharp. For the rest, he was determined to show that hard work and long hours could make up for lack of experience, and he spent the summer of 1957 camped in his office, attempting to serve both his department and his demanding prime minister, scheduling his appointments as late as 10 at night.[54]

Churchill might have been expected to show the same animosity towards Bill Bennett that he displayed to Mitchell Sharp. Bennett, after all, was a direct Liberal appointee, who had come into government years before as Howe's private secretary, a political job, while Sharp was a pure civil servant. But Bennett's political instincts served him better than Sharp's, for as president of Eldorado and AECL he had cultivated opposition members with an interest in their affairs, including Howard Green, the new minister of public works, and John Diefenbaker, whose Saskatchewan constituency lay next to Eldorado's flagship uranium mine. For whatever reason, Churchill decided, or was told, that Bennett was not to be touched. Once that was clear, Churchill gave Bennett his entire confidence. He might not be as knowledgeable or as well-connected as Howe, but nobody would say that he was not co-operative; and nobody did.

Bennett remained the conduit for policy between AECL and the cabinet, and he proposed to use the occasion of the election of a new government and the coincidental redefinition of the reactor program as an opportunity to rethink and redefine Canada's nuclear policy. The exercise began at a meeting of the Board of Directors at Chalk River in July 1957, and it continued until Bennett had worked up a state paper for presentation to Churchill and the cabinet late in the fall.

The first thing to be done was to establish a new division of AECL to direct and co-ordinate the development of NPD and the large reactor of 200 megawatts (electric). It would be located in Toronto for 'more ready access to the manufacturing facilities required in the development programme.' That may well not have been the only reason. As far back as 1954 there was some feeling on the part of Hearn and Gaherty, at least, 'that Chalk River should be reduced in scope and size,' an idea linked to their distrust for Chalk River's engineering capabilities. The latter had been remedied, but now it was time to extract the reactor designers from Chalk River's direct control.[55]

The new division would recruit its staff from the utilities and the manufacturers. Hydro got the ball rolling with a donation of fifteen engineers. To head the division, Harold Smith was appointed to continue his hybrid life as half a servant of Hydro and half an executive of AECL. He and his division symbolized the new relationship that Bennett hoped he was both creating and cementing between designer and consumer; it only remained to work out where the intermediary manufacturer, CGE, now fitted.[56]

The immediate answer to that question was available at the beginning of November. AECL's executive committee was told that a preliminary design study of NPD-2 had concluded that it would indeed be possible to build a 20 megawatt (electric) demonstration reactor using the principles sketched out by Smith and his team for the large reactor in their study, NPG-10. It would take three and a half years; it held promise as the best heavy-water, natural-uranium design for larger reactors; and it would cost $30 million.

Much work had still to be done. Further development of the calandria, the fuel, the fuelling machine, and the pressure tubes was indicated. George Laurence, who was wearing a new hat as chairman of the AECB's Reactor Advisory Safeguards Committee, was worried about safety and, in particular, about the lack of an overall containment shell 'as a last line of defence' in case of an explosion. That question need not be decided immediately – in November; but it would have to be decided in December if the new operational target of mid-1961 was to be met.

That said, the executive committee, with Duncan and some of his officials from Toronto present, as well as AECL's own senior executives, wrestled with an estimate that was 225 per cent up on the original, in the less costly days of 1955. With reluctance shining through their minutes, the committee and Ontario Hydro took the plunge. They would go ahead. Bennett then told them that Churchill had agreed to present the program to cabinet just as soon as it emerged from AECL; that approval was the last hurdle it had to cross.

Bennett's purpose was prescriptive and educational. He had to tell Churchill what AECL was and what it did. In the process he had to safeguard both its role as a pre-eminent centre of pure science in Canada, and its more easily understandable though more complicated mandate to develop a nuclear power system. Pure science, the president told the minister, in words that had been used before and would be used again, 'has as its purpose the establishment and broadening of knowledge about the bases of sciences from which technology is developed.' Then it was the turn of applied science, also represented at Chalk River, and of the power reactor group.

This group had to work out both the technology and the economics of power stations. Power stations were obviously feasible: the British had them and so did the Americans. High capital costs made nuclear stations, at least so far, uneconomic and uncompetitive. Heavy-water, natural-uranium stations had higher capital costs than light-water models which, using enriched uranium, did not have to be as big; if these high capital costs could be reduced, the advantage of nuclear power would increase.

Canada had a program that so far compared not unfavourably with the leaders in nuclear science – Britain, France, the Soviet Union, and the United States. This pre-eminence meant that Canada would sit as a member of the Board of Governors of the new International Atomic Energy Agency – reserved for the most technologically advanced countries. 'Canada has achieved this position,' Bennett explained, 'largely because of the emphasis which has been placed on fundamental research since the inception of the Chalk River project.' There was nothing to be complacent about; keeping ahead or at least abreast meant encouraging more fundamental research.

As for the Canadian line of heavy-water, natural-uranium reactors, there was some gratifying interest abroad. Gratifying, because Canadian industry might soon be capable of manufacturing an entire reactor system inside the country, something that few other Canadian industries could claim. Bennett recommended continuing along the lines already mapped out, in research, pure and applied, and in reactor design and production. To that end, he wished to establish a Nuclear Power Plant Division (NPPD) in Toronto.[57]

AECL's president was alive to the importance of creating an impression that would be both strong and favourable. He was aware that recent announcements by Ontario Hydro of an expanding thermal electricity program might damage his case by demonstrating a lack of interest, and he begged Duncan to write Churchill himself to reaffirm Ontario's interest in nuclear power stations. Duncan obliged.[58]

It was this reasoning that made its way to cabinet early in January 1958. It had proved more difficult to get an appointment to see Churchill than it was to convince him; even before the new program was explained to the minister he approved AECL's 1958–9 estimates, which were based on the assumption that the program would be approved. At a meeting on 16 January, Gordon Churchill summarized Bennett's paper for his colleagues and outlined both what the atomic program consisted of and what they were being asked to pay for it. There would be NPD, to be finished in 1961, and the 200 megawatt (electric) reactor, which would be developed simultaneously; both were to be under the umbrella of the NPPD. This would cost $140 million over the next four years, $25 million a year for Chalk River's basic research program, and $40 million for the two reactors. Of this, NPD's share would be $21 million, a figure which does not seem to have included what had already been spent.

Ministers were urged not to take a narrow view. If nuclear power were not adopted, Ontario would have no choice but to build more thermal stations and import more coal, perhaps as much as 300 million tons every year. Moreover, Canada's uranium industry, which depended on American and British military purchasing, might very well need a civilian customer. And, Churchill added, if the nuclear power plan proved feasible, it might even become an export item itself, in addition to its role as an import substitute.

With some apparent misgivings, the cabinet acquiesced. Some members worried about the lack of foreign co-operation in the Canadian program, and some urged that more be done to seek out foreign partners. But the weight of the $216 million already invested in atomic energy, as well as Canada's obviously prestigious international position in the field, persuaded the cabinet that this was a program that deserved support. After all, 'adverse criticism' might result from a failure to support a program that enhanced

Canada's standing. That was to be borne in mind, for 1958 would almost certainly be an election year, as soon as the minority Diefenbaker government found a way to call one.[59]

Churchill announced his government's support for Canada's expanded nuclear program to the House of Commons on 1 February. The same day parliament was dissolved, and an election set for 31 March. In a small way, atomic energy was part of the Diefenbaker government's promise to the Canadian people that they had the vision and the spirit to get Canada moving.

Completing NPD

It had been not quite a year since the Board of Directors had cancelled NPD-1. The delay was not costless. Money had to be spent, even if at a more reasonable rate in terms of AECL's budget. And the cost included delay. At first it seemed that about eighteen months had been lost; estimates at the end of 1958 placed the completion of construction in late 1960, with full operation early in 1961. The cost of NPD-2 rose, though slowly: it was $22 million in March 1959, with the bulk of expenditure scheduled for fiscal 1960–1.[60]

It did not stay low, and it lasted longer than expected. CGE took over the manufacture of the calandria, which took longer, and cost more, than expected. By November 1960 the anticipated cost of NPD was $35.3 million, with a target for on-power operation of December 1961. The news did not inspire confidence in CGE, whose representatives were plainly told that AECL was dissatisfied with their inability to predict their cost overruns. At least, Gray thought, they could have warned AECL of what was coming.

CGE was protected by a cost-plus contract, and hints that they might wish to assume some of the extra cost fell on stony ground.[61] Other rises were to be anticipated. Finally, in June 1961, the calandria was delivered, and in February 1962 fuel loading began. Criticality was achieved on 11 April 1962. Interestingly, the 1961–2 *Annual Report* listed only the $25 million that AECL contributed to the project; Ontario Hydro's $8 million and CGE's $2 million contribution would have added somewhat to the total.[62]

Another feature of NPD is worth underlining. Its construction occasioned the first official review of reactor safety by the AECB's Reactor Safety Advisory Committee, chaired by George Laurence, who was gradually moving out of the mainstream at Chalk River, and whom C.J. Mackenzie, still chairman of the AECB, had appointed to the task. The committee was made up from members of AECL's Occupational Health and Safety Division, the Department of National Health and Welfare, and the new Nuclear Power Plant Division. They had to examine the consequences for health and safety, both of workers and of the public, and to make recommendations to the AECB before it issued a licence for the new facility.

This was a new wrinkle to the atomic industry in Canada. ZEEP and NRX effectively pre-dated the AECB, and until 1956 the AECB regulated no more than the use of nuclear fuel for reactors. Early in 1957, however, the AECB issued a Nuclear Reactors Order, which stipulated that 'no person shall deal in any nuclear reactor except under and in accordance with an order of The Board.' It then exempted AECL's existing reactors (including NRU) from the order. University reactors, being built at McMaster University and the University of Toronto, were covered, and so was NPD.

Safety at NPD had already given rise to heart-searching; now, at the beginning of 1958, the whole issue was re-examined. The result was the same: NPD provided a very reasonable margin of safety against accidents, and its designers satisfied the Reactor Safety Advisory Committee that their proposals were feasible; later, the safety specifications were checked against performance.[63]

The Nuclear Power Demonstration Reactor had, as we have seen, a variety of sponsors. It was conceived by AECL, sired by CGE, confirmed by the AECB, and, finally, operated by Ontario Hydro. This distribution of responsibilities satisfied Howe and Bennett, and it resembled what was developing in Britain and the United States. The intention plainly was to develop a close relationship between initiator, supplier, and customer, with the idea that the supplier, CGE, would develop the necessary engineering and manufacturing skills to market a reliable reactor to meet the needs of the ultimate customer, Ontario Hydro, and possibly, eventually, other utilities at home and abroad. The overlapping responsibilities, and the novelty of the work, led to unanticipated friction. As a demonstration reactor, NPD showed Gray that AECL had not escaped from cost overruns by farming out its design and construction functions.

Bennett had seen NPD as the first stage in the transfer of an expensive but attractive new technology to industry. It was not possible in a country the size of Canada, with its limited pool of technical manpower and development capital, to develop more than one company, CGE; the choice was a risk, but one not out of line with other Canadian industrial experiments. CGE was to train itself on the reactor, and develop skills that it obviously intended to put to work in making other similar reactors. That, at least, was the theory.[64]

As for AECL's relations with Ontario Hydro, Bennett later observed that NPD was unique. In future contracts, 'I did not think that the NPD pattern should be followed. To be more specific, it was my thought that AECL would continue to supply and pay for the data needed to design a large power reactor. In other words AECL would have somewhat the same role as an outside engineering firm. Ontario Hydro would assume the total cost of building the plant, and of course would operate it.'[65]

When costs ballooned, from $13 million to over $35 million, directors and executives became alarmed and then disgruntled. With the NRU example fresh in their minds, they were not surprised to be disappointed anew. The entire blame was not attached to CGE, but its procedures and procrastinations were the subject of harsh comment and negative conclusions.[66]

There was also the third party, the customer. Ontario Hydro was involved at both the first and the final stages of reactors. Harold Smith and his utility-based team were crucial in the first and second design of NPD. In changing the design, AECL responded to customer demand as well as to technological developments. The result subjected the system, already running at the limit of Canada's engineering capacity, to added strains.

Canada was not alone in this. Reactor design and construction elsewhere were afflicted by delays and cost overruns. That this was so was, however, of only slight comfort. Just as Canada was proceeding with its most familiar design, so were the Americans, the British, and the French. The British were first in the export field, selling reactors to Japan and Italy. The cost to the British taxpayer was considerable, but it would be recovered, so the AEA hoped, as other countries signed on.[67]

As for the Americans, they had not one design but two. General

Electric (GE) developed an unpressurized reactor type, in which the coolant is allowed to boil; the resultant steam drives turbines to generate electricity. Westinghouse, with its naval connection, was the first in the field: its PWR at Shippingport, Pennsylvania, started producing electricity in 1957. The AEC followed with a program to encourage American industry to take up the reactor challenge, and by 1959 GE had done just that. The fact that GE had its own model did not escape notice in Canada. How would the Canadian reactor fare at the hands of General Electric's subsidiary when the parent company was in direct competition? Though the subject was never mentioned, it lingered at the back of Lorne Gray's mind; and by the end of 1958 Gray was the man making the necessary decisions at AECL.[68]

What would become of Howe's and Bennett's industrial strategy was, by 1958, no longer their concern. Howe's departure we have witnessed; Bennett's came in the spring of 1958, just after the Progressive Conservatives' overwhelming victory in a general election. Bennett became president of the British Aluminum Company. His departure from Ottawa had by this point been postponed several times. He had come with Howe in 1935 with the intention of staying for three years. The war changed that. At the end of the war he had completed arrangements to join a large Canadian corporation, but Howe persuaded him to tackle the Eldorado problem. When Howe asked him to become president of AECL in 1953 it was agreed that his main task would be to develop and launch a nuclear power program, and to set up the most effective organization and procedures for carrying out that program. By late 1956 Bennett felt that these objectives had been achieved and he advised Howe that he would like to make plans for leaving Ottawa at the end of 1957. After the election he advised Churchill of his understanding with Howe, and in late November 1957 informed him that he had accepted a position which would mean his leaving Ottawa not later than 31 March 1958. At Churchill's request, he postponed his departure until after the 1958 election. In leaving he recommended Gray as his successor. Together, Gray and Churchill would work out the country's nuclear future.

7

The Flight of the Arrow

The Diefenbaker government enjoyed one of the strongest majorities in Canadian history: 208 out of 265 seats in the House of Commons. Its triumphant re-election on 31 March 1958 meant, to most observers, that the Progressive Conservatives had replaced the Liberals as Canada's party of choice. But just over five years later they were gone.

The Diefenbaker government began to die on 20 February 1959. That day, the prime minister rose in the House of Commons to announce that his government had sent 'formal notice of termination' to the builders of Canada's new supersonic jet fighter, the Avro Arrow. The plane would not be built. The decision was long overdue. As Diefenbaker told the House, the Arrow cost too much; the government had other priorities, also expensive. He did not tell the House – perhaps he did not know – that the St Laurent government would also have scrapped the Arrow. In closing, the prime minister expressed the vain hope that his decision would remain 'above partisan political considerations.'

Diefenbaker knew that his decision would be controversial, but even veteran politicians were astounded at the outrage that boiled up. The government was denounced as positively un-Canadian, turning its back on the nation's future while throwing a highly skilled workforce – 14,000 of them – onto the street. It might not have been so painful had Diefenbaker not campaigned for election with a promise of 'a vision of our nation's destiny, with a positive message of hope and progress' for the nation's ills.[1]

Diefenbaker was the victim of a problem he had not made or planned. But since he had claimed the responsibility of 'hope and progress' he became the goat of disappointment and apparent decline. This situation was rich in irony, not least because of the prime minister's expansive ego; it was doubly ironic because conditions in Canada were not as bad as they were believed to be. Under Diefenbaker national economic output rose, and rose steadily, as did foreign trade and other economic indices. But unluckily for 'Dief the Chief,' so did unemployment; and in February 1959 the government added 14,000 to the unemployment rolls.[2]

The 'positive message of hope and progress' needed just a bit more emphasis under these circumstances. Unfortunately the government faced troubles on other fronts. European recovery was one of the principal aims of Canadian policy after 1945; and by 1957 it was largely achieved. European recovery meant that Canada's uniqueness as a haven of prosperity and security diminished; without any actual decline in Canadian wealth, output, or population, the country found itself for the first time in fifteen years scrambling to redefine its place in the world. In future it would be increasingly necessary to accentuate the positive in dealings with the outside world.

It did not take long for the government to begin counting its assets. The Arrow had been one, psychologically at least. If the Arrow was a failure, was there any place in Canada for skilled designers and highly trained technicians? The departure of some for greener pastures in the United States, part of a highly publicized 'brain drain' from north to south, was duly noted. The end of the Arrow was therefore a lesson, apparently no end of a lesson, for Diefenbaker. Hesitation would be his hallmark for the future. At the same time, it was not at all surprising that his government subsequently favoured plans to improve its shaky standing in the realm of high technology.

Considerations of politics, prestige, and industrial policy were therefore linked to public and political perceptions that the economy was faltering, and must be given a dose of government tonic. But what kind of tonic and where? C.D. Howe had liked to sponsor super-projects, but the trans-Canada natural gas pipeline and the St Lawrence Seaway, both Howe projects, were complete by 1959. The Seaway did afford food for thought. The St Lawrence was southern Ontario's last untapped big river. After 1959 new electricity would have to be thermal, from coal- or oil-fired generating stations. Soon, Ontario Hydro believed, it would be needed.

Ontario's need coincided with Ottawa's desire to prove that it could follow a consistent and progressive industrial policy. The Diefenbaker government, like St Laurent's before it, defined progress in terms of the American and British examples. Their atomic programs were impressive; besides, it was attractive to replace foreign coal with domestic uranium. The government also feared 'adverse criticism,' a phrase from Bennett's memorandum to Churchill of February 1958, if it showed anything less than enthusiasm for its atomic power program.

Churchill lent his ear to AECL. He wanted to learn more about his responsibilities, and he took them very seriously. In 1958 he attended AECL's annual shareholders' meeting, something Howe had never done. It was a formality, since Churchill had little to do other than confirm Bennett's hand-picked successor, Lorne Gray. Gray's appointment took effect at the beginning of May 1958.

Gray was just forty-five years of age in 1958, two years younger than Bennett. A native of Brandon, he had been educated in Winnipeg and Saskatoon, where he attended the University of Saskatchewan and in the process attracted the attention of Dean C.J. Mackenzie. Between spells of hockey in Europe, an expedient to keep his finances afloat, Gray took an MSc, taught briefly at the university, and joined the air force, where he rose to be a wing commander. Plucked from private industry by Mackenzie, he served in NRC from 1947 to 1949, before transferring to Chalk River as chief of administration in 1949. He had been in the company ever since, and as one of the two vice-presidents was in the line of succession.

The other vice-president, W.B. Lewis, would never be president. Bennett admired Lewis, but regarded him as too much of an enthusiast for a task that required an independent and critical judgement. He did not entertain a high opinion of Lewis's ability to handle his superiors or to exercise unchallenged authority over his subordinates. Life with Lewis would be spectacular, but it might also be brief.[3]

Lewis was not especially covetous of the presidency, but he was anxious to keep the company's agenda running on what was largely his own timetable. Gray offered that possibility; he added that in case of any conflict between them, Lewis would enjoy an avenue of appeal to the Board of Directors. Subject to that limitation, Lewis accepted Gray's authority. Only once did he appeal to the board over Gray's head; and when he did he failed, as Gray knew he must.[4]

Gray's presidency was different from Bennett's. Unlike his predecessor, an Ottawa man by avocation but with broad experience with business and businessmen outside the capital, Gray was an engineer and a company man. He had lived in Deep River for ten years; he had shaped the town, and he had come to love it. Gray, not Bennett, was on the firing line with the C.D. Howe Company and CGE, and he had drawn his own conclusions. He admired Bennett's ability to handle politicians, and he strove to imitate it. But Churchill was not Howe, and while he and Gray worked cordially together, Gray understood that AECL's minister was regarded in public and even by Diefenbaker as second-rate. When Churchill was demoted to be minister of veterans' affairs in 1960, he took AECL with him, and he kept it until the Diefenbaker government fell in 1963.

Gray looked back on Churchill with tolerance and affection. Churchill was, he thought, the best minister he served under. It was not that Churchill was actually a great minister, as Howe had been, but he was supportive. In the face of discouraging odds, Churchill soldiered on: against his colleagues, against the Treasury Board, against anyone who did not share his vision or affection for AECL. Yet no matter how much he tried, he could not hope to equal Howe. Howe had the power to impose himself on events; Churchill did not. Not, at any rate, without some considerable reinforcement. Unluckily for Churchill, Diefenbaker did not believe in reinforcing his ministers; the Chief was far too much of a lone wolf for that.

So it was Gray's business to muster the reinforcements. Bennett had had his own circle of friends among the bureaucracy, and Gray knew how useful such a circle could be. Unlike Bennett, Gray lodged, rather than lived, in Ottawa; his home remained in Deep River. Unlike Bennett, he could not draw on twenty years' experience in the capital. But AECL needed Ottawa, because Ottawa supplied it with money ($21 million in 1957–8, $25.7 million in 1958–9). Ottawa for these purposes meant the Treasury Board, but it also meant the Finance Department, the Department of External Affairs, and the senior bureaucrats who worked in each.

Theoretically, the Advisory Panel on Atomic Energy existed to supply the link with the bureaucrats. It was chaired by the most senior civil servant, the clerk of the Privy Council, and its membership included most of the people who counted in the bureaucracy. But Mackenzie and Bennett had ignored the panel; as far as Bennett was concerned, its only use was to facilitate AECL's foreign contacts and to harmonize them with Canada's overall foreign policy. In any case, like most senior committees, the panel had ballooned in size, with troops of members in attendance; its ability to decide matters declined in proportion. Bennett replaced it with a cabinet committee, which met infrequently; though the committee survived, under Diefenbaker it had to concentrate on the vanishing market for Canadian uranium.

Making the panel work was beyond Gray's powers. But if that were too ambitious, why not create an executive committee of its key members? Easier said than done, since deputy ministers were not easy birds to trap and confine for the necessary space of a meeting. Gray needed a meeting: how to stock it?[5]

The key figure on the panel between 1958 and 1963 was the clerk of the Privy Council, R.B. Bryce. An engineer turned economist, Bryce was a veteran of the Finance Department where he had been a major influence in shaping Canada's postwar economic policy. Bryce's competence, assurance, and imperturbability impressed Prime Minister St Laurent, who appointed him clerk of the Privy Council instead of making him, as might have been expected, deputy minister of finance. Bryce's balding, bustling figure was therefore at the centre of things. In Bennett's time he, Jules Léger, and John Deutsch had sat as a senior officials' committee to reinforce the ministers above; Bryce enjoyed the experience. His influence, already pronounced, received a boost when John Diefenbaker became prime minister. At first Diefenbaker looked on Bryce with suspicion, but it did not take long before he 'realized that he had inherited a gem, a man who could be trusted and one whose advice was worth taking.'[6]

Bryce was crucial to Gray's plans, but so were the other deputies. They all had to eat lunch, he reasoned. Lunching over business was an accepted and cheerful practice. It was true that Ottawa restaurants were something of an obstacle either because of the quality of their food or because they presented too many competing attractions – other tables, and problems other than AECL's. There was, however, the Café Henry Burger in Hull, which boasted wood-panelled private rooms upstairs, of just the right size. So Gray issued his first invitation to lunch. The response was gratifying. It was much easier to thrash out the problems of reactor building or Canada's policy at the International Atomic Energy Agency over canard à l'orange. Getting an invitation from Gray soon became an event, and if the price was chatting about Chalk River it was well worth the experience.

The results were pleasing, but, Gray thought, they could be improved on or at least supplemented. He had a largish office and a boardroom, and he installed a kitchen. His secretary's duties, he let it be known, were unusual. They involved selecting the best produce at the market and rendering it into palatable lunches, which were served right at Gray's office. Mrs Parry –who had also been Bennett's secretary – lived up to the job. She was, Gray remembered, 'the best cook in Ottawa,' and her talents were indispensable. The fact that Gray gave his own lunch rather than restaurant or country club meals raised eyebrows among his other staff and among the civil servants who were not included. Some objected to the style; others to the content; some confused the two; and some were merely envious.

As to style, it was obviously indispensable. Style made the lunches an event rather than a chore. Liquor and wine were served, but not by the gallon. It was lunch, after all, and Gray wanted his guests to go back to their offices and implement what had been decided. As far as we know, they went sober.[7] But lunch, wine, cigars, and camaraderie promoted good feeling, not to mention liberation from the dead hand of precedent and departmental self-interest. The deputies could reach decisions as a group rather than as heads of delegations from civil service principalities.

In 1958 or 1959 the civil service was still comparatively small, and the corps of deputy ministers few enough to be collegial. That would not always be the case, as time passed and departments grew. Excluding casual labour, the federal government employed a total of 183,000 personnel in May 1958, when Gray took on the presidency; by 1974, when he left, it had 300,342.[8] While Gray built for his own generation and the needs thereof, the generation beneath would eventually rule the system that he so ably circumvented.

Circumvention, in Gray's mind, made the system work tolerably well. There were enough decisions pending not to waste more time than was necessary, and between 1958 and 1963 the decisions came thick and fast; of these decisions, two will concern us in this chapter: the final decision to proceed with a large reactor; and the decision to build a new and innovative experimental reactor.

Gray's Turning-Point

Sometime during the winter of 1958–9 Herbert Smith, the president of Canadian General Electric, called on Lorne Gray. They had a number of things to discuss. There was NPD, well underway at Rolphton. Completing it would take three years or more; it was, therefore, high time to discuss what came next. What, Smith asked, about the 'large reactor' that NPPD was working on in Toronto? What could CGE expect from that project, and when?

Not much, Gray replied. There would indeed be a large reactor, but CGE would not get the contract to build it. Ontario Hydro objected to being tied down to a single supplier, objected to becoming a captive market. It was accustomed to tendering its projects, and if it were not possible to get a competing Canadian bid – and it was not – then it could at least require tender on the components. Hydro's preference, indeed its demand, was that AECL handle the new large reactor project itself. AECL had no choice but to accede.

'Bullshit,' Smith responded. CGE had not created an entire atomic division and invested $2 million in order to retire from the field after one reactor. It expected better from its atomic partners; surely Gray understood that.

He did. He had not said that CGE would get no more business. He had something in mind; not, to be sure, the large reactor, but something to go on with. AECL was contemplating another reactor, an experimental model that would not use water as a coolant, but an organic substance. He knew CGE had been working on the idea, which had originated in Chalk River, and he knew that there was even some enthusiasm for turning the 'large reactor' into an organic-cooled model. That could not be contemplated, but a small demonstration model might be. If CGE wanted the business, it could be arranged. It would, however, be arranged slightly differently than for NPD. With a reactor behind it, CGE could surely now agree to a fixed price, instead of cost-plus, for the organic-cooled reactor.

Smith, as he had to, grasped at the straw of preservation. Gray, as he wished to, reported to his directors that a new project would soon be launched, and that it would keep CGE on side, even if not completely happy.[9]

The extraction of CGE from the higher direction of Canada's atomic energy business was a major reversal of policy, with profound long-term implications. Howe and Bennett had had in mind off-loading responsibility for building Canada's reactors, for limiting the federal government's responsibilities and future liabilities. It was enough that the government should design an energy system, develop it, and set up a connection between the client – Canadian utilities and especially Ontario Hydro – and the licensed supplier. Thereafter, CGE was obviously expected to be the standard-bearer.

The Bennett-Howe position was designed to be practical, but some disagreed with it. Gray later claimed that he had never liked the arrangement; but it was Gray who had recommended CGE in the first place, back in 1955. By 1958 he may well have decided that the selection of CGE had been a mistake, to be corrected before it was too late. The mistake, as Gray later remembered it, was confirmed by NPD's cost overruns, secretiveness, and subsequent recriminations. He knew that a good estimate could win a contract.

But he also knew that the lower the estimate, the more attractive the estimator. Once committed, a project could be allowed to drift upwards.[10]

CGE's upward drift had been disconcerting. It had provoked fear among the Board of Directors, and then rage. But secretiveness was probably part of CGE's corporate culture. It did not wish to tell the customer, AECL, any more than it absolutely had to. When Gray phoned for information, he did not often get what he wanted. This did not create or sustain a climate of trust between CGE and AECL. For the short term it made no difference, because AECL could not afford to abandon NPD; but once Gray became president, he could bring unhappy memories to bear on the CGE problem.[11]

It did not help that a contingent of AECL veterans stocked the CGE works in Peterborough or that some of the CGE executives were very well regarded by some if not all of their Chalk River counterparts. The problem, Gray explained in July 1957, was that CGE was just not very good in rendering up-to-date estimates to AECL. This was an important consideration, he continued, because cost control and estimating were expected to be a particular strength in dealing with a private firm – presumably to improve on AECL's experience with NRU. If CGE had taken a proper attitude on supplying information to their client, one of Gray's assistants wrote, 'a great deal of stalling and unpleasantness' could have been avoided, for things were not all bad between the Peterborough firm and its governmental client. As it was, CGE was suspected of using NPD 'to obtain information and to embark on development projects which are not in our opinion directly applicable ... thus improving their commercial position and limiting their effective financial contribution toward NPD.'[12]

It was only somewhat later that Gray's feelings towards CGE were recorded; in November 1960 he told his CGE counterparts that he felt 'grave concern' at their delays and cost increases. 'Mr. Gray,' the minutes read, 'assured CGE that they would not receive a further cost-plus contract from AECL unless he had proof of a major change in CGE's method of handling a project of this sort [NPD].' These views were consistent with AECL's earlier sentiments on the subject.[13]

Gray had recommended the CGE contract back in 1955, when

only Harold Smith, among those most directly concerned, was opposed. It was, he argued, unwise if not worse for Ontario Hydro to place the fate of its reactor program in the hands of a single private firm. CGE, moreover, was a firm that could only get going by raiding staff from Chalk River; why not stick with Chalk River or, better yet, go with Ontario Hydro's engineering staff? For Hydro did have its own large engineering department, developed under Hearn. It had gradually replaced the consultants who had worked for earlier Hydro projects, such as the Acres firm in Niagara Falls, and as it grew Hydro came to appreciate the luxury of its own in-house engineering staff. In Smith's opinion it should do what came naturally and participate in Hydro's next great continuing project.

It took time for Smith to convert others to his opinion, but the opportunity was afforded by the rough relations between CGE and AECL. Gray appears to have been an early convert, and then Hearn. Next was James Duncan, Hydro's chairman and Hearn's colleague on the AECL board. Duncan was widely regarded as a lightweight, but against Gordon Churchill he could make his opinion count. By the summer of 1957 Duncan was openly sceptical of CGE's performance.[14] Writing to Bennett in mid-July, the Hydro chairman indicated that he had considered cancelling the CGE contract. He had changed his mind since, but it was a straw in the wind – a chill wind where CGE was concerned.[15]

This was especially so because of NPPD. It had been managing liaison with Peterborough since the spring of 1958, and it was ready to do more. Its large reactor was baptized CANDU, for 'Canadian deuterium-uranium,' a term that communicated, as it was meant to, the idea of Can-Do, an intrepid commitment to surmounting obstacles. One of the first would be CGE.

The original proposal for NPPD contemplated a small, self-sufficient design and development team, whose arrangements mirrored Smith's dual status as an employee of both Ontario Hydro and AECL. AECL would have the principal responsibility, but Hydro would assist in every way imaginable, from lending its research facilities to staff support. In fact, Smith thought that Hydro should furnish all standard engineering functions, on the premise that NPPD was essentially a transitional organization, a project team for the future CANDU. It was unreasonable for AECL to set up an elaborate staff in Toronto when it would presumably not be permanent.[16]

NPPD therefore had a strictly limited mandate. It was to produce a fully fledged design, sometime around the end of 1961. After six months of contemplation and absorption, work could begin on construction in 1962, with completion set for 1965 or 1966. There would be a joint AECL-Hydro arrangement, with Hydro assuming responsibility for operation when complete. The completion date was appropriate because the utility's forecasts showed there would be a need for additions to its baseload power at that time.

These were Bennett's assumptions when he wrote to Churchill in November 1957; they were embodied in the long-term plan adopted by the cabinet and announced by the minister just before the 1958 election. They were prudent, providing for long and careful study as well as an opportunity to assimilate any lessons to be learned from the design and construction of NPD which, it was believed, would be operating by the time ground was broken for the CANDU. As with the commitment to private enterprise, they represented a philosophy that limited government obligations, in this case by spreading them out over time and controlling as best they could the elements of surprise and risk.

The caution did not commend itself to Lewis, either intellectually or emotionally. Intellectually, Lewis was convinced that the CANDU design would work; and he was temperamentally inclined to action, as immediate as possible. Circumstances supported him, because other countries were forging ahead with their own programs and designs; only Canada seemed to be holding back, at a time when competition in design seemed to be a fundamental consideration. As an AECL study put the matter in November 1958, following Bennett's timetable would postpone operation of Canada's first large-scale power station until 1965, if not later. 'In view of the progress being made in the U.S., U.K., France and a few other countries, such a date may be too late. There is a very real danger that the Canadian programme will be overtaken by events and developments in other countries.' Canada must produce a reactor by 1963, at the latest. Lewis did not say what the consequences of failing to meet the deadline would have been; one is tempted to say that they would not have been earth-shattering.[17]

FUEL RODS

CANDU reactors use a uranium-dioxide fuel within a zircaloy cladding. Pencil-sized rods are bound together in a bundle 3 1/4 inches in diameter and 19 1/2 inches long; each fuel bundle weighs roughly 33 pounds. The fuel rods are designed to withstand high heat production and multiple power cycles, without changing dimensions or requiring replacement. The reliable performance of fuel rods is a major factor in the economical operation of reactors, and an important aspect of CANDU's attraction.

Not everyone saw Lewis's point. Harold Smith, admitting that in his opinion the CANDU design was almost as good as proven, nevertheless questioned whether entering a race with the Americans, British, and French was really necessary or even desirable. CANDU, he thought, needed more time. The original proposal at the beginning of 1957 had been worked up very quickly, and he was worried that it might not be properly finished. '[C]ertainly at our present rate we will have much less capacity of nuclear plant than the other three [countries] by 1965,' he wrote in November 1958. 'But is this a suitable criterion for setting the Canadian programme?' Ontario Hydro would derive no benefit from the speed-up; its timetable was already fixed. As for exports, Smith was sceptical. Canada had a high-wage, high-cost economy. 'Can Canadian industry achieve any lasting success in foreign markets in the nuclear field by having advanced know-how when their labour rates are two or three times greater than other highly industrialized nations?'

Nevertheless, Smith sketched a plan for a speed-up. It was possible to choose a site for the CANDU and to do it soon. Given that final approval of the CANDU was virtually certain, it would save time to order its components as soon as possible. When approval was given, as it must, they would be ready to go. Just in case, Smith suggested that CGE's organic-cooled reactor be developed as a parallel if not an alternative to the CANDU.[18]

On 10 November Gray and Duncan called on Gordon Churchill. The board had just met, and it was appropriate to let the minister know what had transpired. Gray told Churchill what he had told the directors, that the British and American programs were 'considerably larger' than Canada's. A visit to France revealed that that country's nuclear effort was 'far advanced,' in fact, eight times larger than Canada's. Under the circumstances, Gray said, 'the question had ... been raised of whether or not the CANDU programme might be speeded up.'

The board thought it should. Gray should ask for tenders on important components in the CANDU system, and place conditional orders to reduce delay. If CANDU was not approved the orders could be paid off, but such an outcome was very unlikely. Site selection was next: let Ontario Hydro decide on a site and go ahead. The question of who would pay was more involved and more difficult than what should be paid. James Duncan thought that the federal government should pick up the tab; Gray believed that the cost should be shared between the federal and provincial jurisdictions. That thorny question the board had left for Gray to discuss with the minister.[19]

Churchill, however, wanted to discuss first principles. Why, he asked, was Ontario getting the reactor? He and his visitors both knew that in a federal country like Canada the distribution of expenditures and projects among the several provinces was always a lively political item, and it was a safe bet that his colleagues would be curious to know the reason why Ontario, the richest province, was getting a very large hand-out from the national treasury. (When Churchill brought the project to cabinet the following spring this was in fact the first question asked.)

Churchill was 'a little anxious' that the sums committed to atomic energy would raise hackles in the House of Commons. 'I told him,' Duncan wrote, 'quite to the contrary ... that if there were any criticism of Canadian policy it might be on the amount of money which the Government was placing at the disposal of A.E.C.L. which was approximately eight times less than France's similar programme.'

AECL's emissaries now went to the heart of the matter. The company would need money, more money than the cabinet had approved only seven months before. They wanted to proceed with site selection, with the pre-ordering of components, and with the development of an organic-cooled reactor – with Smith's list. As for paying, Gray and Duncan differed. Duncan urged that nuclear power until finally proven both feasible and economic must remain a national responsibility; the real issue, he told Churchill, was not Canada's first nuclear power plant, but its second, third, and fourth. Otherwise, there was always 'the necessity of buying coal from the U.S.'

According to Duncan, 'Mr. Churchill was inclined to agree that this was a national responsibility, and expressed his willingness to go along with this viewpoint.' Gray suggested it might be possible for Ontario Hydro just to buy (or 'rent') power from the CANDU; once the plant had proved itself, it could buy it from AECL. Only on a depreciated basis, Duncan interjected, and there the interview seems to have ended.[20]

With Churchill's apparent consent, Gray returned to his office to turn acquiescence into enthusiasm. Large sums were involved, $15 million more for the CANDU and $6.5 million for the organic-cooled job (called OCDRE: Organic-Cooled Deuterium-Moderated Reactor – Experimental). The minister had given Gray a final piece of advice. If an accelerated CANDU went forward, the cabinet would prefer to see results sooner rather than later.[21]

Some of the winter was spent dealing with a bright idea that NPPD had evolved for a vertical rather than horizontal calandria; it would be simpler, certainly, and also 'more advanced' than the horizontal model, 'from an engineering standpoint.' It is probably not too much to say that this was the least favourable moment to bring forward any substantial alteration to the CANDU design. Gray gave the engineers' proposal every consideration, and then politely but firmly told NPPD to forget it. With Lewis standing, figuratively, at his elbow, Gray wrote to Toronto on 13 March to tell Harold Smith that he was to 'proceed with the CANDU programme as outlined' – that is, horizontally.[22]

Churchill and the cabinet never knew what they had been spared. Knowledge of that sort might have given CANDU an even rougher ride than it received from the cabinet when it finally arrived on that body's agenda on 21 April 1959. 'Ontario would be obtaining a good deal under the proposed transaction,' ministers (presumably not from Ontario) argued. Let Ontario pay more for what it was about to receive, or tell the cabinet the reason why. Seeing the CANDU settling below the horizon, Churchill begged his colleagues at least to give him the organic-cooled reactor, but that too was denied. Both questions were put over, as the minutes recorded, 'until an early meeting.'[23]

Gray scrambled to rescue something from the wreck. He generated a special 'supplement' to the submission to cabinet that had gone with the original proposals. Ontario had in fact spent much more than the cabinet thought: $9 or $10 million on NPD and $2 or $3 million more on CANDU, even though the sum was masked by including it under personnel. Though Ontario was the first province to get a reactor, it would not be the last: Manitoba and the Maritimes were likely candidates by the late 1960s. Their utilities knew what was happening and had consented, in the interest of experimentation and development. Besides, only large reactors were at this point economical, and only Ontario could really use one; of course large reactors were only a stage on the road to smaller ones.[24]

Churchill duly repeated this line when next the subject surfaced at the cabinet table, on 16 June. He assured his colleagues that everything possible had been done to check the accuracy of AECL's submissions, adding that Manitoba Hydro, in particular, had approved them. Ontario Hydro was not getting a free ride, he emphasized, and would pay for the plant once it was proven. He added that this work would materially assist Canada's uranium industry which, he had good reason to believe, was facing the loss of its American markets after 1962.

Neither of these points was entirely valid. Ontario Hydro had reserved its position on buying the future atomic plant, while the amount of Canadian uranium that the CANDU could consume was a drop in the bucket compared to the uranium industry's available capacity. But valid or not, the cabinet accepted them, and approved both the CANDU reactor and the organic-cooled experimental model.[25]

What cabinet discussed was important. Equally important was what it did not consider: the fate of the CGE connection. The cabinet was more interested in larger affairs: the distribution of federal investment in an equitable manner across the country.

Once it was satisfied, or at any rate assured, that Ontario was not benefiting unduly, the details were of minor interest. It was a useful lesson in the realities of life and politics in federal Canada, and it was one that AECL did not soon forget.

The Organic Reactor

The morning and evening bus ride between Deep River and the Chalk River laboratory had a number of advantages. It was cheap and it was convenient. Most people took it – not Lewis, to be sure, but division heads and section heads as well as more junior staff. It afforded an occasion to chat, to catch up, and sometimes to talk about something completely different.

So at least Ian MacKay found, when he and his next-door neighbour shared a seat as they trundled back and forth along Highway 17, talking about the idea of a different kind of cooling system for a reactor. The reactor they were shaping in their minds still used heavy water as a moderator, but would be cooled by organic liquids of the terphenyl class, which had been used for some years in engineering experiments at Chalk River. (Organic compounds are the compounds of carbon, which has a unique ability to form immensely complex molecules. Living things are largely made of them, although not all organic compounds are associated with life.) These coolants were much cheaper than heavy water and could be used at higher temperatures, thus making the reactor more efficient. They absorbed more neutrons than heavy water, but not to an intolerable degree.

The idea was hashed over in bull sessions during lunch in MacKay's office, where it was further developed. It was not an entirely novel concept: there were two American organic-cooled reactors operating successfully by the early 1960s. But MacKay and his fellow engineers had other fish to fry, more important ones as far as the laboratory was concerned: fuelling machines, pressure tubes, the myriad details of reactor design. When MacKay left AECL for Peterborough in 1955 he took the idea, still undeveloped, along. In Peterborough he interested his superiors at CGE – enough, at least, for them to allow a study to be made, one that ultimately cost $250,000 of the company's own money.[26]

This study took time to bear fruit, but by November 1958 it had attracted W.B. Lewis's attention. Lewis brought it up as an afterthought, at a meeting of the Board of Directors otherwise directed to CGE's progress on NPD. By this point the board was suspicious of CGE and all its works, and Lewis thought it important to stress that the Peterborough company was working on an organic-cooled reactor entirely on its own time and from its own funds. Organic coolants, Lewis added, were important and interesting, and should not be ruled out for CANDU design. He thought it a good possibility for smaller reactors, less than 150 megawatts (electric) capacity. These held some interest for AECL because they could be established in energy-scarce northern outposts; this was an item that could attract support from another department of government, and at a time when the Diefenbaker government was trumpeting a 'Northern Vision' as part of its political stock in trade. Possibly for that reason the board authorized Lewis to pursue the matter with CGE.[27]

The timing was extraordinarily convenient. Gray wanted to distance CGE from the CANDU program, and an organic-cooled experiment offered a way of compensating CGE's natural disappointment when it was told that its hopes were about to be blasted. CGE, duly informed, duly consented. The course of negotiations was not, however, entirely smooth. CGE at first presented what was in effect a proposal for a rival reactor system. AECL was to commit $4.5 million to a study which was to last thirty months. At that point CGE was to submit a proposal for a 200 megawatt (electric) reactor or, if its proposal was not yet ready, to commit $250,000 more of its own funds. This was too rich for Gray's financial, and Lewis's technical, tastes. Gray was nevertheless attracted by what he termed in February 1959 'a second string to our bow.'[28]

By February 1959 Gray had decided to recommend to the board the organic-cooled experimental reactor, with an eye to eventual use in the Canadian north. In March the board approved a $500,000 study by CGE. CGE was directed to proceed 'as speedily as possible with work on the design and development of OCDRE.[29]

Gray and Lewis knew of the interest shown in organic coolant systems by the AEC and by Euratom, a recently founded combined European atomic agency, whose prospects at the time seemed bright. OCDRE thus became a counter in a longer-term arrangement that promised interchange of ideas and, should the Canadian design succeed, eventual benefits. In the short term, there was hope that some development costs could be offloaded onto Euratom as part of that agency's contribution to a joint investigation of organic cooling.[30]

Not quite all AECL's new experimental eggs found their way into the OCDRE basket. Simultaneously with approval of the OCDRE study, another study, this one by Canadian Westinghouse, CGE's rival, of a small enriched-fuel reactor suitable for the north, was authorized. By the time this study moved out of the executive committee in May, matters were far advanced with CGE, and OCDRE was teetering on the verge of cabinet approval. And thereby hangs another tale.[31]

OCDRE was intended to be a straightforward experimental project which, like its three predecessors, ZEEP, NRX, and NRU, would be located at Chalk River. This made sense in terms of process as well as location, for reactor research and development was also there, and the preliminary work on coolants involved placing an experimental loop in NRU to test the materials. Moreover, C.E. Grinyer, whom Gray had snapped up from A.V. Roe's collapsed Arrow program and made AECL's vice-president, engineering, was at Chalk River; he would handle AECL's end of construction.[32]

Cabinet approved both OCDRE and CANDU in the middle of June 1959. Two months later, at the end of August, a new proposal surfaced, one which, apparently, Gray had not previously discussed with either his board or his executive committee. He had, however, discussed it with Gordon Churchill and with W.B. Lewis.

Lewis, it seems, was not a willing discussant, but after a certain amount of persuasion he swung around to Gray's point of view. What arguments Gray deployed must have been persuasive, for Lewis was not one lightly to concede a point. Gray may, of course, have spoken with the authority of the minister and the cabinet behind him, which the date, late June or early July, tends to confirm. (It was a scant few weeks since the cabinet had made its regional feelings plain.) After further consultations outside the company, Gray drafted a letter to Churchill on 14 July. In it he made a proposal.

It was for a second atomic research and development site. Although the board had not recorded the fact in its minutes, Gray transmitted its conclusion that Chalk River was getting too big. 'It is universally agreed,' he wrote, 'by administrators of research and development programmes that when a research establishment has reached a certain size, further expansion becomes inefficient.' This was because its components tended to become self-contained, and cross-fertilization among its several units declined. This was true at Harwell, or so Gray and the corporate secretary, Donald Watson, argued. For Chalk River, a ceiling of 2500 could be tolerated. Excess staff, including the hundred or so needed for the new reactor, would have to go elsewhere.

Where? Watson later suggested that he had Vancouver or Victoria in mind, in or close to large cities, with an attractive climate and with a university close by; but as the board had learned, the Diefenbaker government took the location of federal facilities very seriously. A quick survey of federal research laboratories by a committee of four (Gray, C.J. Mackenzie, E.W.R. Steacie, and Adam Zimmerman of the Defence Research Board) indicated that three provinces were sadly lacking: Newfoundland, Alberta, and Manitoba. 'Newfoundland,' they decided, 'is [not] a possible site for a major research and development establishment at this time.' Alberta had no need of atomic energy, blessed as it was with abundant oil and gas. So it would be Manitoba, 'somewhere along the Winnipeg River,' the stream that connects the Lake of the Woods with Lake Winnipeg, along the edge of the Canadian Shield. Manitoba was also the home of Gordon Churchill, and Churchill was not displeased.[33]

A preliminary survey went forward under the supervision of Shawinigan Engineering, and Gray journeyed to Manitoba to meet with that province's acting premier (the new Conservative premier, Duff Roblin, was in Europe). His meeting went well, and so did the survey. On 8 November he reported progress to the board: a probable site near the Seven Sisters Falls on the Winnipeg River; an opinion by the CMHC, the federal government's housing agency, that a new townsite would be required; and a preference for Shawinigan Engineering to be the prime engineering consultant, using Winnipeg firms as required. Over all, Gray placed Grinyer, who had quit Chalk River and was moving back to Toronto as a manager in NPPD.[34]

The reactor, as proposed by AECL and as approved by cabinet, was to incorporate a number of novel features. It would permit experimentation with organic coolants, but it was also intended to demonstrate other aspects of power reactor design, such as on-line refuelling and an organic top-shielding device. These apparently superfluous additions would attract and concentrate interest from utilities on organic-cooled designs. But eventually they proved a false start, and were abandoned as likely to hamper the reactor's experimental functions.

The original estimates of OCDRE's cost hovered around $6.5 million. Site costs were minimal, since the device was to be in Chalk River. By February 1959, still at Chalk River, it was up to $11 million; in September of the next year, with moving costs included, it was $18 million.

The story was a familiar one. The design was more complicated than originally allowed for, which upset the timing as well as the cost. There were difficulties in finding a proper organic coolant and, according to a later analysis, 'some areas of the design could only be guessed at, so that contingencies had to be increased to allow for the unknowns.' The number of fuel channels, originally fifty-four, was reduced to thirty-six, but that failed to reduce costs sufficiently.[35]

Negotiations with Manitoba were also complicated. The new research centre would not be costless for the province. It would have to look after some of the infrastructure, such as roads and a bridge across the Winnipeg River, as well as housekeeping details. By the summer of 1960 agreement seemed far away, and Churchill had to be convened to sort out the matter. With Churchill's help an agreement was forthcoming, and on 21 July 1960 cabinet approved the relevant order-in-council.[36]

The research centre and the townsite were both on the edge of Whiteshell Provincial Park about sixty-five miles northeast of Winnipeg. A townsite, later called Pinawa, was selected and designed by Central Mortgage and Housing on unimaginative lines. (First preference for a name was Whiteshell, a postal address that already existed elsewhere.) The name was accepted after the board was told that Pinawa meant 'slow, calm, gentle water.'

Like Deep River, Pinawa had a shopping centre, including a

bank, and a Hudson's Bay Company outlet, and a company hotel (Kelsey House) along the lines of Deep River's Forest Hall. The initial target was 55 houses, rising to 146 by 1963, with hotel accommodation for 70. This was of course smaller than Deep River, although it was not beyond the bounds of possibility that the Manitoba community would eventually reach the same size. Pinawa was much closer to Winnipeg than Deep River was to Ottawa, and there were no intermediate communities of any great size, and so citizens of the new model town found it possible to hit the highway after work, drive to Winnipeg, and make it back in time for bed and work the next morning; the fact that the terrain was flat almost all the way, and that the roads ran in straight lines, helped. As in Deep River, the best recreation was sports: sailing on the river (there was a marina), hockey, and the inevitable curling.[37]

Final agreement was reached on joint facilities, as between AECL and Manitoba, just as the company was reconsidering its commitment to OCDRE. Frank Sayers, who worked as liaison between Gray and NPD and Whiteshell, commented on progress on OCDRE in August 1960. He told Lewis that the initial advantage of OCDRE, the elimination of some expensive heavy water, had 'got lost in the wash,' resulting in 'a proposition only marginally better than CANDU and that only if one can accept the estimates' – in his opinion a major act of faith. (This was, to be fair, something of an exaggeration; besides savings on heavy water, OCDRE was intended to have a higher outlet temperature and therefore a higher power density and high efficiency.)[38]

Andrew Gordon, a chemist from the University of Toronto and a long-standing member of the board, was and remained sceptical about organic reactors. He knew his flame-throwers, he told his fellow directors, and terphenyl was the same stuff.[39] But despite his misgivings he admitted that 'in many respects he had found greater success than he would have expected and had concluded that the use of organic materials in nuclear reactors was not unpromising.' He was pleased that the designers at Chalk River and Peterborough were held in high regard by their American cousins, who considered that the organic-cooled design would 'eliminate many major problems' and was 'imaginative' besides.

Gordon awarded OCDRE a grudging go-ahead, provided it was

kept in bounds and its costs under control. This was a sore point, since AECL was having its costing problems with CGE at the time. Though Grinyer expressed optimism, the board was nervous. There were problems choosing the precise kind of organic coolant, and there were troubles designing sheathing. And though Lewis remained interested in organics, he pointed out to the board that OCDRE was very much an unproven technology, while CANDU was far advanced.[40]

Under the circumstances the board recommended to the cabinet that work on OCDRE proceed in stages, each new stage depending on satisfactory completion of the preceding one. (The cabinet had told Churchill that he must return to cabinet with each new stage before work could proceed.) At least they let the first stage go forward, but it seems that approval was secured by emphasizing how much the British and the Americans thought of organic-cooled reactors and the benefits to be obtained from a 'pooling' of knowledge that would result, a pooling whose scope Churchill seems to have exaggerated.[41]

Indeed, by the spring of 1961 the design still had serious kinks. Grinyer reported in April on problems with the coolant and difficulties with the sheathing material. Lewis criticized the basic engineering, and worried aloud about the possibility of the decomposition of the organic coolant under radiation. More loop experiments were ordered; the development phase was extended for another year and, as the meeting adjourned, Gray raised 'the question of whether any alternative reactor should be considered for construction at Whiteshell.'[42]

Nothing further needed to be said. When Grinyer next appeared before the board in May he proposed a further delay. The board agreed – there was little else that could be done – but the news came as a blow to those working on OCDRE. Some, Grinyer later reported, treated the word from Ottawa as a death blow for their project. That was not the intention, but it did not take long for it to surface. In September F.W. Gilbert, who had been appointed to manage Whiteshell, urged scrapping OCDRE and plumping for an NRX-type reactor, at a cost of $20 million.

NPPD now hatched an alternative. Their idea kept the organic coolant, and it cost half as much as the NRX proposal. It would resemble OCDRE, except that a solid shield would be substituted for a pool tank. It was called simply OTR, Organic Test Reactor. A design would be ready for the start of the construction season in April 1962. It was this proposal the board accepted.[43]

It may be asked why. The political reasons for placing a reactor facility in Manitoba were strong, but not insurmountable if it were demonstrated that the reactor to be placed was excessively costly or persistently impracticable. That had not been demonstrated, but as discouragement mounted, the one thing in the organic's favour was the fact that CANDU had not yet been proven to work. If it did not, there would be only OCDRE, OTR, or some variant. There were still experts in the United States as well as in Canada to say that an organic reactor was preferable, and AECL could not afford to ignore them.[44] Another study from CGE was therefore commissioned. It suggested that, assuming the organic coolant behaved itself, power might be produced at about 5 mills per kilowatt-hour, a figure which AECL considered competitive with competing sources.

A design was produced. The OTR was bounced back up to fifty-five fuel channels (one more than the first OCDRE design), all of which could be detached and connected to separate loops. The fuel, like that at Chalk River (which had been converted in 1962), was uranium oxide slightly enriched by uranium-235 – to about 2 per cent by weight. The designers did not try to resuscitate on-line refuelling but, pursuing the organic idea, they provided two separate cooling systems, each with a different organic coolant. The cost, they estimated, would be $14.25 million.[45]

The cabinet, grumbling, concurred on 5 January 1962. Whiteshell could have its reactor. Better still, CGE could build it, an aspect of the Whiteshell dilemma that had given the board some concern. It is to this aspect that we shall now briefly turn.

CGE's nuclear program had been created by AECL. In return, AECL had sustained CGE's Civilian Atomic Power Department with various contracts, first for NPD and then for OCDRE. Although Gray and his staff were exercised by CGE's cost-accounting practices they were considerably more impressed by what they called 'a highly competent atomic energy team, the best in Canadian industry and perhaps as good as any comparable industrial group in any country.' It would be a pity to destroy such a group, and an even greater pity to do it while people vividly remembered the dismemberment of the Arrow team. Unemployment was higher than it should be, polls showed that the electorate was unsettled, and Peterborough, which should have been reliably Conservative, had kicked over the traces in a 1960 by-election and elected the first member of the New Democratic Party to the House of Commons. It was not surprising that when Lorne Gray discussed CGE with Gordon Churchill, the minister agreed 'that there should be no drastic layoff.'

Gray listed several projects that would keep CGE employed: an up-rated NPD station to produce 70 megawatts (electric); some refinements on the CANDU design; and, last but not least, 'the design of a test reactor for Whiteshell.' Gray knew that CGE was searching for export markets in Brazil and India, but if it were to qualify for them it would still have to be in the nuclear business. For that, AECL would have to tide it over.[46]

With the decision made, work went ahead remarkably fast. A firm offer came in from CGE in April 1962 and it was accepted. 'Firm' in this sense means that the contractor is bound to a specific sum, and must make a profit or loss according to how well he meets the sum. As Ian MacKay observed years later, CGE was pretty good at meeting such objectives in its regular line of work, where it understood the components and controlled their manufacture and supply; and it was good this time too.

Construction began in 1963, just as the electoral tide swept out the Conservative government. Getting the reactor to the sticking-point had taken almost its entire lifespan. Getting the reactor built and then up to criticality took just over two years. On 1 November 1965 WR-1 reached criticality. A month later, it reached full power – a record, according to Ian MacKay.[47]

During its twenty years of active service, it was a most useful research facility, testing experimental fuels, reactor materials, and other coolants. It was the centrepiece of a thriving research community – 567 employees by 1967. Pinawa, their dormitory, reached 1500 population by the same year. As time passed, test loops were added to the reactor; experiments proceeded with the organic coolant, and a variety of expedients were considered – recycling plutonium produced in the CANDU, and new varieties of fuel, such as graphite impregnated with uranium-235. Small reactors, always a part of Whiteshell's mandate, continued to be studied without, it must be admitted, very much in the way of economic results.[48] It would not be the progenitor of a new line of Canadian reactors, nor was it the focus of a new international generation of nuclear reactors.

For that, delays and difficulties in design bore some responsibility. Whiteshell was a known quantity when the time came to choose a reactor type for Hydro-Québec, or so it seemed. What was known was the problems Whiteshell had raised, not how they would eventually be solved. On the basis of what was then known, the organic model was turned down, a decision that Lorne Gray considered his most serious mistake. But opinion, informed opinion, was against the organic reactor. Andy Gordon, who as a chemist advised his fellow board members, harboured dark memories of flame-throwers from war service. By the same token, the boiling-water model which would be chosen for Quebec was not as well understood; rejecting the devil they knew for the one they did not, the board plumped for a boiling-water reactor for the future.[49]

It was not quite the end of the line for organics. The ill-starred career of Douglas Point during the late 1960s kept interest and hope alive. Only success at Pickering in the early 1970s finally decided which model of reactor AECL would favour. There would be no CANDU-organic. But for a critical few years, in the mid to late 1960s, it was the most plausible alternative to the CANDU system, its twin in securing cabinet approval. In a future chapter we shall return to this road not taken.

Errington and Isotopes

AECL's Commercial Products division followed its own bent during the 1950s. Unlike Chalk River, 'CP' or 'CPD' had profit as a short-term goal; unlike Chalk River, it achieved it.

Its lines of activity were well-established by 1953. It harvested isotopes from NRX and, after 1958, from NRU. These it packaged and sold, in competition with British and American products, also produced in government laboratories. The Canadian part of CP's business was limited by the small size of the domestic market, and CP early on learned that if it were to be profitable it had to cultivate the international market.

The rhythm of business was determined by transportation. The bulk of CP's business depended on cobalt-60, a long-lived isotope. Cobalt-60 could therefore be picked up and shipped from Chalk River to Ottawa, and then incorporated in sources for therapy machines or industrial scanners. To do it, Errington bought a warehouse on Ottawa's Laperriere Avenue; there CP established the largest machine shop in the Ottawa Valley to manufacture its 'Eldorados' and 'Theratrons' (the latter utilized the invention of a New York doctor, a rotating unit), or its 'Gammacell' industrial units.

Since there were not enough machinists, CP imported talent from Great Britain. Recruiting missions were sent to Britain, and skilled labour flowed back to Canada. The result was to give the air in the machine shop a distinctly British accent; interestingly, the same was true of Chalk River, which saw a substantial in-migration of highly skilled British labour in this period. The same could be said for the Canadian economy as a whole; as British and other European workers were attracted by the larger opportunities and higher salaries of Canada, Canadians tended to leave for the United States for essentially the same reasons. Without immigration, it is doubtful whether any kind of technological advance could have been sustained for long.[50]

Having assembled a skilled workforce, it remained to find something for them to do. There was a stable, but small, Canadian market to be supplied and serviced. Throughout the 1950s, however, the Canadian portion of CP's business was smaller than its sales to the United States; by 1956 it was only half as large as the American market. The same year sales to other countries surged, but for a variety of reasons sales to overseas markets were not as consistent or reliable as those inside the continent. CP peddled its services as a machine shop to Chalk River, but its ability to perform this service depended on the level of its own sales. If sales went up, outside work had to go down.[51]

Errington had always predicted a profit, and achieved one in 1956; but thereafter CP went back into the red. It was not hard to see why. The Canadian dollar was at a high against the American: good for national pride but bad for exports. Dollars, American or Canadian, were scarce outside North America. Between 1952 and 1954 CP had to buy isotopes from its competitors while NRX was being repaired. To expand, it had to wait for NRU.

Against that, there was a steadily increasing worldwide demand for isotopes, up 1500 per cent between 1955 and 1960. AECL's share of this larger market was higher in 1960 than it had been five years earlier. Using Oak Ridge, always viewed as the prime competitor, as a yardstick, CP was far out in front, selling 80 per cent more isotopes in 1959–60 than its American rival. When NRU came into service, CP was granted second priority on its facilities, behind plutonium. Its ability to meet incoming orders benefited in proportion.

CP's pre-eminence in the isotope market received an unexpected boost when the AEC, which had been faltering in isotope sales, withdrew from the business altogether in 1961. And despite price reductions, AECL passed into profitability in 1959 and stayed there; it was helped by a change in American legislation that admitted therapy machines destined for public hospitals duty-free. It may have been helped even more by a decline in the value of the Canadian dollar; it would remain below the American for most of the 1960s.

As business increased, so did the workforce. Additions were cobbled onto the Tunney's Pasture premises, but in 1964 Errington announced that CP had reached its limit and it would soon be time to build a new industrial plant. That requirement brought CP to the attention of the Board of Directors.

It is worthwhile pausing to consider the nature of CP's relations to the rest of the company. Errington had started as sales manager for Eldorado; when CP joined AECL he became manager and then general manager. He remained so until 1963 when he became vice-president of AECL, in charge of CP. In that capacity he reported first to Bennett and then to Gray on behalf of his division.

Bennett and Errington were especially close. Bennett was Ottawa-based and so was Errington, even though Errington was on the road much of the year. Errington and Bennett shared Eldorado memories, and Errington's ambitious sales philosophy

('All we could get') appealed to the president. 'There is no doubt,' Bennett wrote in 1987, 'that I did consider him as a possible successor to me once I had decided to leave.' Errington, after all, was a trained physicist, at the master's level. 'Second,' Bennett added, 'he had proven his ability as an administrator and operator .... Third, he had a strong commercial sense, somewhat rare in scientists, not to mention engineers. Finally, and perhaps most important, he was decisive and tough in carrying out his decisions once they had been made.'

It was this last quality that made Bennett think twice about Errington. A tough and unfamiliar hand might upset the delicate balance with Lewis and Chalk River. Lewis had to be kept happy and productive. Keeping him happy was a notable feature of Lorne Gray's performance; therefore, Bennett concluded, Gray was the man for the presidency.[52]

Gray never knew what Bennett had been thinking; Errington, who had an inkling, did not want the job; and Lewis was spared the experience of working under a physicist who had not merely left the profession but gone into trade. During Gray's long term as president his relations with Errington were pleasant and straightforward. Except at budget time Errington did not impinge on him, and he tried to make the reverse true as well. It helped that during the 1960s the news from Commercial Products was for the most part bright with profit.

Errington's reports were considered periodically by the Board of Directors. The board, and Gray too, accepted them without many qualms. On the one occasion when the board took the time to tour the facilities at Tunney's Pasture and Laperriere Avenue it was because Canadian Curtiss-Wright, an aircraft company, had expressed an interest in relieving AECL of its commercial burden. In the board's view, expressed in the winter of 1959, commercial products was no burden. It did spend rather a lot on research and development (20 per cent of gross income) but it was soundly managed and should in fact be helped by the rest of the company in its R&D. It was.[53]

There were also connections at a lower level. CP was bound by AECL procedures, for example on expense accounts, and the result was not always happy. AECL was still basically a research organization, with scientists and engineers at the top; CP was a sales company, with salesmen travelling round the world. Eyebrows were raised at head office, and frowns deepened at Tunney's Pasture. The resultant unhappiness never broke out into open warfare. CP and its officers did not want to leave AECL's embrace, and the head office staff did not want them to. While a certain residual snobbishness was expressed at Chalk River, and resented at Tunney's Pasture, it did not affect the essential harmony of the relationship.[54]

Administering Science

Within Gray's first two years at AECL the structure of the company changed almost out of recognition. It was not so much a matter of increased staff as of diversification. Until 1958 there were three units in two places: the scientists and engineers at Chalk River, and the salesmen and head office staff in Ottawa. In 1958 NPPD emerged in Toronto while Whiteshell was created out of the Manitoba wilderness. Head office and commercial products remained in Ottawa.

Power projects (NPPD) perforce grew the most, to 227 staff by 1964. Commercial products was next, up 100 personnel (40 per cent) between the late 1950s and early 1960s. Head office remained minuscule, numbering only ten as late as 1964. Chalk River did not grow at all, while Whiteshell had almost no staff on site until 1963–4.[55]

Gray worried over the physical spread of company activities. Long-distance telephones, company station wagons, frequent flying (at one point a company plane was dreamt of) could compensate for some of the problems of distance, but not enough. His style of management also helped, for Gray believed as much as possible in laissez-faire. He had inherited a strong senior staff – Lewis, Errington, Smith, Foster, and Don Watson, the corporate secretary – and he left them in place. Knowing Lewis's limitations, he tried to bolster Chalk River's administration by sending Charles Grinyer there; when that arrangement collapsed, Gray, like Mackenzie and Bennett before him, searched out a number of personalities to balance Lewis's on the administrative and operations side of the laboratory. But within his bailiwick, Lewis, like Errington and Smith, was left undisturbed.

It was obvious that with the creation of Whiteshell and the establishment of NPPD, Lewis's direct jurisdiction was not quite what it once was. His influence, intellectual and moral, remained great, but he had to rely on others, and eventually on Gray, to command. While he still reached out, as on the matters of vertical reactors or the proper length of fuel bundles, it was not axiomatic that he would prevail.

Lewis had to accept that NPPD and its CANDU must now follow an industrial rather than a purely scientific track; with his contribution limited by distance and circumstance he turned elsewhere, to activities that were simultaneously more general and more specialized. He had for some time been a public figure, gradually supplanting David Keys in the public eye as AECL's scientific spokesman. In public and in private Lewis forwarded the reputation of the Canadian heavy-water reactors; on the side he cultivated the organic model; and in his mind's eye he reserved time and energy for a new generation of Chalk River science: different reactors, to be sure, but perhaps, in the longer term, something very different. There, time would tell.

With a tiny head office, Gray relied on committees to link the separate sections of AECL. At the top, there was the Board of Directors and its executive committee. The executive committee, Gaherty, Gordon, Gray, and Hearn, was unchanged until 1962; when Gaherty, fearing a conflict of interest, retired, he was replaced by Grinyer whom Gray employed as a trouble-shooter. The executive committee exercised great influence; more than once its members held the fate of Whiteshell in their hands, and eventually it was A.R. Gordon who diverted Gray's preference, and the board's, from the organic reactor towards a newer and apparently bolder model, as we have seen.

The other members of the board lent their influence and support, or at least their names, when required. The utilities' interest was expressed through Gordon Shrum of BC Hydro and D.M. Stephens of Manitoba Hydro; Stephens proved his worth in dealing with his provincial government. J.S. Duncan, Ontario Hydro's chairman, was succeeded by Ross Strike; neither man seems to have made much of a mark on AECL's well-established conduits to Toronto and Queen's Park, which moved at the official level through Harold Smith, and at the political level through provincial and federal ministers. Later, in 1963, J.C. Lessard of Quebec Hydro was appointed to the board; as with the other utility appointments, his selection was a harbinger of deeper interest in nuclear power.

Gray inherited from Bennett an Advisory Committee on Atomic Power. The advice, however, was intended to flow from AECL to the committee, whose members paid periodic visits to Chalk River to be told the latest news on nuclear power and its prospects. It was, Bennett thought, an effective way of getting AECL's point of view across where it counted, to utilities, high civil servants, and industrialists. Gray saw no reason to question Bennett's judgement, and the Advisory Committee continued in place.

Inside the company, Gray sponsored two super-committees. The first, the Power Reactor Development Program Evaluation Committee (PRDPEC), was self-descriptive. It received reactor concepts and appraised them for safety, development problems, and, eventually, cost. Chaired by Lewis, it included representatives from both Chalk River and NPPD. Its agenda at its first meeting in 1962 listed four designs to be scrutinized: an improved CANDU, a fog-cooled reactor, the organic model, and a natural-uranium, boiling water-cooled reactor, better known as the SGHW (steam-generating heavy-water). The latter Lewis undertook himself, along with a bright young engineering physicist, George Pon; Pon, however, was also to sponsor the fog-cooled model, of which we shall say more below.[56]

PRDPEC (purd-peck) would have a major influence in setting the company's agenda; in the last resort it was Gray's principal source of technical advice. Gray had the uncomfortable feeling that he needed more, and in 1964 he established a Senior Management Committee; the members were heads of division and met under Gray's chairmanship. Meeting at least every few months in Ottawa or Chalk River (and sometimes in Toronto or Whiteshell), it reviewed administrative details and heard reports. Though it was not strong at making decisions, it was useful to be able to review events from the perspective of each interested group; at the very least it reminded Gray's barons that they were part of a common and much larger enterprise, while to Gray it afforded an opportunity to keep up contact with and, if necessary, to control the several parts of the company. Given his loose management style, the SMC, or something like it, had to be invented. Nowhere else could the president keep his senior officers informed of what he proposed to do, or of the environment in which the company was operating.[57]

The environment was changing. In Bennett's time, connections with the British and the Americans remained strong. Regular conferences and the presence of liaison officers in Chalk River testified to the lingering wartime origins of Canada's nuclear project. In politics as in science the generation that had run the war, and created Chalk River, was passing from the scene. By 1962 Mackenzie, Bennett, and Howe were gone, Mackenzie retired, Bennett in private business, Howe dead; and the same was true in Britain and the United States.

As reactors became increasingly industrial in nature – creatures of a standardized design – rather than experimental, their specifications exited from the sphere of scientific interchange and entered that of intellectual property. That was true at Westinghouse and General Electric; and once the British understood that the Canadians were increasingly unlikely to buy from them, Canadian visitors perceived a cooling in attitude and a diminution in information flow. A change from partnership to rivalry was underway and, although the change might never be complete, every dollar spent on a 'national' design destined for the international market increased AECL's, the AEC's, or the AEA's interest in seeing its design prevail. Since almost every dollar came from government, governments' priorities and preferences played a larger and larger role. This was even more the case because, in 1962, no country had as yet produced nuclear power to compete on equal terms with coal. If the nuclear miracle was to happen, governments, Canada's included, wanted to know when.

AECL's environment, Gray well understood, depended heavily on the success of AECL's major project, the CANDU. Whether the CANDU could actually succeed was, however, an open question.

8

The Burden of Proof: Douglas Point

Ontario was the only part of Canada where there was an immediate need for nuclear power. Ontario was large, it was prosperous, and it consumed a great deal of energy. Its electricity supply flowed from the province's abundant lakes and rivers, symbolized by the magnificent falls at Niagara; stretched along the river below the falls were hydro plants, pumping power to southern Ontario, the 'golden horseshoe' around the west end of Lake Ontario.

The benefits were obvious, in the car plants and steel smelters, the light industries, and the miles of suburbs that stretched out from Toronto and Hamilton and St Catharines. 'Ontario has continually proved itself to be in the forefront of Canadian development,' Richard Hearn told the Royal Commission on Canada's Economic Prospects in January 1956, 'and I can see no reason why this state of affairs should not continue.'

Naturally Hearn saw electric energy as central to his province's pre-eminence. For energy to be truly 'available' it must be domestic; as Hearn put it, 'dependence on outside sources for basic fuel ... constituted a handicap.' Ontario Hydro, by exploiting southern Ontario's water power, helped avert the handicap, but by the end of the Second World War it was evident that it could not cope with expanding demand from domestic electricity supply. The completion of two large coal-fired steam plants in Toronto and Windsor in 1951 signalled a turning away from hydraulic resources. By 1968, Hearn estimated, half of the province's power would come from such stations. There was no escape from conventional fuels, coal or oil, but Hearn hoped to minimize dependence by the construction of a new breed of thermal generating stations: nuclear.

Hearn predicted that nuclear would give Ontario, by 1981, one-third of its generating capacity of 23,600 megawatts (electric). (The actual figure was 25,900 megawatts, a third of it nuclear.) No lesser amount would suffice; in fact, Hearn projected that nuclear power would reach 1531 megawatts (electric) by 1971; building that capacity would cost between $5.8 and $6.9 billion in 1956 dollars.[1]

Hearn was known to be an enthusiast for nuclear power. Admittedly, projections of future capacity did little more than extrapolate in various creative ways on past experience. But the similarity between Hearn's projection and the actual figure suggests either that his speculative powers were unusually accurate, or that he and his successors had an altogether unique ability to get their own way.

The truth probably combines both these factors. Ontario's economic growth remained steady through most of Hearn's twenty-five-year projection. Demand for power increased with economic growth. The Ontario government during the entire period was in the hands of a single party, the Progressive Conservatives, and connections between the government and its electrical utility were close. The government was strongly inclined to favour economic growth; under Leslie Frost, premier from 1949 to 1961, its program amounted to a cornucopia of mega-projects – most of them, interestingly, co-sponsored by the federal government. In the 1955 provincial election Frost ran on a platform of highways, waterways (the St Lawrence Seaway), hydro, and the promise of nuclear power.[2]

Frost won, and the election was effectively a ratification of his government's development policies. In Ontario it paid to think big. It also paid to think of the government as an economic leader. Although conservative in stripe, the provincial government was anything but conservative in its approach to public ownership. It took pride in the success of its government-owned enterprises, Ontario Hydro being foremost. It identified their success with what might be inelegantly called 'Canadian-ness': public enterprise, like Hydro, was an outward sign and an inner bulwark of Canada's separate identity. It directly linked government and citizen. 'This is your chairman speaking,' Bob Saunders intoned as he reported to radio audiences across the province on what the provincial utility had most recently been doing. In the words of Hydro's founder, Sir Adam Beck, this was 'people's power.'

That did not preclude taking a 'fiscally responsible' view of Hydro and its works. Frost and his ministers worried over debt and kept a close eye on Hydro expenditures. The eye blinked at the likely cost of generating stations, but it accepted them. Thermal stations were the cost of doing Ontario's business. If they were coal-fired or oil-fired they were a regrettable necessity, particularly because they involved a perpetual bill for out-of-province fuel. Nuclear stations were another matter entirely. It helped – indeed it helped a great deal – that nuclear fuels came not only from within Canada but from mines in central Ontario. The Ontario government liked the idea of controlling its own fuel supply. Those parts of the government that thought about such things were concerned about questions of foreign exchange and worried that Ontario was placing its energy supply at the whim of foreign governments even if the principal foreign government concerned, the American, was on the whole well disposed. Nuclear power offered a way out of the province's energy conundrum, but there were things to be done first. Nuclear power might be the way of the future, yet it remained unproven and its capital costs were high.

Saunders's hesitation in 1954, Duncan's skittishness about financing the CANDU, and Hydro's insistence that any power derived from it should not be 'base' power showed that in the utility's collective opinion nuclear power remained to be proven. Proof would depend on performance, but it would also be enhanced by participation.

Here we should pause to consider what Ontario Hydro was: a utility, a central agency for electric power supply over the whole province. It was, however, much more. Its employees and even its customers were assumed to be *supporters*, a concept that dated back to Hydro's early origins as a political crusade for people's power. For its supporters and for many of its employees, Hydro was promethean – the bringer of light. What brought light was laudable, and the more light the better.

That it could do so depended on its engineering capabilities. Like any utility of any size, Hydro had its own engineers, but unlike at least some other utilities most of its dams and power plants were designed and built in-house. They were not small projects. The Saunders generating station at Niagara, the Hearn thermal plant in Toronto, and the St Lawrence power development of the mid-1950s were big by any standards. To build them Hydro relied largely on its own staff. If Ontario's goal was a dependable energy supply, independent of outside suppliers, Hydro wanted to be free of outside control. One way of doing it was to rely on its own workforce. By the mid-1950s Hydro was architect, contractor, engineer, and customer for its own projects.

Such a development – the creation of a vast engineering capacity – coincided with the inherent efficiencies of large projects. Central generating stations were, in the opinion of engineers and economists, the most effective and most economical way of providing for Ontario's hydro needs. Given the growth of the sprawling suburbs and satellites of Toronto, it was the *only* way of coping. Large stations required great technical knowledge, repeatable with local variations. The process as well as the product had to be reliable, and the best way of guaranteeing it was to defer to past experience. That, in turn, was the dominion of the engineers, headed by the company's general manager – always, in this period, an engineer. The chairman and the commission turned to the general manager for advice, and the general manager turned to the engineers. Their advice was, frequently, what was done. 'We don't ask people,' one of them reflected. 'We tell people.'[3]

Harold Smith became Hydro's chief engineer in 1959. He had been Hearn's protégé; Duncan had approved his transfer to NPPD. With Rolphton going forward and a large reactor a practical certainty, Smith was the inescapable appointment. He was a man of unusual intellectual ability and great force of character. 'He's an engineer's engineer,' a contemporary suggested, 'brilliant, with an inquisitive nature.' When Smith's team produced NPG-10 they laid the groundwork for what was to come; and Smith proposed, designed, and established NPPD, which would be responsible.

Smith had strong opinions, but opinions that were very much in the Hydro tradition. If anything, Smith's point of view exaggerated what Hearn and the previous generation of senior engineers had believed and then built. Free enterprise, inside Hydro, was just another term for the freedom to have a private monopoly. 'The bracing winds of competition' were nothing of the kind. Terminating CGE as principal contractor for domestic nuclear plants was plain common sense to Smith; it was a question not only of public advantage but of public superiority. It was the Hydro way.

Becoming chief engineer put paid to Smith's direct connection with NPPD. The chief engineer worked in downtown Toronto, with a range of time-consuming responsibilities; it was difficult to put together a day to spend at NPPD in the suburbs. And so Smith gave way to his deputy, John Foster. Foster's experience with AECL dated back to 1953, when he was sent up to Chalk River on behalf of Montreal Engineering, Gaherty's firm, to appraise the aftermath of the NRX accident. When the nuclear power group was set up, Foster was a natural candidate on behalf of Montreal Engineering, and when CGE came to Chalk River in search of staff in 1955, Foster agreed to go to Peterborough. At Smith's behest, Foster moved to NPPD in 1958 as deputy head of the division; when matters had to be explained thereafter to AECL's board or executive committee, the two men turned up together and offered a common front on CANDU matters.

The link between Smith and Foster continued after Smith's formal departure from NPPD, for reasons of business as well as personal confidence. Ontario Hydro was AECL's utility partner, with the Hydro chairmen, one after another, sitting on the board. The two were linked in more than a formal, contractual way. NPPD was stocked with engineers from both companies and although it had, as well, engineers from other interested firms (including Foster from Montreal Engineering) they usually chose to transfer to AECL as their connections with their parent companies dwindled.[4]

NPPD was a microcosm of AECL's political and institutional world. AECL was intended to be closely allied with its customers, and NPPD was the fulfilment of the promise. Close relations were sustained at every level: between president and chairman and general manager, between Foster and Smith, and on a daily basis between AECL's engineers and Hydro's. The customer could shape design, influence specifications, and compute cost – and so could the supplier. By building up its own engineering division, AECL was following the same path as Hydro. Like Hydro, it was tantalized by the idea that it could act to control its environment. With Hydro furnishing the basic business, there was security of employment.

This was no slight concern. The demise of the Avro Arrow highlighted the plight of the skilled Canadian professional, with no better future than to migrate to the fleshpots of American technology. NPPD took over some of the newly unemployed from Avro; and along with the staff it inherited a morale problem. 'Will it last? Will it go the way of the Arrow?' were frequent questions at staff parties in 1959–60.[5]

Not likely. NPPD was for the time being invulnerable to political pressures. It would not shrink but grow, as it had to if it were to meet Gray's commitments to Rolphton, to CANDU, and, especially, Canada's adventure in donating nuclear aid to India, which was at its peak in the late 1950s and early 1960s. At first NPPD had some of the overflow from NPD, under Grinyer, as well as the CANDU team under Foster. There were just thirty staff members, half from Hydro and half from AECL, a few others from different companies, and a metallurgist hired right off the street. There was no clear-cut division of labour in terms of the origins of the staff, but it was apparent that AECL and the 'others' (known as 'attached staff') were clustering around the nuclear end of things, and Hydro's people around the conventional side.[6]

Other projects soon came along, and their proliferation was reflected in the steady increase in numbers of staff:

Year	Professional	Non-professional	Total
1961	71	122	193
1963	74	136	210
1965	101	193	294
1967	202	475	677
1969	252	623	875

As before, some were paid by Hydro and more by AECL; Foster remained on secondment from Montreal Engineering.[7]

How did this arrangement work in practice? Where were the dividing lines between AECL and Hydro in function and personnel? We shall examine this question under two headings: function and form.

Let us take as a point of departure that NPPD performed services for both companies and that it furnished the expertise of each to the other. At the crudest but most essential level, it trained people from both sides to know what button to push when the time came. But it was more than a transfer point for information. It controlled the overall design of the CANDU, both nuclear and conventional. It would design in detail what the nuclear components would be, leaving the non-nuclear side to Hydro, but it also had an expert available to examine Hydro's conceptions of what the turbo-generator should be, and to pronounce on whether the civil engineers had allowed for proper foundations for the reactor building.

Controlling design, NPPD also controlled supply. Manufacturers were encouraged to tender, according to standard Ontario Hydro practice, on specifications drawn up by NPPD. When the supplies were received, they had to be inspected, a function generally known as 'quality control,' which might be performed on the conventional side by Hydro and on the nuclear by NPPD. As we shall see, this was a tall order.

Next came construction. This was a Hydro specialty and to manage it the utility assigned Phil Stratton, one of their more promising construction engineers, 'to be a fifth wheel' at NPPD in 1959. When the time came, Stratton would be second-in-command in the actual building of the CANDU.

After construction, there was commissioning, getting the reactor into working order and starting to produce on-line power. It was common practice for manufacturers to send along commissioning crews with their turbo-generators; in this case, and subsequent ones, it made more sense, and saved time and money, to train the people who would operate the reactor to take over right from the start. And so commissioning and operations were merged into a single process. As for operations, there had never been any question that the reactor, even if owned by AECL, would be operated by Ontario Hydro, just like NPD.[8]

The form of NPPD tended to blur the distinction just made between Hydro and AECL functions. Under Foster there were three principal sections: engineering, under I.L. ('Willy') Wilson, administration under B.P. ('BP') Scull, and development, under M.D. (Mel) Berry. The first, engineering, included reactor design, 'control engineering' which dealt with control systems, applied science (there were a physicist and a metallurgist on the premises), and mechanical engineering, among others. Administration included supply and inspection, while development included in-house and contracted-for experimentation. There was a fourth management slot, for the project manager of the CANDU; Foster assigned that job to himself.[9]

To Foster, a proper design engineering organization would do two things: design components, and put them into a system that was efficient, economical, and safe. It was a large responsibility, but not unprecedented. Since NPPD was not merely the designer, but the customer, it had considerable power. In practice it needed the power because it was not dealing with a routine construction job where the boilers and pipes could be made according to standard and well-understood principles. A mistake, whether by a contractor or a supplier, could have catastrophic consequences.

Running NPPD was not merely a question of space and drawings. It also had a time dimension. There was a deadline. Design must proceed, and before design was finished construction must start. NPPD must run both; if it failed, its future, and AECL's, would be dim. They would know in five years, Gray predicted, when CANDU was built. Foster thought it might take six; his design chief, Willy Wilson, gave it seven. Only if Gray was right would CANDU escape political complications. But Gray was not right.

Douglas Point

The form of the CANDU was already determined. It was to be a 200 megawatt (electric), pressure-tubed, natural-uranium fuelled, horizontal system. Harold Smith would have preferred a 100 megawatt unit to start, but was overruled by a combination of Lewis's enthusiasm and the insistence of other Hydro officials who wanted to see the program keep abreast of developments in conventionally fired units, already with capacities of 500 megawatts or larger. It was to produce power at 5 mills or less. Whatever the figure, Ontario Hydro would pay AECL on the basis of what it would have cost had it been produced at one of the utility's existing thermal plants. Lastly, the new reactor was to be sited somewhere in Ontario.[10]

Where? Obviously not in the north, far from large urban areas and heavy power consumption. But possibly not in the south either, too close to cities where airborne emissions and waste disposal might pose problems. For Canada's first large reactor it was unlikely that the AECB's reactor safety advisory committee would approve any site too close to a large city. Yet it should tie in with the southern Ontario power grid, and there should be room for expansion up to a capacity of 1000 megawatts (electric). It should have good bearing rock, flat, and no more than a few feet above water level. And there should be plenty of cold water.[11]

A site near Toronto was considered and then discarded. Georgian Bay was too far away. Lake Huron seemed a better bet, and engineers and surveyors began to traipse up and down the coastline looking for a suitable spot. Municipalities, hearing that Hydro was on the prowl for a large and lucrative construction site, advertised their special virtues. By the end of June attention was focused on a spot about ten miles north of the town of Kincardine in Bruce County: Douglas Point. Douglas Point was low-lying, close to water (it jutted into the lake), and with a limestone formation at grade along the shore. On 24 June 1959 the Hydro Electric Power Commission decided to proceed with the acquisition of 2300 acres, at a price between $50 and $70 an acre.[12]

Douglas Point is located in Bruce Township, Bruce County, Ontario. Both were named in honour of the family of a colonial governor general, Lord Elgin (Port Elgin is just up the coast). Port Elgin, Kincardine, Walkerton, and Southampton are the principal towns, and agriculture – livestock-raising – the main industry. It was not a particularly prosperous part of the province, rather below the national average in terms of individual purchasing power (83 per cent in 1961), but not so far below to cause government programs to be showered upon it. Hydro's interest was strange, but welcome. 'I don't remember anyone asking if they could come in here,' the township reeve recalled. 'I first read about it in the paper.'[13]

The purchase was closely watched for its impact on a local balance that had endured for generations. One township was Conservative, another Liberal, and the patterns, it was hoped, would be maintained. Unfortunately Douglas Point, and its access road, lay closer to Kincardine than to Port Elgin. Foster was called to a meeting in the reeve's farmhouse to deal with the location of the access road, and had to drive a delegation from Port Elgin up and down concession roads to demonstrate that the road he had chosen – Concession Road Number Two – was the only one possible.[14]

As a crown corporation, Hydro did not pay local rates, instead donating grants in lieu of taxes to host municipalities; this compensated expenditures, for example on roads or schools, resulting from Hydro's presence. In the case of Douglas Point only Bruce Township got the money. Other citizens, however, shared to some extent in the bonanza. Empty farmhouses suddenly became desirable, stores could expand, and shopping centres were contemplated. With more money and new jobs, the occupational structure changed. 'Most of us used to keep a hired hand but that stopped as soon as union wages came in,' Reeve Mackenzie reflected. 'So the farmers, most of whom had a mixed farming operation, had to change their operation to beef farming because it isn't as labour-intensive. And many farmers went to work on the site themselves. In fact, many farms around here were probably saved because of the wages.'[15]

Bruce County stretches along the shore of Lake Huron, with Highway 21 paralleling the lake. At Douglas Point the provincial highway is about three miles inland. In 1960 traffic to the construction site left the paved road at the intersection of Highway 21, a small matter during the summer but a great concern during the winter, for Lake Huron is justly famed as the heart of Ontario's snowbelt. Just inland the annual snowfall reaches over 305 centimetres (120 inches), and on the average during the 1950s there were 80 days out of 365 with measurable snowfall. This is known to climatologists as 'the lake effect,' and occurs in the lee of the Great Lakes as cold winds blow over the lakes towards the shore.[16]

The 'lake effect' was in full force when John Foster paid an early visit to Douglas Point in January 1960 and rammed his car into a snowbank. A party of American wolf hunters tried to rescue the party, and failed; finally Phil Stratton dug them out with a snowplow. Snow or no, construction started the next month.[17]

Hydro construction crews – a nucleus sent out from Toronto, with labour hired both from the area and across the province, and eventually numbering 500 – settled down to clear and excavate the Douglas Point site. The construction trades were heavily unionized, but federal labour policy directed that bids for work should be open to any competent firm, unionized or not. Some difficulties therefore arose when non-union firms got contracts. While those difficulties could be appeased by informal means, such as contributions to union recreation funds, Foster eventually signed a site agreement which bound everyone on site. Labour relations were not entirely smooth – construction was delayed by a six-week strike in 1961 – but there were no serious disruptions.[18]

Design of the reactor went ahead in Toronto; by 1962 design involved sixty men. Because NPD was far from completion, no prior experience could be brought to bear, and the decision to withhold the overall contract from CGE must have had an effect at this point as well. Inexperience, conscious or unconscious, was not the only problem. More important in the long term was the realization that the country that had committed itself to a natural-uranium, heavy-water design – in part because there was so much natural uranium lying around the Canadian Shield – had no heavy-water production of its own. Trail, BC, had been out of production for almost a decade; the nearest reliable source lay in Savannah River in South Carolina.

Given Ontario Hydro's sensitivity to security of supply in the energy field, it might have been expected that the heavy-water problem would have given it pause, but this does not seem to have been the case. Lorne Gray had promised that heavy water would be available, and Gray's word was good. Beyond Gray's word, there was the involvement of the Americans in NPPD, where the AEC had made available $5 million in grants and had assigned liaison officers to Toronto to keep tabs on the Canadian program and see that it lacked for nothing that the Americans could provide – within reason and depending on the availability of a purchase price, of course. It was even possible that Canadian demand for heavy water helped sustain some of the research programs at Savannah River, or so some of the American officials believed.[19]

The supply of heavy water might therefore be considered secure, but it was expensive: $26 a pound. Douglas Point would need tons, not pounds, and it would get it only by the most stringent conservation. And so into a program that depended on the husbanding of every spare neutron there entered a compulsion to save every possible ounce of heavy water. 'We squeezed it down tight,' W.G. Morison stated.[20]

There were two main reasons. Substantial and continuous heavy-water loss would lead eventually to a shut-down. At the same time, with replacement running at $26 a pound and with some loss considered inevitable, leakage above a certain point would raise the operating costs of the reactor to an intolerable level. The design of CANDU had to be adapted to minimizing if not preventing the loss of heavy-water in the heat transfer system; in practice, E.R. Siddall recalled, 'we aimed simply at good standard engineering quality.'[21]

The conservation of heavy water was an *internal* decision by the designers of the reactor – internal, at least, to Hydro, AECL, and NPPD. The considerations of others now played a role in defining its design. NPD had been the first reactor whose characteristics were negotiated between AECL and the AECB's reactor safety advisory committee. At NPD there had been containment, but reliance was placed on a rectangular structure below grade; Douglas Point would be different.

The difference had to rely on experience elsewhere, and especially, according to Morison, on the work of Farmer in the United Kingdom. The Americans had evolved concrete shells to contain any escaping fluids, to keep radiation as low as possible, and Douglas Point therefore had a domed containment building on the American model. The containment shell held the reactor proper and fuelling machines on one level; above that were the steam generators and on top of them a large dousing water tank.

The dousing system would operate in case of a break in the steam lines, de-pressurizing the escaping steam. The option was to make the containment walls so thick that they could resist escaping steam pressure by themselves; the dousing system thus permitted walls a mere three feet thick.

Pouring concrete was a delicate operation. The building shell itself was a lesser problem than the concrete vault in which the calandria would lie. Opportunities for repair, once the reactor went active, would be limited, and it was important to do it right, ensuring that the holes for pipes and struts were correct the first time; a pilot operation, full-size, was devised to ensure a high degree of accuracy.

Control of reactions was to be achieved, as in previous reactors, by raising and lowering the level of heavy-water moderator in the calandria. Experience at Chalk River justified the simplicity of the concept: there were 'a minimum number of components'; at the same time it lent itself to triplication, a concept developed by Ernie Siddall, who had been working on improved control and instrumentation systems at Chalk River in the aftermath of the NRX accident. In this scheme, three parallel channels of control are provided for the most important functions. It is arranged that if one channel disagrees with the other two, an alarm is sounded to draw attention to the problem and control is exercised by the two good channels until repairs can be made.

Douglas Point was much larger than any previous reactor, having 306 fuel channels compared with NPD's 132. It was considered necessary to monitor every channel in order to guard against possible blockage of coolant flow. A more sophisticated control of the neutron flux was essential to obtain maximum output from the different zones of the reactor.

Siddall now took on the problem of extending the control principles developed in NRX, NRU, and NPD to the 200 megawatt (electric) plant. In the AECL program, automation of reactor controls had already received more attention than was the case in other countries, including the United States. The main motive was to combine the immunity from fatigue and monotony of the machine with the human ability to cope with the unexpected – by smelling smoke, in the classic example. Digital electronic technology

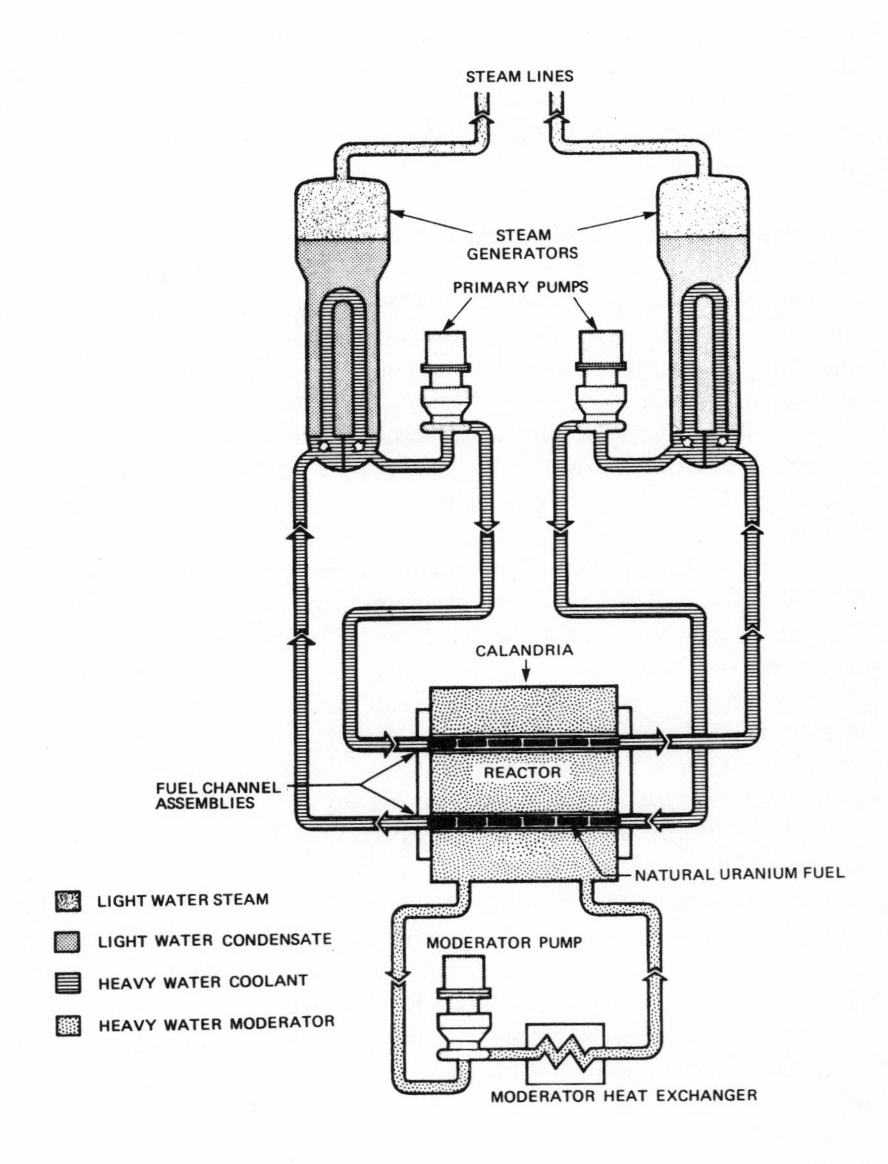

CANDU nuclear steam supply system

was just starting its explosive expansion, and Siddall and his group were able for the first time in a nuclear reactor to use a digital computer not only for monitoring a large number of essential measurements but also, in two applications, to carry out control functions by direct digital methods. The promise for the future was great; the computer could easily carry out sophisticated functions involving many inputs, and could routinely record everything it did and print it out any time the operator or a technical engineer wanted to know.[22]

Contracts for components for Douglas Point were let to 600 Canadian firms and to manufacturers in the United States and the United Kingdom. The calandria would be stainless steel, easier to work with than the Rolphton aluminum tank, and it would be made in Canada, by Dominion Bridge in Montreal. It would, however, be large: 20 feet in diameter, and weighing 60 tons. The fact that Douglas Point was on water was therefore handy in shipping the calandria by barge from Lachine to Kincardine.

CALANDRIA
A tubed vessel or tank that, in a CANDU reactor, contains the heavy-water moderator.

For manufacturing, an elaborate quality control system was devised, after some trial and error. Canadian manufacturers were eager to bid for the business, and the prices they quoted seemed reasonable; but they had not done much in the nuclear line and, as with NRU and NPD, the quality of the products delivered was not always optimal. It almost seemed, in the words of an NPPD study, that suppliers had not paid 'sufficient attention to the specifications in the bidding stage and only [learned] later that we usually require higher standards of workmanship and quality control than accepted by industry generally.'[23]

Phil Stratton took a walk through a set of the larger steam pipes prior to installation and was interested to spot daylight through the presumably solid steel. The welding, he concluded, had to be done again, and inspection at the factory beefed up. This took time. The same problem occurred with a boiler already inspected and shipped from Montreal to Douglas Point; back it went. Another set of pipes came with sleeve welds joining pipe ends; but 'butt ends' were what was required, and the whole shipment had to be re-machined. Seven months were lost. Sardine cans and toothbrushes were found in boilers that had been sent ready for installation.[24]

Experience with manufacturers was not therefore entirely pleasing either to AECL or to Hydro. Canadian manufacturers supplied most of the components, 71 per cent, followed by the British with 17 per cent and the Americans with 12. What followed in installing the components held an element of surprise if not suspense. Foster reported to the Board of Directors in May 1964 that the design of the moderator system was now complete, 'except for a few small details.' Other items, however, had still to be designed – spent-fuel inspection equipment, canning, and underwater storage equipment. According to Stratton, design sometimes took place on the spot. Components were matched and fitted. 'As-built' drawings were then sent to the designer, 'and he'd better like it.' One unintended consequence was that design proceeded system by system, with insufficient regard for the functioning of the reactor as a whole.[25]

Time passed. NPD was completed in 1962, and data from the demonstration reactor began to flow into Chalk River and NPPD. Some of it was good and some, especially where the NPD fuelling machine was concerned, not so good. During an attempt at on-line fuelling in December 1962 the fuelling machine leaked heavy water at its snout and in its body, thereby losing somewhere between one and four tons of heavy water; one ton of the stuff went up the stack and was gone forever. Some was 'downgraded' by ordinary water. It was cleaned up, and the fuelling machine repaired. (The machines would eventually be replaced.) The dousing system operated at NPD as designed; the only problem was that nobody had intended to test it. To get NPD restarted, more heavy water had to be borrowed from Chalk River, a serious consideration at a time when heavy water was in notably short supply. The reactor was losing large amounts of heavy water because of a number of leaks which were difficult to eliminate; the loss was too large for any economy of operation.[26]

Douglas Point lumbered on. One manufacturer after another reported that the contracted delivery date could not be met: the turbine and the calandria were each nine months behind schedule, the end shields ten months, the end fitting bodies fourteen months. It was 'a bit of a struggle,' Bill Morison remembered; 'we were conscious that we were falling behind.' That was partly because of Lorne Gray's optimistic target of five years from approval to completion, meaning 1964, but even Foster had thought that 1965 might be possible. It was not.[27]

Lorne McConnell, who had worked on NPD, was reassigned to Douglas Point. Times were tough, he was told. The pressure was on for a new, larger reactor, and Lorne Gray was unhappy with this endless prototype. Problems had developed with the fuelling system. There was, he recalled, some thought given to pulling the plug on the whole thing. It is, however, difficult to detect the same anxiety and frustration in the board minutes about Douglas Point that had obtained for NRU or NPD. For one thing, Douglas Point was more or less on its budgetary target despite an intervening devaluation of the Canadian dollar; and perhaps where the board was concerned that was the principal consideration.[28]

The economics predicted for Douglas Point when construction was begun did not materialize. It had never been intended that it should be competitive with conventional sources, but the gap between nuclear and conventional turned out to be wider than anticipated. When the reactor entered operation, on 15 November 1966, AECL issued a press release. It stressed that Douglas Point was a landmark, not the culmination of the CANDU design. Ontario would benefit from Douglas Point's power (it would take power into its grid from Douglas Point on 7 January 1967), and pay no more for it than from a conventional source. Optimistically, Douglas Point and its staff of ninety-two would produce power at 6 mills, one up from the 1960 estimate. But by comparison with conventional sources, it was too high.[29]

Douglas Point would operate for twenty years. From the start it attracted criticism. It was frequently 'down,' more than half the time between 1968 and 1971. Its pumps and valves were difficult to get at. Repairs were difficult, time-consuming, and costly. Because vital components were in congested and inaccessible locations,

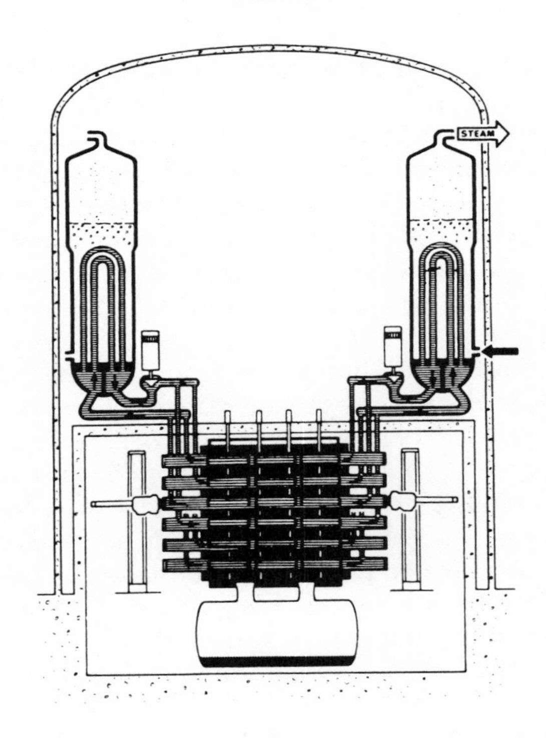

CANDU reactor. *Nuclear Engineering International*

usually because of attempts to minimize the 'hold up' of heavy water in systems, repairs had to be done by remote control or by large teams, each member of which could stay only a short time in the 'active' part of the reactor. In 1972 it lost its heavy-water supply to allow AECL to start up newer reactors, resulting in more 'down time.' In its early stages it leaked prodigious amounts of expensive heavy water. That problem was corrected, and never recurred on later models, but it was costly while it lasted.

Ontario Hydro operated the plant, as agreed, but refused to buy it, and as a result it was still an AECL property when it was shut down for the last time in 1986. There was no better testimony to its lack of success as an operating power reactor.

'Douglas Point, which should have been designed like a dray-horse, was designed like a bloody race-horse,' an NPPD veteran remarked. It was 'delicate.' It shut down too easily and too often.

The lesson, a 1967 analyst argued, was for simplification in future reactors; for that alone designers of future reactors should spend time at Douglas Point, contemplating the mistakes of the recent past. As it happened, they were already doing so: the designers of Douglas Point were by 1967 dividing their time with a new project, the future Pickering A, and applying what they had learned at the earlier reactor. It seems that they learned their lessons well.[30]

Staying Alive

From the moment it was announced, Douglas Point concentrated the hopes, public and private, of Canada's nuclear future. When Lorne Gray appeared before the House of Commons committee on research in 1960, the members accepted what he had to say; but not without some opposition.

The opposition came from Winnett Boyd, one of Canada's more prominent engineers and the C.D. Howe Company's manager for the building of NRU. Boyd had a reputation both distinguished and varied and had experience in the nuclear field. When he announced to the MPS that he did not favour AECL's heavy-water, natural-uranium reactor, there was a considerable stir.

This was not precisely news to AECL. Boyd had at one point sought assistance from the company to develop his reactor concepts – a high-temperature, gas-cooled, graphite-moderated system, first proposed in the United States in 1946 – but it had not been forthcoming. In 1959 he went public, telling the Ottawa branch of the Engineering Institute of Canada that AECL had taken a wrong turn in its quest for a heavy-water, natural-uranium reactor. Graphite, Boyd argued, was better than heavy water; natural uranium reactors were too low in temperature to drive turbines efficiently; the real prospects for the heavy-water reactor could be discerned from the fact that no other country was actively pursuing a heavy-water system. There were, besides, real questions of safety and reliability in the CANDU design. Although he mentioned several aspects of CANDU, Boyd concentrated his fire on the pressure tubes which, he suggested, might not be able to bear sufficient stress: 'if one of them was to burst,' he said, 'the reactor might possibly run away and explode.'

Boyd exaggerated, as Lorne Gray was quick to point out. Gray spoke after Boyd at the engineering forum. With Boyd's text before him, he set about to destroy his arguments. Gray argued that Boyd was misinformed and out of date where AECL was concerned; as for his own system, it was at best unproven and at worst unsafe. The suggestion that the CANDU's pressure tubes could not stand any conceivable stress was ridiculous, Gray added. 'Even major ruptures of pressure tubes' would mean nothing more than replacing them. Boyd's argument to the contrary indicated 'ignorance of the real facts or a direct attempt to mislead.'

Gray devoted most of his speech to defending the heavy-water system's international credentials. It was not true that nobody else was considering it. AECL had a string of international connections: with the British, the Americans, and the European atomic agency, Euratom. The British were actively investigating heavy water, and AECL had a team in Britain to help. The Americans showed their interest by posting personnel to Peterborough and Chalk River, as well as to NPPD. And so it went, before the engineers, before nuclear forums, before politicians.

Gray was not alone in refuting Boyd. Lewis of course did his bit, and so did C.J. Mackenzie, who told the MPs when he appeared before them that if Boyd entered the room he, Mackenzie, would leave. The politicians were impressed enough to take Mackenzie's word for it; Boyd's testimony to the contrary was not considered important or interesting enough to merit inclusion, or even mention, in the annual unofficial summary of Canadians news, *The Canadian Annual Review*; instead, the *Review* faithfully repeated the testimony of AECL's officials.[31]

Whether Boyd's criticism of CANDU was true, only time would tell, though every successive delay at Douglas Point lent added credibility to his arguments as far as the general public was concerned. He had, however, touched a nerve when he underlined the uniqueness of the heavy water system. What Gray said in response was correct, at the time, but in the longer run Boyd was the truer prophet. Though it could not be predicted in 1960, other countries' heavy water projects folded up one by one: Swedish, American, British. It took a very long while, but eventually only CANDU was left to carry the heavy water banner.

The burden of proof that lay on Douglas Point was accordingly very great. Its early failures were proportionately discouraging, if not devastating. It cannot be said that Douglas Point's performance helped to destroy any other nation's heavy-water program, but it certainly did not help. At the point when it was most needed, in the mid to late 1960s, Douglas Point proved to be a feeble reed.

When Gray, Mackenzie, and Lewis had confronted Boyd in the early 1960s, their prestige was high and their authority great. Nothing they said was untrue, but what was at stake was not so much truth as plausibility. Parliamentarians and press were unskilled at uncovering the former, but they had a fair instinct for the latter. Mackenzie's high-handed refusal to confront Boyd directly could be, and was, swallowed in 1960; it would be another matter in 1970.

Reflecting on the experience, Ernie Siddall, who was eventually sent in to Douglas Point with a mandate to straighten it out, commented that the real defect of the first CANDU was hubris, the overweening self-confidence that made NPPD's designers, himself included, think that they could bypass their modest 20 megawatt, privately designed cousin, NPD. By the time NPD, or most of it, was ticking over, Douglas Point was floundering, and publicly floundering.[32]

The case for nuclear power, Canadian-style, still rested on faith. In 1963–4, faith would get another, and possibly final, workout.

9

Pickering

The world turned at the end of 1962. The Diefenbaker government was on its last legs, after an inconclusive election. Ironically, as the government staggered towards its demise, conditions in Canada were improving. Investment was up, unemployment was down, and incomes were rising. Canada was launched on another boom.

In April 1963 a general election tossed out the Progressive Conservatives, and Diefenbaker, and returned the Liberals under Lester B. Pearson to power. The 'Pearson team,' as his slogans called it, was made up of strong personalities; one of them later quipped, 'We looked down the cabinet table [at the prime minister] and said to ourselves, "He's here because we're here."'

The Pearson government had its problems. Despite two elections, it never gained a majority in the House of Commons; but it was so close that replacing it with another party was inconceivable. It had to deal with a worsening world scene: war in Viet Nam, post-colonial conflicts in Africa, and the grandiose politics of French president Charles De Gaulle. It had its troubles at home, too: growing nationalism on both the French and English sides of the country. The English version was expressed in terms intended to distance Canada from the United States; the French kind tended to weaken if not dissolve the connection between Quebec and the rest of the country. As time passed, the 'Quebec problem' overshadowed all the others; few indeed were the federal agencies and policies that were not affected by it.

AECL got a new minister. C.M. 'Bud' Drury was a wartime brigadier, a former deputy minister of national defence as well as a member of a very prominent Montreal family. Though his roots were Conservative he had gone political as a Liberal in 1962, and had been elected in a wealthy constituency on the English-speaking side of Montreal. Drury at first received the minor job of minister of defence production, one of C.D. Howe's relics from the 1950s. In keeping with one of the Liberals' electoral pledges, he was promised the job of 'minister of industry' as soon as an industry department could be carved out of the old trade and commerce portfolio. AECL would be part of 'industry,' and so from the first it went to Drury.

Lorne Gray was not pleased. He had worked well with Churchill, and he had got used to his minister's unvarying and sometimes uncritical support. Drury was a smoother customer altogether, much more knowledgeable about Ottawa and its ways. He knew the deputy ministers as well as Gray did; he may have thought that during Churchill's term of office the AECL president had held his minister in tutelage. What is certain is that the temperature in the ministerial office dropped considerably when Gray and Drury conferred. From Gray's point of view, Drury gave him insufficient support before the cabinet; Drury's view is not recorded. This was a pity, because during the 1960s AECL was a frequent item on the cabinet's agenda.

When his department of industry was finally established in 1964 Drury took aboard as his deputy minister Simon Reisman, a rising star in the finance department and a well-known trade expert. Reisman was influential and wished to be more influential; he was also active and energetic. Some of his energies would, inevitably, rub off on AECL. The energies were appropriate to an activist government that believed in an expansion in governmental direction of the economy and society.

Three aspects of Pearson's policies were particularly important for AECL. The first was economic nationalism, with its implicit commitment to energy self-sufficiency. The second was styled 'le fait français,' giving more prominence in national affairs to more and better French Canadians. The third was a commitment to regional development, especially in the Atlantic provinces. The

Pearson administration, more than its predecessor, was philosophically inclined to dabble in AECL's affairs, or to second-guess its initiatives.

AECL was first touched when J.C. Lessard was appointed to the Board of Directors in the spring of 1963. Lessard had been federal deputy minister of transport; he was, since 1960, president of Quebec Hydro. His appointment suggested that the federal government expected to see AECL active in promoting nuclear power to Quebec; consent by the provincial government – a Liberal one, headed by Jean Lesage – implied that Quebec would be receptive.

If the Ottawa barometer read 'variable,' the Ontario weather was fair – very fair. There the government was in the hands of the amiable and politically canny John Robarts, the fourth in succession in a Tory dynasty that would last forty-two years and six premiers. Robarts understood politics; he had made a name for himself as education minister at the end of the 1950s; and he stood on the moderately progressive end of his party. His style, avuncular but apparently vigorous, suited Ontario. Some suspected that where the economy was concerned he preferred to leave the initiative to others.

Robarts came to Rolphton to open NPD in 1962, and performed admirably. He sounded both optimistic and committed, and his audience, which included Gordon Churchill, was suitably impressed. After the ceremony, Robarts turned the conversation to the local high schools, his subject of preference.[1] Atomic energy, which during his premiership would absorb hundreds of millions of dollars, and which he had just seen operating for the first time, interested him not at all. He was, in other words, perfectly adapted to serve as a figurehead for a decade of social turmoil and economic change. Robarts's solidity gave his party something to lean against while practising the delicate art of political adaptation in a time of prosperous uncertainty.

It was a considerable bonus for the Ontario government that Simon Reisman negotiated, in 1964, an 'Autopact' with the US government. The Autopact provided for a limited kind of free trade in automobiles and automobile parts across the Canadian-American border. The Canadian end of production was hedged about with safeguards which lent a certain stability to Canadian automakers and their suppliers; these automakers were almost entirely located in southern Ontario.

The Autopact was a harbinger not merely of better times but of good, very good, economic times for Ontario. Prosperity for Ontario could mean expansion for Hydro; but as recently as 1947–8 good economic times had been matched by brown-outs and reduced service as Hydro failed to provide the power necessary to support the province's industrial expansion. Hydro's peak load reached 4909 megawatts in 1956, 6157 in 1960, and 6969 in 1963. It was for that reason that Hydro hoped to bring into service, by 1969, 485 megawatts (electric) of hydro projects, and 4400 megawatts (electric) of thermal stations. Its planners worried that this might not be enough.[2]

A new government in Ottawa and prosperity in Toronto were two necessary ingredients to the change that AECL was about to experience. There was a third, and it had been present since 1958: NPPD. The Nuclear Power Plant Division was still growing. It grew to satisfy the need for supervision at NPD and the needs of design and construction at Douglas Point. By 1962 its engineering and design effort at Douglas Point was peaking, and by 1963 it would be on the downward slope. At least it would be unless something else was found for it to do.

The timing was off. Douglas Point, whether it failed or worked, was four or five years from completion. Hydro's need was immediate: to accommodate the booming economy, decisions must be made now: 1962 or 1963. NPPD did not mind much; it had to justify its future existence, or get out of the business. The conjunction of these events was a coincidence. As it turned out, it could not have been better contrived.

'The future of NPPD' appeared on the agenda of AECL's executive committee on 18 March 1963. The end at Douglas Point was still distant, but by summer it would no longer require as many staff as at present. 'Mr. Foster,' the minutes said, 'was anxious that additional work be authorized for NPPD so that key staff would not be lost owing to uncertainty of the future of the division.'

The committee, still Gray, Gordon, and Hearn, with the addition of Grinyer, who had joined the board the previous year, lent a sympathetic ear. There was money in the estimates to keep NPPD afloat at full strength until the spring of 1964. Admittedly there were no firm prospects, but it seemed reasonable to have them work on a 500 megawatt (electric) station for Ontario Hydro – the size of station that, according to Lorne McConnell, the Hydro planners had wanted for Douglas Point in the first place. Subject to carefully drafted cancellation clauses, contracts for such a station could be placed with suppliers. 'It is a project like this,' Lorne Gray later wrote, 'rather than a repetition of [Douglas Point] that is needed if we are to keep this highly trained and excellent group of designers together.'[3]

Fortuitously or not, Robert Macaulay, Ontario's minister of energy resources, invaded the airwaves and the public prints to demand large nuclear power stations. In light of his pronouncements, AECL nudged Hydro to discover whether, and when, some formal approach might be expected. With Hydro's chairman, Ross Strike, sitting on the AECL board (he had succeeded Duncan in 1961), an answer was expected no later than the next board meeting, in Douglas Point in June.

It was a positive answer. Ontario Hydro was interested in 'units' of 500 megawatt (electric) reactors, though how many units Strike did not say. He did, however, state that they were in the 'national' interest, a sure sign that Hydro believed that the national government should kick in something towards their creation. Gray did not jib; he and Strike had already worked out a formula on the ratio 100 (Hydro): 54 (the province): 80 (the federal government). The total came to $234 per kilowatt of capacity. The Hydro portion was based on what it would have invested in a similarly sized conventional station. The two governmental contributors could lend the money and get it back from future power sales.[4]

It could easily be done, Gray told the board. NPPD existed for just such an eventuality as this, and unless it took this opportunity it might be dispersed. Nor was it NPPD alone; industry also depended on new orders, and this was a highly desirable one. The board agreed, laying down terms that closely resembled those for Douglas Point: a joint engineering and design effort, each side paying its own way as required. A final settlement would be embodied in the price of the plant. The board expected that not one but two units would be built, and in that hope proposed that a detailed agreement be ready by October.

It then turned its attention to the construction of a heavy-water plant in Canada, the last component in the construction of a fully autonomous nuclear industry in Canada and a subject much in the minds of the board since the NPD spill the previous December. The connection between the heavy-water project and the 500 megawatt (electric) units is obvious; but like the coincidence between Whiteshell and Douglas Point, it would exact a price. How large and how long a price remained at that point to be seen. For the moment it was merely a good idea whose time seemed to have come.[5]

The Price of Pickering

An exchange of letters in June 1963 between Gray and Strike (wearing his Hydro hat) set in motion the design of a 500 megawatt (electric) station. The federal government reacted positively. Simon Reisman, then still assistant deputy minister of finance, regarded the 500 megawatt unit as 'the next logical step in nuclear design,' one that promised 'to be competitive with a thermal station of the same capacity.' Equally important, it would keep Canada abreast in 'the technological race and should result in commensurate advantages to Canadian industry and our competitive position in general.' Reisman had his doubts about *two* units, but Bryce had none: 'I would be inclined to go for it,' the deputy minister wrote in mid-July, 'given the cost savings and the close margin over coal.'[6]

There was never any doubt that the federal government would support a 500 megawatt design for Ontario. There was, however, debate over the terms on which it would do so. At first Gray, and then Drury, favoured a 'functional' approach, whereby AECL would contribute the nuclear parts of the reactor. The finance department found the idea appealing; the Ontario government and Ontario Hydro did not. The reason was that under the 'functional' approach Ottawa would end up paying less and Ontario more; Ontario's negotiators therefore sought a solution that would reduce Ontario's share and increase that of the federal government.[7] The result, inevitably, was a compromise, but it was not reached until the summer of 1964. Before we examine it, we shall pause to consider the context of the negotiations.

The Ontario reactors were not the only subject on Gray's agenda over the winter of 1963–4. He had Canada's nuclear program in India and a heavy-water plant to forward as well, and he seems to have convinced his minister. In November 1963 Drury, in describing the Indian project to his colleagues, even argued that the success of the Canadian project hinged on the development of the Indian program.[8]

For Foster and NPPD, it was business not quite as usual. An extra forty personnel appeared in his staff estimates for 1964–5. With the 500 megawatt (electric) reactor in prospect, it was time to provide. Forty seemed like a safe bet; to get that many, AECL approached other Canadian atomic firms for the loan of staff. The terms were 50 per cent of salary, plus experience to be accumulated as part of the design team for the next generation of reactors.[9]

It should not be thought that relations between AECL and private suppliers were all one way. The whisper of a 500 megawatt station, not to mention Macaulay's subsequent bellows, attracted interest. CGE, as AECL drily observed, was 'known' to be 'anxious to get work in this area of activity.' Not to be outdone, Hawker Siddeley, the British industrial and engineering conglomerate responsible for the Arrow, wanted into this new field, while in 1965 Montreal Engineering delicately suggested that they might be willing to incorporate NPPD lock, stock, and reactor if AECL ever gave thought to getting out of the business.[10]

The suggestion, once made, floundered. The possibility of a contract with India and another with Hydro brought AECL up against the requirements of partners who were simultaneously customers. Neither India nor Hydro wanted to deal with anybody other than AECL, and Hydro's attitude was very well known and largely (though not universally) shared inside AECL. Les Haywood, who as vice-president, engineering, directed those parts of Chalk River that had not come under Lewis, argued the 'temporary' nature of the existing situation. AECL might be on the verge of drifting into a permanent organization that would be competitive and not complementary to other firms' efforts. What if reactors became standardized to the point where they could be bought, as it were, off the rack? Where would AECL's advantage be then?

To Foster it was purely a practical matter. A design and development team existed. It was expert, but few in numbers. 'He felt,' he told other AECL executives in a meeting in December 1963, 'that in Canada the work would be done by the same experts regardless of the organization chosen to handle the design; in other words those engineers who [had] obtained experience in this field would tend to follow the work wherever it might be done.' It would not by that token be done better elsewhere.[11]

Gray's decision, when it came, followed Foster's reasoning. After sounding out certain government officials, who were found to be favourable, he told his senior management committee that he believed AECL had little choice. Its existing clients, Ontario Hydro and India's Department of Atomic Energy, would deal only with another governmental organ. And in Canada there were, unfortunately, no other clients. While it was theoretically possible for CGE to forward its NPD design (it was doing so, eventually successfully, with Pakistan) it was not likely to find takers and it might not do so even if it had an effective monopoly, because of purchasers' dislike of paying monopoly prices. Besides, although CGE could offer a fixed price package, it could not offer what AECL could: in-depth scientific background as well as a consulting engineering capacity that would permit if not encourage competitive tendering. Nor was it simply a question of designing new reactors and supervising their construction. Maintaining the reactors meant, in effect, preserving the organization that had designed and built them.

As Gray put it, 'this all suggests very strongly that AECL is not going to get out of the nuclear consulting engineering field for several years – certainly not less than ten and probably much longer in a reduced role related to new systems.' Doubtless sooner or later private consulting firms would take over the business, but only when it *was* business and not a fully supported or guaranteed offshoot of the federal government. When that happened, he thought that Montreal Engineering, then working on an Indian contract with AECL, would be the first in the field.[12]

Gray told his executives that he intended to discuss his recommendations with Drury at an early date. First, he ran them past the Board of Directors, who met the next day, 14 January 1964. They approved, and they also approved the creation of a sales force of three or four, so that AECL might operate 'as an aggressive commercial consulting nuclear engineer.' The board sent their recommendation on in support of Gray's, to be used when he talked with Drury.[13] But if and when he did so – the records are sparse – it was in the context of larger issues: the negotiations with Ontario Hydro, discussions with India, and the question of domestic heavy-water production. Meanwhile, planning for a 500 megawatt (electric) unit went forward, and another cycle of estimates came and went. Even inertia forwarded the prospects of the new reactor, since until a decision was made the capability to respond positively to a government directive had to be preserved.

It is important to emphasize the larger picture that confronted AECL – a picture in which the 500 megawatt project was the largest part, but nevertheless only a part. The government was, from its point of view, being asked to make a series of decisions linked by the common factor of atomic energy. It justified its reactions, which were favourable, in terms of the national interest. Energy self-sufficiency was promoted both by the pact with Hydro and by the creation of an independent Canadian capacity to manufacture heavy water; and the Canadian atomic program was strengthened in the eyes of the cabinet and the Ottawa bureaucracy by the fact that India, a friendly Commonwealth partner that bulked large in Canadian schemes to pacify the world, was interested in the CANDU design.

These developments had a domestic economic dimension. Even though Canadian loans financed the reactor for India, the construction of such a reactor would be good for sectors of the Canadian economy on which the Pearson government, like the Diefenbaker and St Laurent governments before it, looked with great favour. These sectors were not only generic but geographic, for the Pearson government, again like Diefenbaker's, preferred to spread the national wealth into regions where high technology was only a word. The linkage of an Ontario project with a more diversified interpretation of the national economic interest was therefore extremely useful, not to say essential to the prospects of the CANDU. Those prospects were bound up with the issue of the negotiations with Ontario, and we now return to that subject.

The federal cabinet authorized negotiations on 12 September 1963. On the federal side, Bryce and Reisman took the lead, with

Drury, the minister responsible, and Walter Gordon, the minister of finance, informed on the sidelines. (Gordon by coincidence was Drury's brother-in-law.) The federal side argued for a division of risk that allowed Ontario Hydro to escape with the cost of a conventional plant ($100 million), while dividing the rest between the governments of Ontario and Canada. Anything over and above the operating costs of the station would be paid back to the investors on a strict percentage ratio.[14]

The final agreement was not quite so straightforward. The total was different, up $30 million from the first figure. Ontario Hydro's contribution was higher too. The residue was split between Ontario and Canada, but the percentages were different, with the federal government contributing 54 per cent and Ontario 46 per cent. Net revenue from the station (to consist of four 500 megawatt units) would be divided according to investment and paid out over a thirty-year amortization period; should the investment (plus interest) not be repaid by the end of thirty years, payments would flow until it was. Finally, the Ontario government became pressing; though the federal cabinet still hesitated, Drury prevailed. Ontario would get its big reactor and the federal government would help make it possible.[15]

A prime consideration in all of this, as it had been for Diefenbaker's civil servants four years before, was how Canada's industrial aims and foreign policies could be merged to advantage with the wonderful new generation of reactors that AECL dreamed of building. They included an organic model, a boiling light-water model, and a 'fog-cooled' reactor, all to be made available to AECL's next generation of customers, the most promising of whom was the province of Quebec.

By mid-1964 it seemed a foregone conclusion that some new reactor would be built, so AECL authorized the NPPD on a more permanent basis. Hydro wanted its Manby site, where NPPD had been squatting since 1958, back. AECL decided that a move was overdue, especially in view of the extra staff that it was hiring. By 1967 NPPD would reach its 'peak' of 400 staff, and by then a new building would definitely be needed. The move took time – getting a good design took time – but eventually it was made, west to Toronto's suburbs, to the Ontario Research Foundation industrial complex at Sheridan Park in what is now Mississauga.[16]

The announcement that Pickering would proceed could not have been much of a surprise by the time it was made, and its audience on a hot Ottawa August day was presumably pleased at this further evidence of Canada's industrial progress. But that was not the only meaning that we may ascribe to the ratification of Pickering by the politicians. As the announcement was made in the House of Commons on 20 August 1964, Gray, vastly pleased, sat in the gallery. He heard the words that confirmed AECL's next reactor. He also heard confirmed the status of AECL as a major engineering firm.[17]

Building Pickering

Pickering would be different from Douglas Point. It would be owned by Ontario Hydro, and would feed into its basic power supply. To build it, Hydro turned to AECL, through NPPD, to act as its consulting engineer. AECL would charge Ontario Hydro the cost of its services, but it would be Hydro that would supply the site engineer and the construction expertise.[18] Pickering was a small township east of Toronto on the shore of Lake Ontario, in direct line with the metropolitan area's expanding suburbs. The station was on the lakeshore, accessible by water, not far from the railway main lines parallel to the lake.

Pickering, or some other lakeshore location close to Toronto, was the original choice for the 200 megawatt station that was eventually built on Lake Huron. The high cost of long-distance transmission, in terms of power lost over distance as well as the cost of the lines themselves, was a major economic factor in the choice of site.[19] Inexperience and prudence – the latter enforced by the AECB – had suggested that distance from a metropolitan centre was more desirable than proximity and economy in transmission of electricity to customers. The electricity that would be drawn from Douglas Point was in any case marginal, or 'surplus.' 'Surplus,' however, was not what was now required. Now the necessity for competitive and plentiful electricity supply was drawing attention back to the Toronto area. The decision to place two 500 megawatt reactors at Pickering was a risky one, but not for the most apparent reason, location. There were risks attendant on the fact that only one prototype reactor, NPD, was actually in operation. There were risks in the size of the proposed station, which would represent a substantial proportion of Ontario's electricity generation. An outage of any duration would harm the fortunes of Hydro, not to mention its effect on the provincial economy.

Economically, bigger could only be better as far as Hydro was concerned. Large demand meant large stations, whether conventional or nuclear, a view that Lewis enthusiastically espoused. Writing in March 1963, he claimed that a 500 megawatt unit would cut the cost of power generation; doubling that unit would cut costs even more. He emphasized that a nuclear system's ongoing costs were and would be less than for a conventional facility of the same size and relative sophistication. It was, of course, still true that a conventional plant would initially cost less than its nuclear rival, but by endless computations involving the cost and availability of money, projections of the price of coal, and assumptions about the efficiency and capacity of generating plants, he arrived at the conclusion that a 500 megawatt nuclear station would save $49 million per year over its coal-fired competition.[20]

Even if Lewis's estimates of costs were correct – and many, inside the company and out, had their doubts about his ability as an economist – an expanded nuclear program could only be justified if demand increased. Future demand was extrapolated from previous experience, and previous experience suggested that demand between 1962 and 1982 would grow by 5.5 per cent to 7 per cent a year. This meant that Canada's generating capacity would grow from 26,000 megawatts in 1962 to between 75,000 and 100,000 megawatts by 1982. It is interesting to compare these speculations with the growth rates and capacity actually achieved. Between 1963 and 1973 the annual rate of increase in generating capacity was 7.9 per cent; these figures diminished in the 1970s to 6.5 per cent.

Reactor type and reactor economics were studied by PRDPEC, which we have already met. PRDPEC had a double function. It contemplated the development of power reactors, but its mandate was not confined to the CANDU. Indeed, CANDU (heavy-water cooled) was just one of several designs that PRDPEC had before it; of the others we have already encountered the organic-cooled model then being built at Whiteshell. There were fog-cooled and water-cooled models; the moderator remained heavy water.[21]

PRDPEC was a forum for arguing about concepts of reactor design, but it also served to ventilate Lewis's worries about new and more advanced reactor designs, whether organic, light water, or fog cooled. As Pickering began, it was obvious that Lewis was concerned about the time that any substantial redesign of the CANDU model might take, not because it postponed the construction of a 500 megawatt station but because it detracted from the time available for the other reactor research projects.

Yet changes had to be made. A 500 megawatt reactor was bigger than either Douglas Point or NPD at Rolphton. But bigger in what way? Should it have more or bigger fuel channels and pressure tubes? Should its fuel 'rating' (expressed in terms of power derived per centimetre of fuel) be higher, and if so by how much?

This question, or set of questions, exercised Lewis, Les Haywood (who had taken over operations at Chalk River), and Foster for the first few months of 1964. Foster, speaking for NPPD and for Hydro, urged the adoption of a 4 inch standard for pressure tubes; Lewis and Haywood resisted, preferring the older 3.25 inch measure and arguing that the eventual economies to be derived from the newer and more advanced designs far outweighed any benefit that was likely from the redesign of the pressure tubes. It was a classic confrontation between reactor physics (3.25 inches was better) and down-to-earth engineering. One way or another, whether more tubes or bigger tubes went in – or both – the fuelling machines and fuel channels had to be redesigned.[22]

The issue afflicted the board as well as senior management. Since his executives were unable to agree, Gray in January 1964 had the board assign the problem to the executive committee: Gray, Gordon, Hearn, and Grinyer. The executive committee flinched. Two meetings passed without a decision until finally, in April, Lewis and Haywood threw in the sponge, and agreed 'that the convictions of the design team (shared by Hydro) should prevail,' and that 4 inch tubes should become the standard for the 'CANDU-500.'[23]

Chalk River's concerns did not end with the size of the pressure tubes. There was also the question of the effect of irradiation on the fuel and the pressure tubes holding the fuel, and the related problem of the fuel rating, which Chalk River preferred to keep at 40 watts/cm rather than the 52 watts/cm urged by some. It had been learned that vibration of the fuel channels was a problem and it was hard to predict how serious it might be in the longer tube. The most serious difficulty was the rate of creep (creep is the slow stretching of metal when it is under stress at high temperature, and it is made worse by nuclear radiation), which in Haywood's opinion risked having to replace the pressure tubes within the lifetime of the reactor. The cost of spare parts would be considerable, but the heaviest loss would be in terms of down-time on the reactor – six months' worth of lost power.[24]

It may be worthwhile anticipating events at this point. 'Creep' did occur at Pickering; when it did, it took not Haywood's projected six months to repair, but two years. Nor did cost remain constant: it was more expensive to repair than to replace the pressure tubes. But one thing turned out to be correct, or even slightly on the pessimistic side: the fifteen-year life projected for the tubing.

Foster again took a practical view. Haywood might well be right; planning assumed that he was. The life of a reactor was thought to be thirty years. If a pressure tube gave out after fifteen years, a single replacement operation would be necessary, at presumably the same cost. The difficulty was keeping the zircaloy tubes from stretching and sagging, and then hitting the bottom of their calandria tubes which are cool and creep very little.

Testing began at Manby and at Chalk River, where zircaloy was stretched on a mechanical rack in an NRX loop to discover what would happen. Meanwhile, it was decided to use 'garter springs' to keep the pressure tube centred within the calandria tube. The garter springs were made of a zirconium-niobium alloy, a stronger substance than zircaloy, developed in the USSR and first used on the Soviet atomic icebreaker *Lenin*. In the longer term, Foster wrote in 1964, it was hoped that this new alloy could be used for the coolant tubes; because of its increased strength it might permit a reduction in thickness and hence a decrease in neutron absorption. The whole question of zirconium and its uses remained lively, and there was apparently no conclusion. Endless experiments were in prospect. As Foster later admitted, no better expedient could be found. There would be, at any rate, no surprise when difficulties later developed.[25]

Control of such a large reactor was also a challenge, but lurking within the challenge was an opportunity. Moving control rods using cobalt as a neutron absorber were provided for coarse adjustment of power. A copious supply of cobalt-60, Commercial Products' staple isotope, would thus be made available as a by-product. Fine tuning of output and *zone control*, which ensured that all parts of the large core made their proper contribution to the output, was achieved by fixed control tubes containing variable amounts of light water with a helium cover gas, an idea credited to Ernie Siddall, continuing his responsibility for control system and instrumentation design in NPPD.

IBM 1800 computers with added drum memories were selected as the principal items of hardware for the automatic control systems, two sets for each reactor to give full back-up. The programming (software) was achieved by AECL's own designers, with the unique advantage that one of the computer systems for the last reactor was obtained first and used in the design office so that the programs could be thoroughly tested before the first reactor started up. The digital computer was a major advance in automation. Much attention was paid to communications between the computer and the human operator, the two complementing each other. CANDU nuclear reactors commonly operate steadily at full power for *months* at a time, twenty-four hours a day, with almost no need of attention. The computer keeps a tireless watch and remembers complicated procedures; the operator uses his eyes and ears (and perhaps his nose), but above all his brain. In principle, the computer keeps a log of every important happening, and when things happen fast, it can write about 10,000 times faster than a human being could.[26]

Large nuclear reactors could be operated with older and simpler systems of automation and even, to a large extent, under manual control. However, CANDU reactors would lead the world in reliability and continuity of output, and computer control would contribute something to their success. But that is to anticipate: in 1964 success was a desideratum, not an achievement.

The preliminary design was ready to go before Hydro's commission (board) in the spring of 1964. After passing the commission's scrutiny it travelled to the AECB, a major hurdle, since it had been decided to locate the reactor right at the site the board had rejected for the CANDU-200 five years before. Without the board's licence it was back to the drawing-board, or back to Lake Huron.

The AECB's basic safety requirements had evolved over many years as the CANDU design had evolved. They were mainly based on what was referred to as the 'single-dual failure criterion,' developed principally by George Laurence while he was chairman of the reactor safety advisory committee. This called for independence between three major divisions of the plant, (a) the process system, which was everything involved in the normal range of operations; (b) the protective system, which used signals from a wide variety of sensing instruments to operate shut-down devices or to produce other automatic remedial actions in the event that trouble was signalled; and (c) the containment system, which was intended to prevent radioactivity which had escaped from the process system by any means whatever from reaching the public. The AECB required that the failure of any one of these systems should not occur more frequently than a specified figure and should result in only very minor specified radiation exposure to the public. They also required that the simultaneous failure of any two of the systems should be much more infrequent; any resultant public exposure must be less than a specified, though higher, limit.

Although these requirements are on the face of it clearly and unambiguously stated, it has proved very difficult to decide whether a particular design does in fact comply with them. This uncertainty, duplicated in every country in which nuclear power stations have been built, has led to apparently endless correspondence, negotiation, and bickering between reactor designers and operators on the one hand, and regulators on the other. This process got well under way as the Pickering project progressed. The AECB did not consider the designers' proposal to be adequate and asked them to give it further thought.[27]

When they returned to the board in November 1964 they had worked out a refinement. This was a vacuum building attached to the reactor. One of its designers, Bill Morison, described the vacuum building as based 'on fundamental engineering principles.

It is passive; it has no moving parts; and it sucks leaks inwards, not out.' The principles were not novel, but their application to an emergency system was.[28]

Foster's initial estimate was that the Pickering station could be built for $266 million over six years. He believed that given certain assumptions – as to the cost of uranium fuel – it would be competitive with coal-fired stations. He did not need to assume very much about its operation, at least for the present. NPD, as far as it went, showed that it would work, and probably work well. To work it would need fuel, a Canadian product, but it would also need heavy water – a great deal of heavy water. It was not at all clear where it would get it, and to that question Gray now had to turn his attention.

Heavy Water

Canada was well launched into heavy-water reactors; but no heavy water was produced in Canada. The facility at Trail had been superseded by large American plants at Dana, Indiana, and Savannah River; each plant could make about two hundred tons of heavy water a year. Canada bought the product, which was admittedly in surplus. There was so much of a surplus that the AEC closed the Dana plant, reduced the price, and still had heavy water to spare for Canada's research reactors. (The price reduction made even the cost-conscious Hinton take note: perhaps heavy water could be substituted for graphite after all.) It helped balance the plutonium account: at $28.50 a pound, heavy water was a sizeable item in the cost of running a reactor.

Making heavy water was a prolonged, even exotic process. Ordinary water contains heavy water in the ratio 1:7000. The AEC, in building heavy-water reactors at Savannah River, relied on a process devised by Jerome Spevack, a veteran of the Manhattan Project, for its supply. By mixing water with hydrogen sulphide, and pulsing the water through hot and cold zones in multiple columns, heavy water (deuterium) is gradually concentrated. It worked well enough; when the AEC decided to declassify the process, Spevack thought it worthwhile to sue the commission to obtain patent protection for his ideas. He won, and a patent was duly issued in 1959.[29]

Douglas Point in winter, 1966

John Foster examines a model of the reactor at Douglas Point.

Delivering the calandria at Douglas Point, 1961

OPPOSITE

The calandria installed: the reactor face at Douglas Point

Operators bring the Douglas Point reactor to 'first critical' in 1966.

Pickering under construction

Pickering A completed, 1971

Inaugurating Pickering. Dignitaries are, left to right, George Gathercole (Ontario Hydro), James Auld (minister of natural resources, Ontario), J.L. Gray (AECL), William Davis (premier of Ontario), and Donald Macdonald (minister of energy, mines, and resources)

Turbine room at Pickering A

Bruce nuclear power station, mid-1980s. Douglas Point is left centre.

Gentilly 1

The reactor vessel arrives on site for Gentilly 1.

Glace Bay heavy-water plant

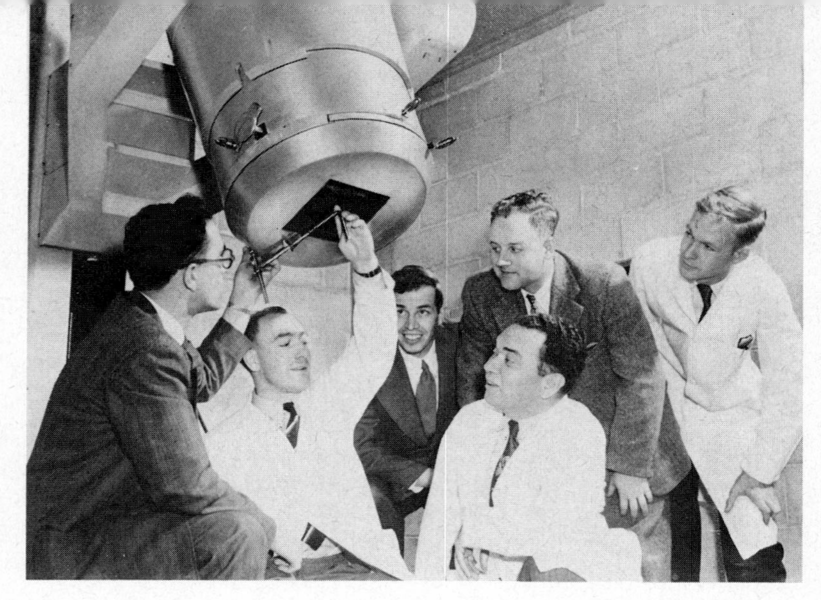

Inaugurating the Eldorado therapy unit, 1951. Roy Errington is standing second from right.

Chou En-lai and Roy Errington at a trade fair in Beijing. Mitchell Sharp is standing next to Errington.

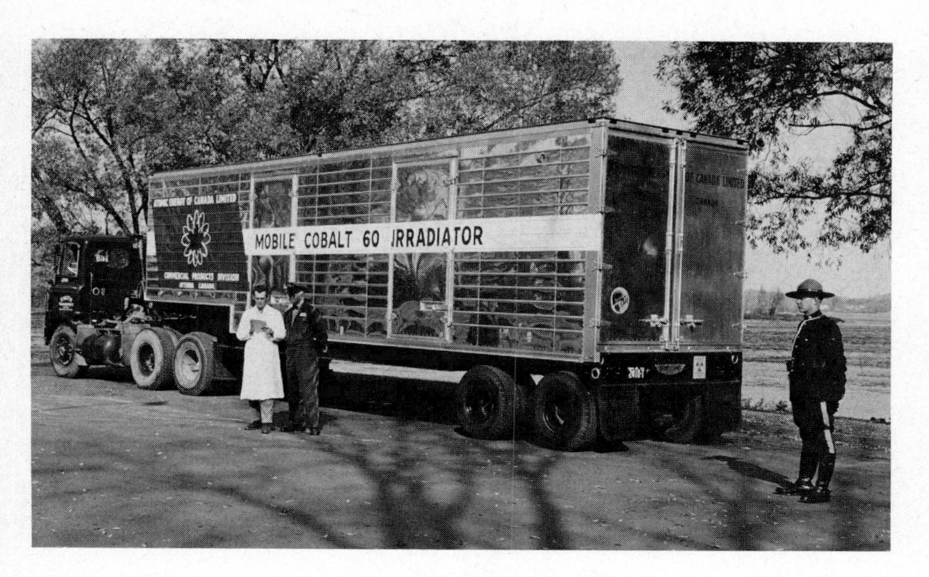

The Cobalt-60 mobile unit

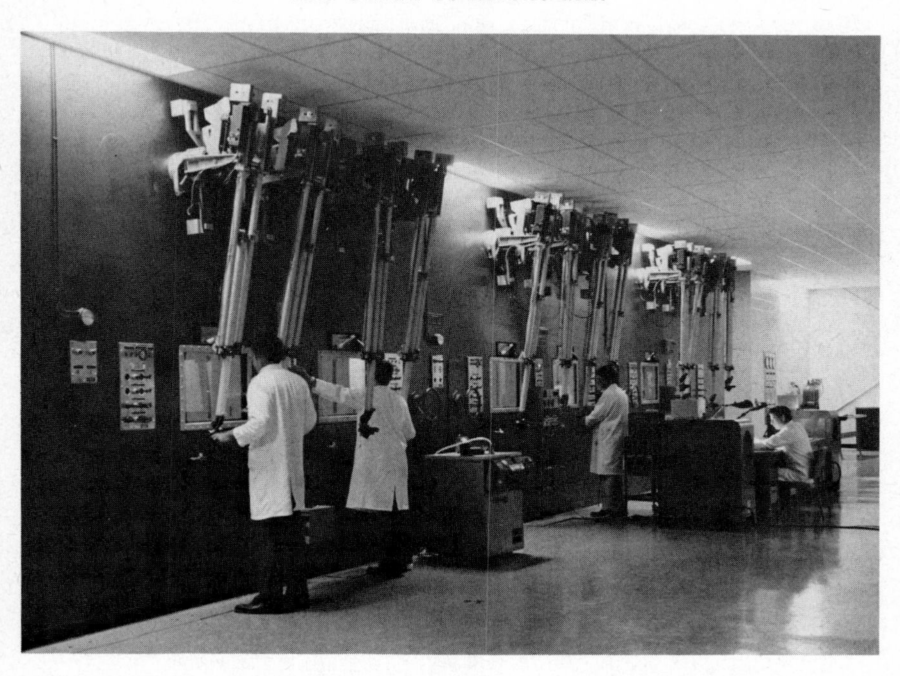

Processing isotopes in hot cells at the AECL Kanata plant

The Radiochemical company's Kanata plant

As early as February 1957 Spevack approached AECL, offering to build a heavy-water plant in Canada. It would be to Canada's advantage, he wrote: using his process, AECL could get heavy water as low as $10 a pound. The time was not ripe. AECL had enough financial commitments to go on with, and was facing cancellation costs for its pressure vessel.

Four years later, however, circumstances had changed. The construction of NPD, of a test reactor in India, and then of Douglas Point strained Canadian supplies. Designers at Douglas Point acted to economize heavy water in every way consistent with a heavy-water moderated *and* cooled design; but Douglas Point required 200 tons of heavy water even so. The prospective Rajasthan reactor in India, discussed in a later chapter, would also need heavy water. When and if the next generation of CANDUs came into operation, they would thirst for hundreds of tons more. At $28 a pound, heavy water for just one Pickering reactor cost $28 million. It was not surprising under the circumstances that Whiteshell's OTR should be organic-cooled, and that by 1963 very serious consideration was being given to a boiling-water reactor, still using a heavy-water moderator but with light water as a coolant.[30]

DEUTERIUM

Deuterium is a constituent of heavy water and is one of the three isotopes of hydrogen, which include ordinary hydrogen, heavy hydrogen (deuterium), and tritium. Deuterium occurs in ordinary water in the ratio of about 1:7000.

Heavy water is of interest in nuclear reactions because of its great efficiency as a moderator. The 'moderating ratio' of ordinary water (used in American PWRS) is 72, that of beryllium is 159, of graphite 170, and of heavy water 12,000. (The moderating ratio is a measure of efficiency.)

Ontario Hydro, having begun with a dual-purpose heavy-water reactor, preferred to stick with what it knew. There was no question that Pickering and any successors would also use heavy water for moderator and coolant, as long as the utility remained committed to the CANDU design. That was, however, an important condition, for it was by no means axiomatic that Hydro would stay committed to CANDU or even to reactors if there was no reliable source of supply. And in Hydro's lexicon, 'reliable' meant Canadian.

It is unlikely that Lorne Gray had Spevack in mind when he first mooted the subject of heavy water to his directors in 1962. He had, he informed his executive committee in June, done some exploring, and had found 'considerable interest' from four companies; the interest was further stimulated by a report by du Pont for the AEC that suggested that 'a modern' heavy-water plant 'combined with fractional distillation of water could produce heavy water for less than $20 per pound.' The AEC had assigned a patent to AECL in 1955; the company owned two other patents on heavy-water production. Gray thought the company could assign these on a non-exclusive basis, for a five-year period, to anyone willing to start up a heavy-water business in Canada.[31]

News of AECL's interest travelled fast. It travelled to New York, where Jerome Spevack resided. He established a Canadian subsidiary, Deuterium of Canada Limited, which had a mailing address in the Bank of Commerce building in downtown Toronto; the subsidiary made itself known to Gray. It would build him a plant producing heavy water at the rate of 50 tons a year for a price of $28 per pound. Gray noted the offer to his executive committee at the end of August, but also indicated that after consultation with the Department of Trade and Commerce he intended to invite offers from the four companies already approached. They were to submit proposals by 1 October.[32]

Proposals flowed in. The AEC gave thought to reducing its price, or to modernizing its heavy-water plant. BA Oil suggested a 200 ton per year plant, to produce heavy water at $28 per pound for the first year and $22 per pound for four years thereafter. BA suggested a four-month engineering study to come up with capital cost estimates. But even as BA's proposal lay on the table, Spevack asked for time to come up with a new and better deal.[33]

Time passed. The AEC lowered its price – to $24 from $28. (The exchange rate was $0.92 US = $1 Cdn.) An internal AECL study reviewed the available options and concluded that an economical plant would require, first, to have a minimum capacity of 200 tons per year; second, a guaranteed purchase of not less than 1000 tons; and third, a location near an abundant source of cheap fuel, since heavy-water plants consumed a great deal of energy. Gray's advisers thought the gas fields of Alberta ideal. On balance, they

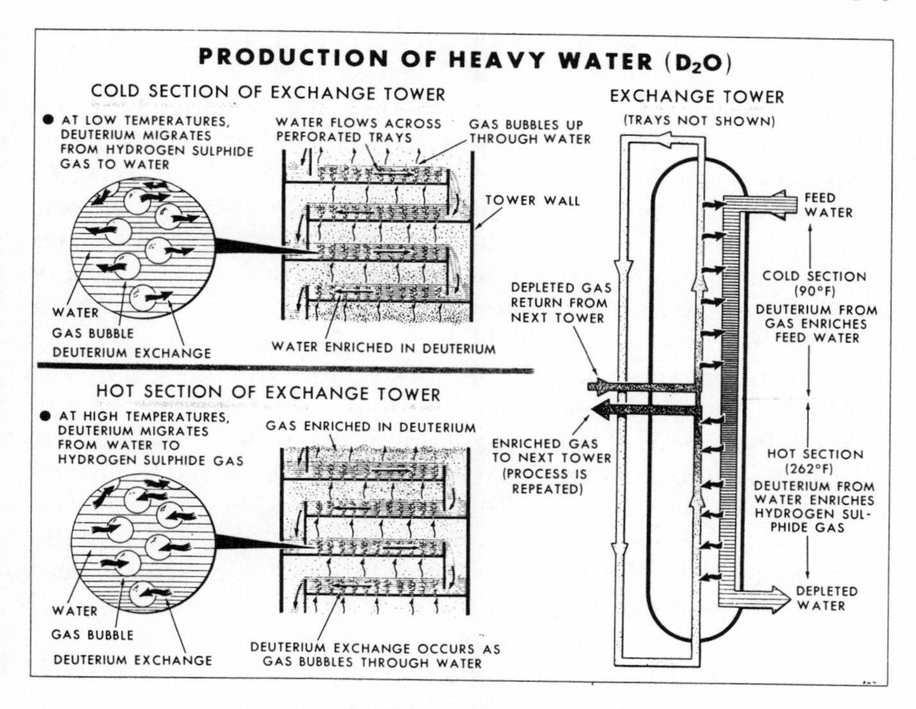

Production of heavy water

judged heavy water to be at least an acceptable risk and even a desirable one.[34]

In December 1962 Gray approached the advisory panel on atomic energy, as a first step towards consideration by the cabinet. Would the government of Canada back the risk? The panel thought it might; the day after it met, 18 December, Gray sent on his proposals to the cabinet.[35]

The Diefenbaker government was by Christmas in its last spasms. As a result, two different ministers of trade and commerce presented the heavy-water project to the cabinet; on the second occasion, 14 February 1963, the government had already been defeated in the House of Commons and an election campaign was on. Before ministers left Ottawa for their constituencies, they held a housekeeping meeting for business that could not wait until after the election. The advantages were great and the risk slight, the cabinet was told. At worst, the government might have to carry an inventory of perhaps 500 tons 'for a short number of years,' at a cost of about $1 million a year; against this were set foreign exchange savings of between $15 million and $30 million between 1965, when the proposed plant would enter production, and 1970, when AECL's purchase guarantee would expire. There were, in addition, export possibilities, and the establishment of 'a new industry processing Canadian natural resources and creating 100 new permanent jobs.' The commitment of a heavy-water facility would also show that the government backed CANDU and would enhance the attractiveness of the heavy-water system abroad.

The cabinet approved, while postponing the deadline for proposals from interested firms to 1 June 1963 (later amended to 31 May). It committed to purchase up to 1000 tons of heavy water, at $22 a pound or less, the heavy water to be delivered in annual batches of 200 tons. A successful proposal would first have to pass by AECL, the Department of Finance, the Treasury Board, and the Department of Trade and Commerce: agencies whose chiefs were members in good standing of Lorne Gray's lunch club. The heavy-water plant would be, as recommended, in western Canada.[36]

The cabinet then dispersed; when AECL and heavy water next appeared on its agenda a new set of ministers were sitting around the table. Drury, who would represent AECL's interests until 1966, we have already met. Gray afterwards retained a lasting and bitter memory of Drury's management of the heavy-water conundrum. Drury was a respected member of the cabinet, and though not its most senior or prestigious minister he nevertheless had more standing with his colleagues than Churchill had had with his.

AECL therefore received a more balanced but not necessarily less effective representation than Gray was used to. It was as an instrument rather than as a subject that the company appeared in cabinet discussions, for the Pearson government's industrial plans (it is too much to call them a strategy) had several purposes. They would enhance Canadian industry and expand Canadian exports; at the same time, they would develop Canada's have-not regions, outside the large cities and the booming Windsor-Quebec corridor. The Liberals had not tended to do well in such places, where the Conservatives had traded on the Diefenbaker government's knack for striking the proper balance between central prosperity and regional patronage.

The Pearson cabinet was correspondingly anxious to do well by regional interests. It favoured regional economic development, and held as an article of faith that such development, properly stimulated, could reverse history and alter geography, if only the right industrial alchemy could be devised. It was not alone in this; nor was Canada unique in attempting to balance the equation between its relatively impoverished or disadvantaged regions and the prosperous industrial heartland.

Pearson wished to improve relations between the central government and those of the ten provinces: this was called 'co-operative federalism.' Provincial governments were also sensitive, none more so than the government of Nova Scotia, whose Conservative government under Robert L. Stanfield had established an industrial development corporation, Industrial Estates Limited, back in 1957. In 1961 it had considered an abortive proposal from du Pont of Canada to build a heavy-water plant in Cape Breton. Nothing had come of it, but the idea was intriguing. It required government aid, but that was appropriate, for the Stanfield government, like other Canadian governments, believed in the efficacy of state intervention and direction in the marketplace.[37]

Nova Scotia had not been in contention for Gray's heavy-water plant. He favoured the west, with its abundant energy. When the bids came in to AECL at the end of May 1963, all four, Imperial Oil, Dynamic Power Corporation of Calgary, Deuterium of Canada Limited, and Western Deuterium Limited, proposed the prairies. Two bids were soon discarded, leaving Western Deuterium and Deuterium of Canada. Deuterium of Canada was still in the race, but just barely, because it had no firm financing. Gray preferred Western Deuterium, which would settle in Estevan, Saskatchewan. When he saw Drury, on 7 June and again on 22 July, that was his recommendation.[38]

Estevan had sufficient energy and plenty of fresh water. But Drury introduced a new factor: the possibility of choosing the winning company not on the basis of energy supply alone but on location, the winning location to be in the maritimes and not the west. In the meantime Deuterium of Canada had changed its venue, and its affections, from Alberta to Nova Scotia. Industrial Estates Limited (IEL) and Spevack's company agreed to co-operate; the plant that they proposed to build, if the federal government agreed, would be at Glace Bay, on Cape Breton Island.[39]

Stanfield now intervened. Writing to Pearson on 14 August, he urged the case for Nova Scotia, for Industrial Estates, and for Deuterium of Canada. Pearson referred the matter to Cape Breton Island's Allan J. MacEachen, the minister of labour. MacEachen was the Liberal party's Nova Scotia chief; he and Stanfield competed for the political affections of its citizens.[40]

Cape Breton Island had a long history of industrial decline; from a major coal and steel centre the island had gradually declined until, in the early 1960s, it seemed that its population had no alternative but emigration to the mainland and even to central Canada. The choice seemed excessively harsh; because it *was* harsh, softer policies should be tried first to see whether Cape Breton could not be turned around, and to turn its liabilities, such as low-grade, high-cost coal, into assets. A heavy-water plant in Cape Breton could draw on coal-fired power; local industry could benefit from making components. The idea had a natural attraction, and MacEachen was attracted.

Other ministers with Atlantic interests, such as J.W. Pickersgill, the senior and powerful government leader in the House of Commons, were enlisted in support. According to Paul Hellyer, then minister of national defence, the choice was 'a private decision taken by [Pearson] and [MacEachen]. It defied economics. It was a political decision, pure and simple.'

There is another version. According to Pickersgill, he and MacEachen combined to move the project from the west to the east. Pickersgill asked Gray if Spevack's process would work. It was a good question, to which Gray did not know the answer; AECL had not specialized in heavy water and had no hands-on knowledge of making it. But Spevack's process had worked at Savannah River, and so Gray replied affirmatively. As far as Pickersgill and other ministers were concerned, that settled the matter. Knowing that the Deuterium of Canada bid was competitive, Pickersgill joined MacEachen to insist that the plant be placed in Nova Scotia. After 'a vigorous internal struggle in the cabinet,' Spevack's project got the nod. It was rumoured that the nationalistic minister of finance, Walter Gordon, had insisted on a majority interest in Deuterium of Canada for IEL; finance officials later denied this, explaining that all they had done was give 'advice ... to Industrial Estates Limited ... from a purely business point of view.' The advice was badly needed, in their opinion, since IEL's representatives were 'rather unexperienced and naive in the negotiation.' Spevack remained president.[41]

Gray was dismayed. Against the pressure from the cabinet he mobilized his board to tell Drury that it would not recommend a maritime location. Such a site, where energy was expensive, could only be established and sustained by subsidy. If the government wished to equalize the terms by subsidizing the cost of energy, then it would have to instruct rather than consult AECL. To the board Gray remarked that any coal-burning station built to service Spevack's heavy-water plant would require double the existing subsidy paid for maritime coal used for power production in the maritimes.[42]

Gray made his views known around Ottawa, including the cabinet committee on economic policy that was sifting through the heavy-water perplex. The cabinet had Pearson's regional strategy on its mind; it was, its members believed, dealing with a larger issue than AECL's site preferences. Some of its members saw a saving to Canada in locating a coal-devouring industry in the maritimes; the heavy-water development could be made to consume coal whose shipment to central Canada would otherwise have to be subsidized under existing legislation. When Gray persisted in his objections, the politicians' patience snapped.

'Are you a member of this committee, Mr. Gray?' A minister demanded. 'Because if not, I think you should shut up.'

'That's good advice,' murmured Drury to Gray. Gray obliged.

Gray protested that he had not been 'meddling in affairs that are of no concern to AECL,' namely the location and economics of the heavy-water plant. He had not broadcast what he thought, or so he claimed, but he made no secret of it either.

Caught between factions in the cabinet – one minister was reportedly ready to resign on the issue – Gray bowed to his political masters, as and when they came to make up their collective mind.[43] That took some while longer. Jack Davis, AECL's former economic consultant, who had been reborn as a BC Liberal MP and parliamentary secretary to the prime minister, circulated a memorandum questioning the economic rationale of the Deuterium project: 'a new $30 million heavy water plant would only employ 120 men ... equivalent to one employee for each $250,000 invested in new plant and equipment.' Plenty of heavy water was stored in the United States – 5000 tons at present. The government should delay until the situation became clearer. Davis had no effect.[44]

On 21 November Drury asked the cabinet to approve the Deuterium of Canada bid. Spevack had made an offer that, on the face of it, promised heavy water at $20.50 a pound, and that figure was indubitably the lowest. There might even be savings in subventions to Cape Breton: an increase in Cape Breton coal production would mean a decrease in the unit cost of subsidies.[45]

There were misgivings. Nova Scotia was playing a mug's game, it was suggested. Deuterium of Canada was only getting the deal because Nova Scotia was lending it a hand. What if other provinces did the same? They did not get the chance, because the federal government moved quickly. On 2 December 1963 it announced that it had approved a negotiation between IEL and Deuterium of Canada to build a $30 million heavy-water plant near Glace Bay, a depressed mining town in Cape Breton. The scheme was presented as a means of helping a depressed area, with the possibility that if it worked, and a subsidized power facility started up, it would assist other industry that might wish to move to the vicinity. It would use seawater as part of its process, seawater being abundant in Glace Bay. Drury anticipated Gray's disapproval by leaving him to read it in the newspapers.[46]

The Nova Scotia Power Commission gave Seaboard Power, a subsidiary of the Dominion Steel and Coal Corporation (Dosco), $12.5 million to install 36 megawatts of new generating capacity; then, in 1966, it bought all of Seaboard for $5.2 million. The province followed up in 1968 by buying Dosco; now it could mine the coal for power and smelt the steel for the plant all by itself.[47]

So far, as Philip Mathias has pointed out, the risk was entirely Nova Scotia's. AECL was committed only to buying 1000 tons of heavy water at $20.50 a pound, admittedly a good price. Difficulties remained. A detailed contract had to be negotiated, but Gray left it behind and flew off to promote CANDUS abroad. The contract was considered 'satisfactory' by the company's executives but, as the board minutes put it, 'having regard to the events that led up to the decision to award the contract,' the finished document would be placed before Drury for his personal imprimatur. It was signed on 7 February 1964.[48]

The contract was well received for, as a contract, it was good business. But once politics had been invoked to put the heavy water plant in Nova Scotia, who could doubt that politics would keep it there, even if Deuterium failed to meet the $20.50 price? Robert Thompson, leader of the Social Credit Party and an Alberta MP, 'publicly repeated a rumour that Spevack had boasted openly he would negotiate a higher price once the plant was a fait accompli.'[49]

The future price of the heavy water was at least in the realm of possibilities. Gray had problems enough in reality, not the least of which was coping with the consequences, otherwise happy enough, of Ontario Hydro's decision to commit itself to Pickering and to its first two 500 megawatt (electric) reactors. Even if Spevack produced the entire 1000 tons on time, there would not be enough heavy water to go round especially if demand for heavy water emerged abroad. Gray said as much to Drury within a month of the signature of the contract with Deuterium of Canada; he thought that by September enough new demand would be manifest to force him to reopen the question of heavy-water supply. The Board of Directors decided in May 1964 that they would have to ask for more – provided Ontario Hydro's commitment to Pickering was firm.[50]

Gray told Drury that he would be back at the trough in September 1964, and he kept his appointment. There would be a deficiency in heavy water, assuming demand from Quebec and foreign customers as well as from Ontario Hydro, but even if demand were limited to Hydro alone there would be a problem. So Drury came back to the cabinet committee on finance and economic policy on 8 October, and, with its approval, back to the full cabinet on 22 October 1964.

The cabinet approved a search for further capacity, with bids due in January 1965; as before, it was Deuterium of Canada that appeared to have an advantage, thanks to its partnership with Industrial Estates, which conferred a privileged access to capital at low rates. As the cabinet noted in December, Cape Breton had become a 'designated area,' an economically depressed region that deserved and even demanded official aid at concessionary rates. 'Designated areas' represented a new level of official intervention designed to retain and retrain the local population.[51]

A final judgement is not yet in on the performance of 'designated areas' in Canada in the 1960s and 1970s, but while there may have been individual successes in the program, they have not succeeded in reversing the imbalance in prosperity between the maritimes and central Canada. Some evidence suggests that in their eagerness to encourage 'high tech' industry in neglected or impoverished areas the governments of the day plunged too quickly into ventures whose practical ramifications they only imperfectly understood; when problems occurred their response was, simply, to persist or, in the words of one senior government economist, 'to throw good money after bad.' The story of Deuterium of Canada adds to that evidence.[52]

Good money set out to create a heavy-water facility in Cape Breton in 1964. Optimism was the keynote. According to Spevack, 'Canada is going to have the world's most efficient and lowest-cost heavy water plant. We will be in a very strong position to meet competition from anywhere.' Admittedly, Spevack was starting from a very weak position, for the $12 million that IEL contributed to Deuterium as its share was nowhere near enough to meet the $30 million cost of the plant.

By 1966 the cost was no longer $30 million. A sales tax here, a strike there, and a rise in interest rates added an extra $10 million. Spevack remained optimistic. Its maritime home gave his factory plenty of cheap raw materials. It would still lower the world price, not raise it. The plant would be finished, not on time – its due date was 31 July 1966 – but not scandalously late either. And despite strikes and confusions, the Glace Bay plant would be ready to go in the spring of 1967.[53]

There was only one slight problem. Deuterium had raised enough money to go on with, but it could not raise any more. When the federal government decided to plump for an expanded heavy-water capacity, Deuterium could promise but it could not perform. Nor, apparently, could anyone else. A bid from Western Deuterium had no issue. Its process would cost a mere $14.65 a pound; but perfecting its process would take more time. When the matter went before the cabinet it reacted cautiously, requiring Gray to satisfy himself that the Western Deuterium process really was practicable. If it was, then Western Deuterium should get the contract and, in return, the federal government would undertake to subsidize the first 1500 tons produced from its facility.[54]

By the time a contract had been drafted, Western Deuterium had had second thoughts and withdrew its proposal. The cabinet, which was monitoring the situation both because of the cost involved and because of its perception that heavy water was a useful tool in its program for regional development, reacted by directing Drury to notify the Saskatchewan government of the failure, and to notify Nova Scotia to stand by.

Gray concluded that part of the problem was that AECL had no independent capacity to assess heavy-water technology. He appointed Ara Mooradian, head of engineering at Chalk River, to investigate the problem with an eye to establishing 'a substantial development program.' But while Mooradian investigated, Gray sounded out other companies on the matter; for AECL to become directly involved in heavy-water production could strain the company's resources, both managerial and technical. Mooradian and his team investigated a French process, but though the French syndicate showed some interest in taking on the business, the cabinet did not reciprocate; instead, on Gray's recommendation, it investigated the next highest bid, from Dynamic Power.[55] A bid from the western-based Dynamic Power Corporation failed, despite the active interest of the government of Saskatchewan. A cabinet direction to try Deuterium of Canada did no good; Deuterium's negotiators said that they would have to delay production at Glace Bay for six months.[56]

In June 1965 the AEC commissioners toured Canada's atomic facilities and Gray took the opportunity to discuss the possibility of buying a large quantity – 1000 tons – of American heavy water from them. Such a large amount, he suggested, should receive a special price. The reaction was favourable, so favourable that Gray reported to Drury that he felt 'quite confident' that the Americans would supply their product for $17.00 (Cdn) a pound, a figure that was certainly higher than two of the bids received for a second heavy-water plant but substantially lower than the existing purchase arrangement with Spevack. Presumably because the figure was higher than the bids currently on the table, the deal was not pursued; when Gray next approached the Americans the price would be higher.[57]

This too failed to pan out. Imperial Oil was confident that it could take on the job, but not at an acceptable price. And so eyes turned to CGE, which was recognized as at least capable of managing a large construction project. CGE, however, was both interested and interesting. With the authority of the cabinet in hand, Gray began negotiations with CGE before the end of 1965 for a plant capable of producing 400 tons a year; in return, the government would underwrite 400 tons of sales per annum.[58]

In informing his directors of the state of affairs in December 1965, Gray added that there had, of course, been further approaches from Deuterium of Canada, proposing an expansion of their facilities. The trouble was that he (and presumably his advisers) had no confidence in Deuterium's ability to raise money or to get its existing design to work, because of its reliance on seawater in its extraction process. This factor, which Gray then brought up for the first time, would loom larger and larger over the next few years.[59]

Despite the AECL president's misgivings, the cabinet authorized him to negotiate a doubling of deliveries from Spevack's plant in January 1966; then, in March, the cabinet agreed to a firm contract with CGE to supply a further 5000 tons of heavy water. The contract brought Deuterium's troubles to a head. The result was a change in Deuterium's status. The government of Nova Scotia had appealed to the legislature that March for a new draft of funds for Deuterium. Pushed by the company's financial difficulties, the Nova Scotia government in August acquired a 100 per cent interest in Deuterium of Canada; it was now a wholly government-owned corporation, though still under the umbrella of Industrial Estates Limited. The change made negotiations easier, since it eliminated the possibility of royalties flowing to Deuterium Corporation, Spevack's American company, and at the same time reassured AECL as to the company's ability to finance what it promised.[60]

Nova Scotia had become involved with Deuterium in the first place not because of its financial status, but because of the Spevack process; its subsequent grant of subsidies and assumption of an equity position in Deuterium were predicated on the overall economic benefits expected from the heavy water plant, benefits that included technical glamour. Whatever the inventor's deficiencies as a fund-raiser, he was still the government's best hope for technical advice. His services were therefore retained, and AECL's misgivings continued unabated.

With production put off until the spring of 1967, Gray had to look for heavy water where he could find it. Douglas Point would soon go active, and the AEC was happy to make its product available, at a price of $24.50 (US) a pound. They were happy to supply more, Gray told the advisory panel in November 1966, 300 to 500 tons more, if AECL placed the order in time for delivery in 1969–70. Meanwhile, Ontario Hydro was consulting its own interests and investigating the possibility of building a heavy-water plant of its own, using steam from Douglas Point.[61]

CGE meanwhile started construction of its Port Hawkesbury works, a survey having revealed that the ocean waters around Cape Breton were peculiarly rich in the deuterium isotope and that the island was after all a specially favourable site for a plant extracting it from the waters. As construction on the CGE plant proceeded, Deuterium's great day finally arrived.[62]

On 1 May 1967 Premier Robert Stanfield and federal health minister Allan MacEachen presided over the opening of the Glace Bay plant. 'I like to think of this great heavy water plant as a foot in the door of the future,' Stanfield orated, 'the foot which has kept the door from blowing shut and keeping the future open when the collieries were threatened with closure.' Allan MacEachen was more prosaic, reminding his audience that the federal contribution was $200 million, the price of the heavy water it had promised to buy. 'This is money which will come into Cape Breton,' he stated, 'and much of it will remain here.'[63]

Opening the plant was not the same as getting it into production. That would take another year, according to company officials. It took longer, and with every passing month Deuterium of Canada lost money in terms of sales forgone and interest payable on the investment in the plant. AECL gazed at the spectacle from a distance; as Gray explained to his directors, he did not wish to arouse suspicions that AECL was poaching on Spevack's patents. Given Spevack's litigious past this was no small concern. The stories of what was afoot at Glace Bay kept coming. Spevack was mostly in New York; his general manager operated out of Halifax. The engineering on site was reported to be of doubtful quality and labour trouble was frequent and lengthy.[64]

In October 1968 the province finally lost patience with Spevack, who was replaced. An appeal went out to AECL for help, and help was sent in the form of Ara Mooradian. There was indeed a problem, and of a perplexing kind. If the AECL advisory staff stayed around, 'they would end up by virtually managing the operation.' Mooradian reported that the situation was worse than he had imagined; as far as he could tell, there was no discernible management at the project and the project itself was in very bad condition. It was not surprising that the province was now $73.2 million in the hole, and the meter was still running. Tests on the process started, and production was anxiously anticipated before the new year. Gray flew to Halifax to inform the provincial government of the dismal state of affairs.[65]

Instead of a rush of heavy water, there was a trickle of announcements. The estimated cost of Deuterium of Canada rose to $100 million. The government of Nova Scotia removed Deuterium from Industrial Estates and set it up as a provincial crown corporation, a virtual admission that the factory was not expected to make money in the foreseeable future. There were technical difficulties. The plant was not quite ready. Consultants were brought in. The price was up again: $130 million, and it seemed that $30 million more must be spent.

Seawater corrodes metal. It corrodes metal very effectively if left to sit in a pipe, and at Glace Bay that was what happened next. When Spevack's successor started up the plant by introducing hydrogen sulphide into the system, it leaked out through the corroded pipes. Hydrogen sulphide gas is highly toxic; luckily for Glace Bay the wind blew it out over the Atlantic. No one knew what or whom to blame: sabotage, incompetence, or simple bad luck. The plant was effectively inoperable.

At the beginning of 1970 the province sued four companies involved in the construction of the Glace Bay plant, while from the sidelines Spevack attributed every failure to departures from his own true process. There was a provincial election in 1970, in which the 'Deuterium issue' featured. Robert Stanfield had left provincial for federal politics, and it was his luckless successor, Ike Smith, who bore the popular wrath that saw the Conservatives defeated after fourteen years in office. A new Liberal government took power – not that its advent made much of a difference to the Deuterium situation.[66]

The stalemating of Deuterium of Canada and its resident genius left AECL with two options. It could, as it had done before, import heavy water, or it could rely on CGE to finish a plant that would work. CGE, AECL, and Ontario Hydro turned to the source of successful heavy-water technology, Savannah River, whose denizens advised the Canadians formally and informally on how to extract isotopes in quantity from ordinary water. Eventually, in 1971, AECL would be forced to move in and rescue the Deuterium plant; but that is a story for a later chapter. Foreknowledge of this event would not have made the company management any happier in 1968 or 1969.[67]

That was slight consolation. The future might still look rosy, but the present was black. Not only Pickering but a new reactor (Gentilly) in Quebec was under construction; when complete it too would need heavy water. To go on with, Douglas Point's heavy water was purchased from the Americans for $10 million in the winter of 1967, seven months after Glace Bay had been scheduled to start up. In October, Gray predicted a severe shortage during 1968, meanwhile begging whatever the Americans could supply from Savannah River; 500 tons were purchased outright, and another 200 tons leased. In March 1968 he sent out an appeal to the USSR, which had its own heavy-water reactors, but was told that there was no heavy water to be had.[68]

'It seemed clear,' Gray told the board in January 1968, 'that another domestic production plant is required.' Ontario Hydro was looking into it, and would probably build it themselves. But it would take years, and before it was finished the opening of Pickering would be upon them. The answer to the question of how

Pickering was to get its heavy water was no more obvious in 1968 than it had been in 1962 when Gray had started to pursue the stuff; and the very name of heavy water was becoming synonymous with futility and waste.

Energie Atomique

There remains one piece to be fitted into this tale of expansion. The province of Quebec was not involved as such in the development of nuclear power in the period down to 1963. This was due, in part, to the particular institutions of the province, and to the ideological and political views of its governments; it was also due to the fact that Quebec enjoyed an abundance of water power, much of it remote and untapped but available when the market might require it.

When AECL appeared on the scene, Quebec had a dual utility structure. There was, on the one hand, the Quebec Hydro-Electric Commission, created in 1944 by the nationalization of the largest private utility in the province, serving Montreal and its metropolitan area; on the other, there were the remaining private utilities, Shawinigan Water and Power, Saguenay Electric, Gatineau Power, and so on. In 1962 the provincial Liberal government won an election on the issue of nationalizing the remaining companies; the amalgamation of these units into the provincially owned system occurred on 1 May 1963.

In consolidating provincial control over its electrical energy supply Quebec was only following where others, including Ontario, had long since trodden. The politics of electricity in Quebec nevertheless had a dimension that was lacking in other provinces, for there the provincial utility spoke French while the private companies spoke English. Their disappearance was regarded as a blow to the economic domination of English business over a French society, and, in retrospect, this perception seems exact. Its larger significance, disputed at the time, remains cloudy. It was seen by many as simply catching up, while simultaneously opening economic roles for the 80 per cent of the province who spoke French and not English as their language of preference; others, however, saw it as a first step to political emancipation. This perception deeply concerned the Pearson government and helped shape its political agenda. From 1963 on Ottawa became alert to the need to share the benefits of Canada with the country's French-speaking citizens.

Hydro-Québec, as the expanded company was commonly called, sat on the richest hydro resources in Canada; its developed hydro power resources accounted in 1966 for 48 per cent of the Canadian total. It would shortly need more, and although more was under development along the Manicouagan and Outardes rivers in north-central Quebec, the province had followed Ontario into thermal power generation, with a 300 megawatt plant at Tracy near Sorel. This nevertheless represented a small fraction of Quebec's electrical generation.[69]

It may seem strange that Quebec, with its abundant water power, should have worried about its electrical future. It was, however, true that in certain localities thermal power was beginning to seem more attractive than the long-distance transmission of hydroelectricity; looming larger, though in the distance, was the prospect that by the 1990s Quebec would find itself where Ontario already was in the 1950s and 1960s. For that reason alone it was an appropriate moment to bring Canada's largest electrical utility (as it became in 1963) into contact with nuclear power; but there were other reasons too.[70]

As we have seen, Gray moved in 1963 to involve Hydro-Québec in AECL's affairs by appointing J.-C. Lessard, its president, to the board, thereby ending a boycott by the provincial utility that had lasted since 1954. Such an appointment reflected the federal government's priorities. It was followed in 1967 by two other francophone appointments: Claude Geoffrion and Marcel Caron in 1967. To those who made the nominations as well as to the politicians who were ultimately responsible, it was important to make an attempt to share nuclear power with a third of the country that had hitherto hardly participated in its development.[71]

The symbolism had a political basis. Lesage was concerned to secure federal benefits for Quebec, either directly, in cash, or indirectly through services. If nuclear energy had a place, though possibly a small one, in Ottawa's desire to provide benefits to Quebec, it also had a place in the debate over Quebec's own energy strategy. Through 1964 the source and the price of electricity were the subject of a highly charged and intensely political debate inside and outside the provincial government; the price and availability of nuclear power were used as a benchmark against which were measured the feasibility and desirability of hydroelectric projects in the northern wilderness. There was as well the feeling that nuclear power was, in the words of two recent analysts, 'a new technology, effective and reliable.' It was essential to 'master the new technology.'[72]

It was assumed in certain circles that nuclear energy would be cheaper than the northern alternative: Jacques Parizeau, a prominent economist with a developing taste for politics, was the standard-bearer of this line of argument. But others agreed, and it was because nuclear power was regarded as the appropriate form of cheap power for the future that the provincial cabinet began to consider the possibility of building a nuclear reactor.[73]

The company minutes suggest that the first approach came from Quebec, in the summer of 1964, and that Gray then committed AECL to produce 'a report on a 250 MW plant – an uprated Douglas Point station' – by May 1965. If Hydro-Québec was anxious to approach, AECL was anxious to reciprocate; yet only in the most general sense did this fit into AECL's plans and then largely because of the political and economic symmetry it represented in terms of the Canadian political balance. It is hard to find what has been described as 'heavy pressures on Hydro-Québec and the Quebec government to have the province launch an ambitious capital program that would utilize Canadian CANDU technology.'[74]

Indeed, Quebec's first interest appears to have been in a tried and true method, and CGE was the first company to enter the marketplace with its own design, which included a 'water-immersed calandria.' Gray and his staff were not impressed. If AECL were to be consultant or prime contractor – another Quebec requirement – then it would not be with such a calandria nor, by preference, would it be with CGE.[75]

Quebec's interest in nuclear power, led by Lessard and sustained by administrator-publicists like Parizeau, coincided with both the flowering of the CANDU in Ontario and with the perceived shortage of heavy water in the mid to late 1960s. It overlapped, too, with the emergence of variant reactor designs of the kind that we have seen were preoccupying PRDPEC from 1962 on. It was natural that one or more of these designs would be considered for Quebec's first nuclear station, and natural too that one should be adopted. And here the conditions for financing such a station reared their head.

Hydro-Québec wanted federal participation in the project; but federal funding even for Ontario Hydro was based on the principle that Ottawa was funding pilot plants, even very large pilot plants like Pickering. How could or should Ottawa contribute to a design that had already been tried? In Gray's opinion Hydro-Québec should either go up to 500 megawatts (electric), or over to another reactor model – for example, 'the installation of a boiling light water type.'[76]

This proposal, Gray later reflected, was one of his worst, since it effectively put paid to the organic reactor design, already downgraded (the minutes speak of 'limitation' and 'termination'), and substituted for it a design that was certainly untried and, obviously, unproven. It had the tangential effect of further undercutting CGE's CAPD, which had built the Whiteshell organic test reactor, and, therefore, once again frustrated that company's desire for a piece of the domestic action as a foundation for whatever overseas business it might be able to attract. For CGE, Whiteshell became a one-time-only affair.[77]

The reasoning behind the decision to terminate the organic experiment has been reviewed elsewhere. It is, however, undeniable that OTR worked well during its twenty-odd years of operation, and plain that the experience of building and running it would have been useful in working up to a 250 megawatt reactor for Quebec.[78]

While AECL reviewed its options and began to make its initial dispositions for accommodating a new reactor program, the bureaucrats and politicians had been working out terms between the federal government and the government of Quebec. Lessard had evidently had an impact on Premier Lesage, who decided that a forthcoming meeting of the Canadian Nuclear Association in Quebec City in May would afford a splendid opportunity to announce that Quebec too was signing on to nuclear technology. There were admittedly a few snags. AECL had been over-optimistic about the time it would take to produce a firm BLW design; in the spring of 1965 the relevant cabinet committee was told that a final design and construction project were still eighteen months to two years away. They could also have been told that the AECL board was withholding approval until the design was in a clearer form.[79]

Lesage attempted to precipitate matters by writing to Pearson on 6 April 1965. He wanted the project to move forward. AECL had suggested a model; Quebec Hydro had a site, at St-Edouard de Gentilly. The other details could wait until after the announcement. But, of course, any announcement should be preceded by an agreement on financing, and it was that small matter that the premier wanted to have settled. He wanted essentially the same terms as Ontario Hydro had got for Douglas Point. The cost would be more because of inflation – say $100 million, of which the federal government's advisers thought between $50 million and $70 million could eventually be recovered.

The Privy Council Committee on Scientific and Industrial Research debated the matter on 30 April 1965. Following advice from AECL it was cautious – more cautious than Lesage wanted. AECL would proceed with its BLW study. If the BLW looked good it would be located in Quebec and, in the meantime, Hydro-Québec engineers would be welcome at Chalk River for a co-operative study. Most important, the terms would be the same for Quebec as for Ontario.[80]

The cabinet, when it discussed the matter on 3 May, remained cautious on some themes while showing a certain generosity on other matters. Any agreement with Quebec should be carefully limited to process, rather than expanded to cover model and site. Mitchell Sharp, the minister of trade and commerce, was enthusiastic, arguing that the Quebec venture was 'likely to lead to substantial sales of this type of reactor in the future.' At the same time Maurice Sauvé, the minister of forestry and Lesage's conduit into and out of the federal cabinet, argued that it should be recognized that Quebec did not need a nuclear reactor or, indeed, any thermal-generated power. The standard against which power charges from the proposed reactor should be set might be those of a hydro station rather than the Tracy thermal unit. The cabinet did not immediately come to any conclusion on financial arrangements, since there was no firm decision to do anything other than make co-operative gestures. Despite Lesage's speech (it was only a week away) nothing more could be done until AECL was ready with a design and, just as important, until the AECB was ready to approve the site.[81]

AECL's preparations for a Quebec site nevertheless received a considerable fillip from the excitement surrounding Lesage's letter. When the executive committee next met, in May, it took the BLW question as one of its highest priorities. A new organization was established to work on a boiling-water design. It came under Les Haywood, which is to say that it reported to engineering at Chalk River rather than to NPPD. The minutes are, perhaps unintentionally, revealing on the subject: 'The [new] Division will be organized on the basis of four or five branches – similar to CGE and Power Projects organizations,' the executive committee was told in May 1965. George Pon was placed in charge, and his staff, Haywood reported, would eventually total sixty. His unit was located in Toronto and French-language staff were sent there for training. Haywood believed that estimates and 'design conclusions' would be ready by the middle of 1966, and that if a decision to build were made at that time the new reactor would be delivering on-line power by the middle of 1971.[82]

The decision to build was made in several stages. According to Lesage's biographer, the Quebec cabinet gave its consent to both project and site on 11 February 1966. This was over three months before AECL's own report was ready and its directors gave their consent to the project, sending their recommendation to the federal cabinet for appropriate immediate action. With the Quebec decision already made, Haywood nudged Gray in March and again in early May for money to start excavations in September. The executive committee declined the opportunity on both occasions; it would return in final form at the Board of Directors meeting held at Pinawa on 31 May.[83]

When it did, Haywood finally prevailed. His study, he told the directors, showed that the CANDU-BLW was attractive economically and held some advantages over the CANDU-PHW that Ontario Hydro was building. Design and cost estimates were ready, and the board should now bite the bullet. The board did, and an appropriate recommendation went on to the cabinet.[84]

The cabinet was told that while Canada's nuclear program had been successful to date, it was necessary to keep improving on past performance 'in order to maintain a competitive position.' One such improvement was the proposed BLW reactor which employed a coolant that was arguably superior to the heavy water used in older models, 'since its cost is negligible compared to heavy water, it is a fluid with which electric utility operators are familiar, and, furthermore, steam produced in the reactor may be passed directly to the turbine.' This had not always been true, for light water admittedly absorbed more neutrons than did deuterium, but 'extensive research and development' had negated that problem: 'by the mixture of light water and light water steam, the neutron absorption is less than with light water alone.'

It would prove to be cheaper than Douglas Point, not, it is true, in nominal dollars, but in sums adjusted to reflect their differing size and the effect of inflation between the early and late 1960s. And its efficiency would be greater, so that its 'inherent advantages' would become even more apparent as time passed. Like Douglas Point, Gentilly would ultimately be purchased by Hydro-Québec, for something like $80 million less depreciation. The difference should be written off as a federal contribution to research and development.[85]

In the interval between the board's approval of the BLW reactor and the transmission of its advice to the cabinet on 11 July 1966, the Lesage government had gone down to defeat; but Lesage's political demise and the succession of Daniel Johnson and his Union Nationale government made no special difference. Lessard remained as president of Hydro, and went on to become a dining and fishing companion of the new premier; no one in the new government seems to have thought to question Gentilly. Nor did the federal cabinet; in what must have seemed a gesture that combined all the best features of its policies of encouragement to technology and industry and national unity, the Pearson cabinet on 20 July approved construction of Gentilly on the Douglas Point formula.[86]

The course of true engagement did not run entirely smooth, as AECL and Hydro-Québec squabbled over the proper interpretation of the Douglas Point terms, but on the whole the early stages of

Gentilly passed without much worry. There were plenty of other things to be worried about: heavy water first and foremost, the early and dismal operating results at Douglas Point, where the reactor spent more time down than on; and the continuing travails of CGE as it searched for a market for its expertise. AECL was afflicted, during 1966–7, with labour worries of its own, but they were dwarfed by Gentilly, which, according to Haywood, lost five months of construction time thanks to work stoppages; but while it might not be ready on time he still believed that it would finish within its budget.[87]

Haywood was not wrong. In the spring of 1970 Gray reported happily that Gentilly was at least on schedule, strike notwithstanding, and would be ready for operation some time in 1971. Though not quite on time, it was a creditable performance under the circumstances. But it also had an ironic undercurrent, for even as Gray issued his predictions in the AECL *Annual Report*, the board learned that Gentilly might have a brief operating record, since the supply of heavy water was far from assured. Even if Gentilly opened, it had second priority to Pickering. However embarrassing that might be, Lessard told the directors, 'Hydro-Québec was prepared to face up to the consequences.' The prime minister – by this time Pierre Elliott Trudeau – had asked the chairmen of Ontario Hydro and Hydro-Québec to work out a reasonable solution, and with that impetus they did.[88]

Gentilly went critical early in 1971 and underwent a detailed and deliberate commissioning program through the year. It had its problems, but this was mainly because everyone knew that it would shortly be shut down so as to send its heavy water to Ontario for use in Pickering. Gentilly's real trial would, so it seemed, come later.

Expansion and its Consequences

The decade of the 1960s saw a vast expansion in AECL. It moved from prototype to power, from demonstration reactor at 20 megawatts to power-producer at 200 megawatts and finally to a million-kilowatt (1000 megawatts) station at Pickering. The sums invested in nuclear power increased almost exponentially, and where at the beginning of the decade they all came from Ottawa, by

1970 there was a substantial provincial component. To make it possible the company created a heavy-water capacity, and consequently became an economic factor in distant Nova Scotia – and a political one as well.

AECL's involvement in politics was not new. Atomic energy was a political creation. Politicians and senior bureaucrats perceived that atomic energy and its offspring the atomic bomb involved nothing less than the future of the world, a future in which Canada must play a part. What part, they did not know, but as long as Canada had its Chalk River reactors, as long as its plutonium flowed south and American heavy water flowed north and its scientists and engineers met in regular conclaves with their British and American counterparts, then Canada had 'a seat at the table' when the weaponry that might put an end to the world came under discussion.

Atomic energy therefore had a political cachet that was difficult to gainsay. But it also had glamour – the glamour, ultimately, of innovative science. That was embodied in Chalk River, impressive behind its security booths, with its up-to-date laboratories and highly educated workforce, but also in the emergence of the CANDU, bright, shiny, clean, and Canadian. The glamour was expensive, in Canadian terms – but not in international ones. And there was the rub.

The United States was richer than Canada by a factor of twelve or more, and Great Britain, while no longer richer on a per capita basis, was more than twice as big. So was France, while the Soviet Union demonstrated that even a country impoverished and devastated by war and the results of its own peculiar political system could mobilize enough resources to be, in this period, competitive with the Americans.

Canada could not be, except at the margins and by concentration of effort. The effort went into the CANDU, the heavy-water moderated, natural-uranium-fuelled reactor. The concentration saved AECL from squandering limited resources of manpower and saved it from running up its budget to the point where politicians and bureaucrats would be tempted to regard it with a jaundiced eye, and to mutter to themselves about champagne appetites and beer budgets.

It was Lorne Gray's job to manage things so that they would not mutter, or at least mutter too much. The job was relatively easy in fiscal 1958–9, when AECL's budget was $32 million; it was less so in 1968–9 when the budget was $73 million plus $10.7 million in capital expenditures, of which $69 million had to come from parliamentary grants. These may not seem vast sums, or significant when set against the federal government's other commitments: $6115 million in expenditures in 1959 and $12,976 million in 1969. But they were significant at times of budget squeezes which, when the fiscal system was struggling to absorb the Pearson government's new social programs, were taken very seriously. When set against the federal government's expenditure on capital projects, AECL was more significant still.[89]

There was, therefore, a constant struggle for the budget in which the company's revenues, whether from plutonium sales or from Commercial Products, could not play a very large or decisive part. It was possible, however, to bring into play the sums the federal government was prepared to commit for its other objectives, foreign aid, regional economic development, or the fostering of national unity. Government intervention was seen as the best, and often the sole, means of achieving a change that would make a real difference. In the case of the maritimes, for example, where, as Pearson's policy adviser Tom Kent later wrote, 'the poor will go on getting relatively poorer unless the Federal Government intervenes to finance provincial activity on a highly discriminatory basis.' Not everyone agreed. To some, especially in the federal Finance Department, such spending raised the question whether, as the saying goes, the federal government was 'trying to pump milk into the cow.' But the trend of the times was with Kent, and not with the older generation of bureaucrats. It was, however, precisely this younger generation with whom Gray and AECL found themselves allied, rather to their surprise.[90]

It made sense to link AECL's interests with those of others. But some of the linkages carried a price. The link with Ontario exacted a cost in regional jealousy and interprovincial rivalry; with Quebec, a high priority with the Pearson regime, the price was expansion of AECL's engineering arm and a scramble for more heavy water; the Nova Scotia link the company would just as soon have foregone, because of the Deuterium of Canada nightmare that came with it. But it was better than nothing, and it guaranteed a cushion of regional interest that in the Canadian federal system was not insignificant.

In the struggle, the intermediaries were important. The most important, potentially, were the ministers. AECL had its own, to whom the president reported directly, an inheritance from the relationship between Bennett and Howe. The deputy ministers of trade and commerce and defence production, the cognate ministries, did not count for much in the equation although other deputies – in finance and external affairs – did. Relations between AECL and its minister from 1957 to 1963, Gordon Churchill, were close. Churchill believed in the company and atomic energy. So did his colleagues, especially Howard Green, the minister of external affairs. Because of the chaotic nature of the Diefenbaker cabinet this did not matter as much as it might have, but it did count for something.

Gray cultivated Churchill, confident that through his minister he could at least get AECL on the cabinet's agenda; and he did, through the frequent revisions of the company's priorities and programs that led from NPD-1 to NPD-2 and from OCDRE to OTR and from NPD to CANDU. To get AECL successfully *off* the cabinet's agenda required a more delicate balancing act, where Gray was dealing with individuals who were technically his equals – deputy ministers – but whose power was considerably greater, either because they served in more powerful departments or because they had the ear of more influential ministers.

Influence did not always suffice. It did not save AECL from locating the organic test reactor at Whiteshell, after recriminations in the Diefenbaker cabinet over the 'favouritism' shown Ontario Hydro; but the contrary tendencies were not yet strong enough to detract from the company's priorities.

These trends were stronger under Pearson. Encouragement for industrial development was a splendid idea, but its measure was job creation, a cruder index, and not necessarily the best one. A desire for high technology was all very well, but when combined with regional economic expansion it could mutate into programs for regional economic greatness. Applied to the maritimes, it held out hope for a prosperity lost since the days of sail. Deuterium of Canada was the consequence.

It is a matter of debate among social scientists what the motivation of such programs may be. According to some, politicians act to maximize their votes, while bureaucrats act to maximize their personal influence and protect their departmental empires. These considerations were obviously not absent from AECL's history in the 1960s. They were not, however, the only or even the dominant considerations that the company had to contend with. The alleviation of regional disparity, the fostering of national unity, or, at a more basic level, the beneficent features of state intervention, were high national objectives. To say that they were at bottom vote-getting mechanisms is to reduce the proposition to a tautology or, better still, to a nonsense.[91]

Like other crown agencies before it, AECL could be made to transfer its spending where the politicians thought it would do most good. Gray was ill-equipped to resist. He believed in expansion, and his programs offered an opportunity. When confronted with the Pearson government's assumptions about appropriate public investment – underlining the fact that it was *public* – his only response was to lean on the broken reed of his Board of Directors, themselves public appointees bound to follow political direction, or to resign. They did not resign, nor did Gray.

It was true that the cabinet's, and Nova Scotia's, choice of Deuterium of Canada as an instrument of industrial renewal was unfortunate and mistaken. Not for the last time large programs and high principles afforded opportunities to the opportunistic; unluckily in this case the opportunistic were also incompetent. Even more unfortunate, AECL did not have the technical capacity to evaluate the Deuterium bid; when Gray was asked by ministers whether it would work, he assured them that it would. Nobody in Canada in fact knew whether it would work or not, but the probabilities – experience at Savannah River – suggested that it would. Deuterium, after all, was headed by an expert, who had helped make Savannah River; Savannah River worked; and Spevack had the patents. The 'wizard principle' was strongly embedded in Canadians; there is no reason to be surprised that IEL adopted it. But to the question, why did it work in Savannah

River and not in Glace Bay, a final answer may never be forthcoming.

Gray's relations with his minister, C.M. Drury, seem never to have recovered from the minister's refusal to support the company's desire to stay away from Cape Breton and from Deuterium of Canada. More important was the fact that the choice of Deuterium affected AECL's ability to deliver on its CANDU promises. It forced recurrent visits to the cabinet table to deal with one heavy-water crisis after another, although later visits were met with increasing resignation from the politicians. There was a sense in which the politicians had contributed, and contributed hugely, to Canada's deuterium dilemma.

The theory of compensating for regional underdevelopment was not, therefore, a roaring success, at least as far as national priorities were concerned. As Kent admitted in 1974, doubtless with the Deuterium experience in mind, 'It might be said that the record of the provinces in entrepreneurship is not uniformly encouraging.' There had been 'conspicuous misadventures,' through the employment of 'methods ... designed to make it as easy as possible to throw good money after bad.' Kent was writing from the perspective of Cape Breton, where he presided over the federal government's latest attempt to tilt the economic balance towards the disadvantaged island. He was the latest, though not the last, of a long series of missionaries in the field.[92]

It is an interesting question whether the loss on Deuterium of Canada hurt Nova Scotia very much. In the sense that it distorted provincial priorities and prevented the province from spending on worthier causes, probably it did. But a study of the activities of Industrial Estates Limited – activities that included another notable disaster, Clairtone – concludes that on balance the province benefited from IEL's investment activities. That is not, however, to say that its performance could not have been better, as the study admits. The experience of Deuterium offers an instance in which that would have been so.[93]

Deuterium of Canada extended Lorne Gray's corporate empire to the Atlantic coast. Though Gray was not pleased by the experience, there is a sense in which the erection of Deuterium of Canada, followed by the CGE heavy-water factory, followed by

Ontario Hydro's decision to produce its own heavy-water supply at the Bruce site on Lake Huron, was a microcosm of AECL's corporate dilemma. Facing a shortage of a rare product, heavy water, AECL sought to create its own production capacity. When it eventually succeeded, it faced a contrasting problem: oversupply. The decision to build a heavy-water plant thus aided neither AECL nor, in the longer run, Nova Scotia; and it encouraged Ontario Hydro to consult its own interests.

That could not have been foreseen back in 1962. But the project envisaged at that time was very different from the one that eventually emerged, years late and hundreds of millions of dollars later. It might nevertheless have served some purpose, had it been able to serve a market outside the country. That, however, it did not do. How this was so is the topic of our next chapter.

10

Experience Abroad

When Canada embarked on its nuclear power program in 1955, its choice of a heavy-water, natural-uranium system was experimental, one option among several prototypes. That the Canadian system was different from other models was perceived as an advantage: it afforded an independent research path and avoided the duplication of others' efforts. That the Canadian system did not rely on enriched uranium was another advantage, since it avoided the construction of costly enrichment plants, and avoided, as well, dependence on American facilities. Independence in energy was a facet of political independence; Canadian leaders were concerned to keep it so.

Other countries pursued similar policies. The British were first: unlike Canada they had a direct military interest, and designed reactors that were primarily plutonium producers. Known from their fuel cladding as 'magnox' (magnesium oxide) reactors, their prototype was Calder Hall, opened with great fanfare in October 1956. The magnox reactors, like their Canadian cousins, used natural uranium fuel, enclosed in a pressure vessel, as with NPD-1; they used graphite rather than heavy water as a moderator; unlike NRX they used gas as a coolant. The gas was then piped out to heat exchangers – 'a fancy word for boilers,' as Walter Patterson observes. The resultant steam powered turbines, which generated electricity.[1]

There would eventually be twenty-eight magnox stations, twenty-six in the United Kingdom and two abroad, in Italy and Japan. These last two are of special interest, for they mark the arrival of reactors as commodities in international trade; the fact that there are only two also indicates the relative lack of success of the magnox as a reactor type, in large part because of its relative inefficiency as a heat and therefore a power producer. A second generation of reactors, the 'advanced gas-cooled' variety or AGR, was already under development in the late 1950s; the first station of this kind was roughly eight years late when brought into operation. Whether because of its early problems or because of the consequent loss of time, the British failed to place any AGRs abroad.

The magnox reactor and the AGR faced intense competition. The competition came from the United States, whose reactor program had received a tremendous boost from government investment and military spending. The American reactors used light (ordinary) water for moderating and cooling; since light water absorbs so many neutrons, reactors of this kind need to use enriched fuel, containing about 3 per cent of uranium-235 instead of the natural 0.7 per cent. The water was pressurized to prevent boiling; to hold it in, large pressure vessels were constructed. The whole system was usually called the pressurized light-water reactor, or PWR; it was initially designed for use on Admiral Rickover's nuclear submarines. Its land prototype was Shippingport, which opened in 1957.

Shippingport was not economical. It produced power at far greater cost than conventional stations. It was, however, heavily subsidized, because the AEC wished to demonstrate the feasibility of nuclear power; the Westinghouse company, Shippingport's progenitor, prepared to build a much larger model, also with the help of subsidy.[2]

While Westinghouse pursued its PWR model, its great corporate rival, General Electric, had become fascinated with a boiling-water reactor, in which water would be allowed to boil inside the reactor. This eliminated several steps between the generation of heat from the nuclear fuel and the steam-powered turbine; the simplicity and elegance of the concept appealed to General Electric, which emerged as a rival nuclear design company with its BWRs. Like the

PWR, the BWR used light water as moderator and coolant; it also used enriched uranium fuel. Unfortunately, as often happens in technology, the simplicity of the BWR proved to be offset by a number of practical problems in other areas, so that it had no overall advantage over the PWR.[3]

The American companies promptly began scouting for customers. They found some modest success at home, and hoped to supplement it with success abroad. To succeed in the export market, they discovered they must first overcome other countries' sentimental preferences for home-grown products; with that obstacle, progress was at first slow.

The US government was prepared to help, by declassifying data on the operation of power plants, by concluding bilateral agreements with friendly countries (Canada's was the first and Belgium's a close second), and then by exporting enriched uranium, a *sine qua non* to the spread of American reactor models. When the Europeans proved unable to reach agreement on a European reactor model, the Americans were ready to step in. In 1958 they concluded an agreement with Euratom, the European Atomic Energy Community that paralleled the six-nation economic community, then in its first year of existence. The agreement loosened restrictions and offered advantages, including a guarantee of reprocessing fuel; nevertheless, only two plants were undertaken under its auspices between 1958 and 1962.[4]

The slow development of American light-water exports mirrored uncertainty on the part of those governments interested in nuclear power. There was uncertainty, to be sure, about the economics of nuclear power; there was uncertainty about the 'sovereign' implications of nuclear technology, whether, in effect, possession of an autonomous nuclear system was an indispensable sovereign attribute. There was also uncertainty as to which system would finally prevail. As long as doubt persisted, the future saleability of the heavy-water technology did not appear worse than that of its equally untried rivals. Unlike the British or the American, the Canadian effort was not dissipated among various systems; all Canadian energies were directed towards just one.

It was in the spirit of adventurous competition that Canada first set out to sell reactors.

Institutions

Atomic energy in Canada always had international implications. These were dealt with by co-ordination among several Canadian government agencies. As a legacy of the war and as a function of Canada's alliances, these rested squarely in C.D. Howe's hands. Howe's predominance and Bennett's reputation furnished such reinforcement and validation as might, from time to time, be required. But it was Bennett and not Arnold Heeney, Canada's ambassador to Washington, who negotiated the Canadian-American atomic bilateral agreement in 1955. Heeney got to sit in on the signature, but only after vigorous protests.

This does not mean that there was any real contradiction of policy between AECL and other Ottawa agencies and departments. The company had a scientific purpose, generally accepted; it had an energy future, largely unquestioned; and it played a role in defence and security through its contribution to the US atomic weapons program. People like Bennett, Howe, or Lester Pearson had, as they saw it, no qualms and no illusions about the desirability of Canadian contributions to the Western nuclear deterrent.[5]

This second aspect of Canada's nuclear energy policy had an economic dimension. The development of some of Chalk River's facilities was paid for, in part, by American money. The connection between Canada's own atomic development and the international market, even the severely restricted one of the 1950s, was obvious. At a time when exports and foreign investment were considered to be crucial to economic growth and prosperity, atomic energy was assumed to benefit in the same way.

The national interest resided in five directing organs, two of which functioned. There was the vestigial Privy Council Committee, chaired by Howe. It had virtually disappeared by the mid-1950s. So had its mirror committee of senior officials, the Advisory Panel on Atomic Energy; it was in any case far too big to function effectively. Replacing the older Privy Council Committee was a Uranium Committee, which brooded over Canada's uranium sales and atomic connections; its mirror committee, consisting of Bennett, Jules Léger from External Affairs, and John Deutsch from Finance, was active and effective. There was also the Atomic

Energy Control Board, under C.J. Mackenzie; it played little or no role in shaping Canada's external atomic policy.

While Bennett habitually ran matters of commercial policy past his board, he kept diplomacy to himself, for discussion with Howe or with his friends in the senior civil service. That included the UN conferences on atomic energy, and the question of reactors as a tool in foreign aid.

That reactors could be considered for foreign aid reflected a change in the international climate in the mid-1950s. The death of Stalin, the end of the Korean war, the achievement of a temporary settlement in Indo-China, and, not least, the Atoms for Peace adventure initiated a time of lessening tensions. 'Peaceful competition' overshadowed, though it did not replace, the armed confrontation of the early 1950s.

Canadian policy aimed to encourage this peaceable trend. As a 'middle power' (not great, but not small either) Canada offered its services as an honest broker attempting to bridge the gaps between east and west, and increasingly between the 'have' and 'have-not' nations of the world. Pearson and St Laurent hoped thereby to offer the developing nations something more attractive than what some considered to be the crass American policy of alliances and military bases around the periphery of the Soviet Union.

The policy was identified with Pearson and his Department of External Affairs, but most members of the cabinet sympathized, as did their senior civil servants. Nationalist by inclination, liberal by temperament and conviction, these men regarded Canada as an enviable compromise between political and economic liberties and state activism – a mixed economy where efficiency and social concern could flourish side by side. The best place to start, they reasoned, was with the countries that shared a common past with Canada: the ex-colonies of the British Empire, Canada's partners in the British Commonwealth of Nations.

The Indian Experiment

Canadian policy selected the largest of the 'new' Commonwealth states as a field for an experiment in foreign policy. India had been independent since 1947. When it changed its government from a monarchy to a republic in 1949, Canada helped keep it in the Commonwealth. It was the world's largest democracy. Its prime minister, Jawaharlal Nehru, was an eloquent spokesman for peace and international understanding. Its governing political elite was imbued with British liberal values and culture.[6]

It seemed altogether appropriate to establish links with the Indians, with their imperial past and liberal present. Sentiment did not, however, completely outweigh practicalities. As one of Canada's senior diplomats wrote in 1947, there was a danger that 'the western powers may have the great majority of the colonial and coloured peoples hostile or unfriendly to them,' in the event of an east-west war; creative diplomacy with India was a means to avoid that unhappy fate. Pearson for his part had no doubt that 'India is a great power and is subject to the temptations of a great power.' Canada, whose standing in India he assessed as high and even exaggerated, could help to direct Indian energies into friendly channels – friendly, that is, to Canada and the West.[7]

In the early 1950s the British Commonwealth of Nations enjoyed great prestige; it seemed to be the embodiment of a hopeful future rather than the legacy of an outmoded past. Commonwealth sentiment counted for something, the Commonwealth enjoyed public support, and Canada was no exception. Canadians hoped that Canada could, through its Commonwealth links, explain and assist the United States in parts of the world where American views by themselves were considered too extreme to obtain a fair hearing.[8]

The decision to establish atomic relations can be dated quite precisely to early 1955. The climate in which it occurred goes back further, to the Cavendish Laboratory where two promising young scientists, W.B. Lewis and Homi Bhabha, got to know one another. Bhabha was one of a large number of Indians at Cambridge; he came from a privileged background at home, where he was linked to the steelmaking Tata family. He took his PhD at Cambridge, but he also picked up a taste for Western music and opera; Lewis and Bill Bennett shared Bhabha's liking for music. After returning to India, Bhabha headed a number of scientific institutes, and after independence became head of its Atomic Energy Commission. In

1954 he was appointed permanent secretary of the newly formed Department of Atomic Energy. Soon India began to build its own experimental reactor; it was scheduled to go critical sometime in 1956.

Bhabha's savoir-faire and culture attracted the prime minister, Nehru; the two men enjoyed an unusually close relationship. Nehru undoubtedly looked on Bhabha as an advertisement for India abroad, an Indian who could hold his own with the best that other countries could produce. When Bhabha gave the keynote speech in Geneva in 1956, it was even more apt. He and others like him would show that India was no longer another country's colony but a self-respecting modern nation.[9]

Bhabha, like Lewis, took an interest in thorium reactors; India had considerable reserves of thorium. He believed, and proclaimed, that atomic power was the best way for underdeveloped countries to advance from misery to prosperity; Lewis felt much the same way. Indeed, for Lewis his convictions in this field were so strong that they lent a missionary aura to his enterprise.[10]

The idea of a Canadian reactor for India arose in the mind of Nik Cavell, the administrator of Canadian aid to the Colombo Plan, one of Canada's major international aid projects. The Colombo Plan was premised on the relationship between misery and poverty and communism. Cavell, 'a highly intelligent and charming person' in Bennett's memory, had the task of selling foreign aid to a suspicious and economically conservative Canadian public.

What his aid program lacked, Cavell decided, was flair. Canada was giving worthy help: aluminum, for example, and wheat. These commodities were not likely to stand out in the public mind, however, or in the minds of their recipients. Over lunch one day, Bennett later wrote, 'he inquired of me whether it would be possible for AECL to build a nuclear power plant in India.'

Bennett responded cautiously. NPD was on the verge of approval, and he had few resources to spare. But he would ask Lewis, and it was Lewis who provided the solution, what he called a carbon copy of NRX. NRX's design had been recently declassified, so no security was involved. AECL could easily act as consulting engineer on the project, beside a Canadian construction company.[11]

Bennett consulted A.E. Ritchie of External Affairs' economic division. Ritchie knew that the government of India assigned the highest priority to industrialization, and would welcome Canadian aid to the energy sector. Canada was building hydro projects in both India and Pakistan, and on the eve of the Geneva conference what better gesture could Canada make? The undersecretary, Jules Léger, was similarly impressed. The symbolic value of a peaceful, research-oriented gift was very considerable, he minuted; at the same time, Canada must not forget the commercial impact that such a gift could have. Let us not, he wrote, 'lag behind.'

Léger did enter one caveat. 'There might,' he warned, 'also be some problems regarding control over the plutonium produced by any reactor which we might supply, but this presumably could be surmounted, especially if we assume that a country like India will acquire a reactor from some source (friendly or otherwise) and will be producing this material.' It was too bad that no international agreement existed for the export of nuclear technology, and it was unlikely that an international atomic energy agency would be constituted 'for some time.' In the meantime it was every country for itself. Pearson should now raise the matter in cabinet, and secure the necessary Colombo Plan funds.[12]

Pearson thought well of the idea. A week after Howe brought NPD before the cabinet, Pearson submitted his own reactor project on 30 March. Howe was not present but sent his full support. The gift of a reactor to India, Pearson advised, would be 'politically ... a most important gesture, the effects of which might be very great indeed.' The cabinet agreed, and authorized an unenthusiastic Finance Department to come up with the funds. Now, finally, the Indians could be told.[13]

Escott Reid, the Canadian high commissioner in New Delhi, made the necessary formal approach; informally, W.B. Lewis wrote to Homi Bhabha. Formally, the Indians were told that unlike the Americans the Canadians attached no strings to their aid. What was meant was that no political payoff was expected in return; but the idea of strings could (and does) have another and more constructive meaning, and it was not emphasized.

The Indians were not sure how to respond. Would Canada's total aid package be reduced by the estimated $7 million cost of a reactor? Should the Canadian model be preferred to the British or the American? The latter point was a matter for internal debate, but Canada had to supply the answer to the first question. With Howe's approval, Pearson plumped for an additional $7 million. This offer, with no strings attached, was sent to India just before the Geneva conference convened.[14]

At the beginning of August Nehru sent his thanks to St Laurent. He appreciated what had been done to meet his concerns, but he had some new ones. He had consulted with his scientists, who had expressed a preference for an NRU model over NRX. This, Bennett snorted, was impossible; Canada could only contemplate NRU because of the experience accumulated on NRX. And an NRU model would be impossible to cost – how impossible, he did not yet know. In any case, most of NRU was classified. This he undertook to explain to Bhabha when they met in Geneva. The meeting seems to have gone well, since the end of the summer produced a grateful and final acceptance in a letter from Nehru to St Laurent.[15]

The Indian version of events is rather different. Bhabha, one of his biographers asserts, actually would have preferred an enriched uranium, heavy water plant similar to the Dido reactor at Harwell. As a concession to his colleagues he agreed that India should have both the enriched and the natural uranium varieties; only after 'heated' discussions did he admit that India's existing resources were not after all sufficient to maintain two high-flux reactors, and that the Canadian variety was acceptable on its own merits.

Bhabha went to Geneva with a negotiating brief but without complete authority. As in Canada, the sums of money necessary to build a reactor required political authority, authority which he sought on 29 August and which was granted, by telegram, by Nehru on 1 September. His feelings on the subject must have been ambivalent, since while he was pleased to acquire a research reactor, he also preferred to have a reactor built by Indians out of Indian plans and resources. As one biographer observes, 'He pointed out the tendency, widely prevalent whether in the private or in the public sector, to think immediately of finding a foreign consultant whenever some new plant had to be established.'[16]

By the time Nehru's letter was in, the Geneva conference had made its triumphant impression on public opinion, and the spread of Canadian technology to India acquired a greater symbolic as well as practical importance. The prospects for an international atomic agency were also rated higher than they had been in the spring, and Canadian officials were giving thought to the desiderata such an agency might hold for them. Their thoughts were concentrated by the unrestricted gift of a nuclear reactor to India, without the Indians' committing themselves to the safe disposition of the plutonium in the spent fuel rods that an NRX-type reactor would produce. Before the ink was dry on the announcement, qualms were being expressed about the unqualified nature of Canada's transfer of a technology that, after all, had originally been designed as a prototype plutonium factory. But the Indians had accepted, and with gratitude. How, then, to get out of an impasse that was of the Canadians' own making?[17]

A long process of constantly deferred gratification got underway. Hopes were held out for an approach to Bhabha, who passed through Ottawa en route to India from Geneva in September, but he proved to be utterly inflexible. His choice to represent India and Nehru was not, after all, accidental.[18]

The first encounter was typical of many to follow. There is some disagreement as to exactly what position Bhabha sustained inside the Indian government. According to Bertrand Goldschmidt, who knew Bhabha well, the Indian scientist actually argued the case for a public renunciation of nuclear weapons to his prime minister in 1955, presumably after receiving an approach from the Canadians on the subject. Nehru replied 'that they should discuss it again on the day when India was ready to produce one.' Neither man, as it happened, survived to see that day.[19]

Bhabha's speeches and formal contacts with Canadian representatives gave little reason for optimism that India would voluntarily accept controls or safeguards over nuclear materials produced in the Canada-India Reactor, or CIR as it was inevitably called. There was an agreement of sorts in April 1956 on an audit of Canadian-supplied fuel, but it was recognized that it did not do much to guarantee the fully peaceful use of the reactor. At New York in 1956 and at Vienna in 1958, 1959, and subsequent years Bhabha and his delegation matched the Soviets' efforts to block any meaningful inspection and safeguards role for the IAEA. He even suggested that perhaps the agency would not last long enough to be worth worrying about: 'it was a weakling that might not survive.' Sceptical Canadian diplomats took that with a grain of salt; the Indians had been so active in securing positions in Vienna for their own nationals that there was legitimate doubt that they would ever want to give up on the agency.[20]

The frostiness of Canadian diplomatic contacts with the Indians over atomic energy was not duplicated within AECL, or between AECL and the Indian Department of Atomic Energy. There was no single view within the company, but to the extent that there was, it was realistic: if the Indians did not get their reactor from Canada they would get it somewhere else.[21]

The conversations at Chalk River in September 1955 may have disappointed External Affairs, but where AECL alone was concerned they were fruitful. The Canadians would supply the hardware, the supervisory engineers, and the contractor. The Indians would pay for local costs, while the Canadians assumed responsibility for training, in Canada, the Indians who would operate the reactor. Bennett also undertook to approach the Americans for a guarantee of heavy water supplies. 'I recall discussing this with Lewis Strauss on several occasions,' he wrote (Strauss was head of the AEC). 'He was anxious to have the US participate in the project, but was also greatly concerned about the possibility of India getting into a weapons program.' Strauss evidently overcame his anxieties: the American contribution also went forward.[22]

The reasons originally advanced for building a Canadian reactor in India remained cogent ones, even if the contrary side now looked more powerful than it had once done. An uneasy cycle developed. After each refusal Canadian hopes were pinned to the next Indian visitor: Bhabha, an Indian minister, or, in 1964, Nehru's successor, Lal Bahadur Shastri, the Indian prime minister. The words spoken were soothing, sometimes very much so, but however reassuring they were, they were never committed to writing.[23] As luck would have it, negotiations for international control were postponed in December 1955, making possible consideration by the cabinet of the Indian reactor on its own merits and without a direct linkage to the questions of safeguards and controls.

CIR meanwhile went ahead. It was built at Trombay, outside Bombay, Bhabha's native city and centre of operations. AECL received money from Canada's Colombo Plan program, and disbursed it to Shawinigan Engineering, which sent a supervisory engineering staff to India. The Foundation Company, which was building NRU, did the actual construction. AECL, as consulting engineer, kept an eye on proceedings, sending Lorne Gray on the first of many trips to India to inspect construction in October 1957. Gray did not find everything to his satisfaction, only to find that Bhabha supported the crew on the spot; the crew consisted of 1200 Indians and 30 Canadians.[24]

It was a problem of slow progress, poorly trained workmen, and a lack of trained supervisory staff; that necessitated sending another consultant to India at the end of 1958 to try to speed things along. Whether for that reason or not, progress began to look up; reports were more encouraging and it began to appear that the reactor would be finished ahead of schedule – early in 1960. The Indians had been trying to make their own fuel rods, but Gray was not confident that they 'would be good enough not to break down during the commissioning period.' It would be advisable to send over fuel rods from Canada. Since there was still no agreement on safeguards – the Indians explicitly refused to accept them – Gray had yet another discussion pending with Bhabha.[25]

The discussion ended in a partial success. There would be safeguards, watered down but acceptable to External Affairs. Thus there would be fuel rods – 100 of them, shipped to India with another 100 on consignment in case of need.[26] A last-minute glitch was surmounted, apparently in a way satisfying to Indian self-esteem: 'the Canadians were unable to identify the cause of the trouble and they left the reactor to Indian experts to set right.' With the reactor set right and with Fred Gilbert of the Chalk River staff looking on, CIR went critical on 10 July 1960.[27]

Relations with India had survived and flourished. From the standpoint of the Department of External Affairs, however, they raised more troublesome questions than they solved. Yet even in the early 1960s, relations with India were sufficiently important to give pause to any who might wish to disrupt them. India was still the key nation in Canada's contacts with the non-aligned world, and the two countries were collaborating in a number of forums, not least a control commission supervising the short-lived peace in Viet Nam following the Geneva accords of 1954.

Thus, when Bhabha announced in 1957 that India would build its own plutonium reprocessing plant, there was no special protest. Although the plant was to reprocess fuel taken from the research reactor, its actual design and construction were Indian, and there was little or nothing that Canada could do to retard it; a minor consolation was that it would not be finished until 1965. Of greater importance was the argument that, like Canada's early efforts in plutonium extraction, the plant's product aimed at a thorium breeding process, eminently peaceful and appropriate to India's abundant thorium supplies.

From AECL's standpoint, India's political prestige was useful in validating its arrangements with the Indian Department of Atomic Energy. CIR was, as of 1961, the only Canadian-designed reactor outside of Canada. It was established to assist Canada's foreign policy, but also for AECL's more particular purposes. It showed that Canada was in a small way an exporter of nuclear technology, like the British and the Americans. Canada was therefore competitive, and was still an atomic power by this new measure introduced in the mid-1950s. The competition had a purpose: the encouragement of long-term commercial sales. The reactor was proudly known as CIR, the Canada-India reactor; it was later extended to be CIRUS, apparently in memory of Cyrus, shah of ancient Persia.[28]

With the completion of CIRUS and without any spectacular disagreements between Canada and India, or between AECL and the Department of Atomic Energy, the Indians were ready to try again. That by itself seemed to justify flexibility in diplomacy, through imprecision over safeguards as political instruments of commercial diplomacy.

Reactors for Rajasthan

The CIR reactor was barely complete before a new gust of interest blew from India into AECL. The imperatives that had determined Indian policy in 1955 were still the same, and the same cast of characters was running the Indian atomic energy program. They wanted, as far as possible, a reactor that could depend on indigenous supplies, meaning natural uranium. There was, if possible, even more urgency because relations between India and two of its neighbours, Pakistan and China, were drastically worsening. Actual war would break out with China along the Tibetan frontier in 1961, and with Pakistan in 1965. The Chinese experience was humiliating for the Indian government; it stimulated thoughts both of revenge and of emulation; emulation involved an even greater commitment to economic growth. And economic growth was closely linked to the atomic program. So was the eventuality of putting atomic research to military use; for China certainly intended to make itself an arsenal of nuclear weapons.[29]

The uncertainties of India's external relations added an acerbic element to the annual debate at the IAEA in September 1960. Canada had helped form the so-called 'Ottawa Group' – including the United States, Britain, Australia, and South Africa, a major uranium producer – to support safeguards through the IAEA in March 1959. As a result, Canada's tactics at Vienna were concerted with Britain and the United States. India, which was assumed to be hostile on the subject, was not consulted or informed in advance. As far as the IAEA was concerned, this was accurate; but there had been contradictory signals from the Indian high commission in Ottawa and the Canadian high commission in New Delhi as to whether India would or would not accept safeguards on fuel. There were also reports that India had approached both Britain and the USSR about the possibilities of natural uranium reactors, even as Bhabha hinted that he had not given up on Canada.[30]

The timing of the Indian approach coincided with the revival of the question of nuclear safeguards. Bhabha visited Canada for a week in November 1960. He arrived with an interest in a small organic-cooled reactor; he also came with a firm and unchanged position on safeguards. Gray did not try to dispute the safeguards in any serious way; instead, visions of an NPD were dangled in front of Bhabha. In the end Bhabha became intrigued by Douglas Point which might, he suggested, be installed next to a Canadian-sponsored hydro project in Madras state; since Montreal Engineering was the consulting engineer there, Gaherty's attention was captured and focused: focused in such a way as to direct Montreal

Engineering's interest towards nuclear reactors. With his company in the business, Gaherty's services would soon be lost to the board; in an effort to avoid such a conflict, AECL itself undertook a preliminary engineering study.[31]

Nothing had transpired before the IAEA, after ferocious debate, produced a model for safeguards. (Safeguards have been well defined as 'accounting checks ... to detect the diversion of nuclear material to nuclear explosives production.'[32]) It would apply to reactors over 100 megawatts, but not under. Canada immediately signed on; among those who did not was India, whose delegation resisted to the last. The cabinet informed Canadian government departments and agencies that henceforth it would refuse to sanction the export of Canadian reactors, reactor components, or 'natural' uranium (over ten tons) unless a safeguard agreement was signed first. Of course, Canada would observe safeguards only as long as its allies did, a reasonable precaution given the competitive nature of the market in nuclear products that was developing.[33]

Negotiations bogged down. Finance was an issue, since India's ability to build reactors depended on establishing international lines of credit, and some bankers, especially at the World Bank, were dubious of the appropriateness of such costly capital projects for a country in India's relatively undeveloped state. There were, however, options and avenues open in negotiating finance. Safeguards were another matter. The respective positions on safeguards were clear, and clearly opposed, from the start. Gray predicted that Bhabha would hold out for a simple governmental assurance, if such were required; he also predicted something similar to the CIR agreement. Since American heavy water would be needed, the Americans would also have a say. Gray consulted his directors, as well as his senior lunch group at Madame Burger's restaurant in Hull, and came away with permission to go ahead, 'in view of the political importance of continued cooperation between India and Canada in the atomic field and the possible economic advantages related to opportunities to supply equipment and uranium to India.' Once again a rigid insistence on safeguards was postponed until a later and presumably better day, when the advantages of the Canadian reactor would be manifest to the Indians, or until the Indians were so far committed to the

Canadians and so far behind in negotiations with anyone else that they would swallow the safeguards without wincing.[34]

The Indians wanted and needed World Bank funding, which required a considerable expenditure of effort to present a plausible proposal to bank officials in Washington. It was not until July 1961 that, with the World Bank's approval secured, the Indians formally requested the survey that AECL had approved back in November. It would look at two possible sites, determine the time it would take to build, assess costs, and divide them into appropriate national shares.[35]

The study was only part of a comprehensive Canadian package that Gray set about assembling. A canvass of the Finance Department produced agreement to consider either Colombo Plan financing or a long-term loan for the $30-odd million that would be the Canadian share (a figure that the study confirmed). John Foster took the report to Bhabha in the winter of 1962, with the news that the total cost of a 100 megawatt reactor would be around $75 million, with Canada's share at $27 million, lower than expected. 'The principal problem now facing India is financing,' Gray stated in March 1962; in the face of economic adversity, Bhabha remained 'optimistic,' according to Foster.[36]

Bhabha was still optimistic when he wrote to Gray on 16 March that he proposed to start that year on a station at Tarapur near Delhi, finance permitting. Gray was prepared to guarantee the Indians against unusual losses of heavy water, a point of concern to Bhabha; it was hoped that this guarantee, added to Bhabha's satisfaction with the Canadian preliminary study, would swing the reactor to CANDU and to Canada.[37]

In all this negotiation, safeguards went unmentioned. But they were very much involved in the decision by the Indian government to choose both a US-designed system and a Canadian one. The US candidate, a General Electric BWR, was chosen for the site that Gray had expected to build at Tarapur; instead of two 200 megawatt stations Canada would be asked to build a single 200 megawatt reactor at Kota in Rajasthan. There was a silver lining in the cloud; to get the American reactor the Indians agreed to 'rigorous safeguards,' even though this was explained away as a special case.[38]

There was another sense in which the choice of GE was a special case. GE offered extraordinarily favourable terms on its first reactor sales, hoping thereby to firm up its market position. The terms offered India were simply too good to refuse, at a time of cash flow problems and foreign exchange scarcity. But the Indian government had already decided that the proper route to take was that of natural uranium and heavy water. For its next atomic project it did not even call for tenders.[39]

In August 1962 the Indian government informed its parliament that it proposed to build a Canadian-designed heavy-water reactor in Rajasthan near Delhi, on the conclusion of a satisfactory agreement. To the Indian government this meant more money from Canada; but to the Canadian government it also meant, as Gray's minutes state, 'the question of safeguards and the protection of commercial rights on CANDU,' so that the design could not be pirated and reproduced by the customer.

Nevertheless, AECL's management was very pleased with the Indian announcement. Bhabha was a world-class expert; his selection of the CANDU 'above the systems being developed in other countries would have a considerable impact on world thinking.' Gray naturally recommended acceptance of the Indian terms, and urged an immediate conference with the Indians with AECL present beside External Affairs at the table. Churchill took the matter under advisement, but would bring it before the cabinet as soon as he could get it on the agenda.[40]

The cabinet gave a green light to negotiations. These took place in Ottawa for their first stage. They went well, although, as is customary in financial dealings, figuring out the ways and means took time. AECL did not negotiate in a vacuum; besides the concerns of other departments and agencies the company had to take the American promise that they would not conclude an agreement with the Indians for US-designed systems unless and until a satisfactory agreement was reached on safeguards. As for safeguards, Gray explained to the Finance Department that he believed the time was ripe to strike a deal that would be even better than CIR.[41]

A trip by Gray to India in November 1962 kept the pot boiling. India's recent war with China had impressed responsible officials and made them determined to have nuclear power as soon as they could. The Indian finance department approved Canada's financial terms, but it had to be admitted that safeguards, 'handled in Canada by External Affairs,' were a stumbling-block. The Indians agreed to submit statements on fuel burn-up and to co-operate in an annual audit of fuel; they also promised to proclaim their intention to use any plutonium produced solely for peaceful purposes. This could be effectively guaranteed by keeping the fuel in the reactor for an unsuitably long time for a military-grade product.[42]

The Canadians doing the negotiating and the ministers to whom they reported were alive to the favourable implications such a deal held for Canada's exports, which by then might include heavy water from the plant Gray was hoping to build. It was unfortunately true that any deal struck would not concern the Indians alone but the Pakistanis too; if advantages were conceded to the Indians they would have to be given to the Pakistanis also, adding up to a higher price for the Indian reactor than purely Indian costs would justify.

It looked good enough for a mission to leave for New Delhi, bearing with it permission for $30 million worth of credits, plus $7.5 million for Pakistan. A satisfactory conclusion was anticipated, but it remained subject to agreement on safeguards and fuel supply.[43]

The agreement did not come easily, or for some considerable time. There was, to begin with, some disagreement within the Canadian delegation to Delhi. Some at least of the diplomatic staff wanted to use the Indian desire for a new reactor to lever proper safeguards for the first reactor, CIRUS. 'We knew that reactor was naked,' one of them asserted in 1974. 'Here was a chance to do something about it. But the commercial people kept saying that if we didn't give the Indians what they wanted, they'd get it elsewhere.'[44]

The financial problems were resolved as the result of the Canadian mission to India. Safeguards remained. It took time to deal with them, especially because, as the safeguards issue developed, the Canadian general election of 1963 was called.[45]

Bhabha arrived in Ottawa to greet the new Pearson government and try his luck with a cabinet that had got into power on a promise to accept nuclear arms – American-owned and produced arms – for Canada's armed forces. He dangled the bait of 'several more Douglas Point type stations' in front of the Canadians. These would be a matter of prestige rather than profit, for the Indians intended to build these themselves, using the know-how acquired from building their first CANDU. (Canadian rights would be covered by a lump-sum, once-only payment, and by an Indian agreement not to pass the technology on to anyone else.) Bhabha was ready to start discussions with Montreal Engineering, who would act as Canadian engineer on the first CANDU; Geoff Gaherty, who did not believe in losing time, had already suggested going ahead with construction, leaving the diplomats to catch up later.[46]

Gaherty was premature. External Affairs and AECL were deadlocked. It was Bhabha who gradually gave way. He allowed Canadian inspection of Canadian fuel elements. The Canadians wanted more: the inspection of all fuel elements. Bhabha took that question home for further consultations. The consultations were fruitful, and productive of a promise that peaceful usage would be maintained, through a promise of reciprocal inspection of analogous Canadian facilities.[47]

It seemed promising, and of a piece with better news internationally. At the end of 1962 the Russians and Americans exchanged letters expressing the possibility of a ban on nuclear weapons testing in space, in the atmosphere, and underwater. Because there was no agreement on detecting tests underground, such tests were excluded from the limited test-ban treaty that was subsequently signed at Moscow. This development was an immediate relief for those concerned about radioactive pollution in the atmosphere as a result of nuclear tests – no small concern – and it seemed to promise easier east-west relations. The treaty was initialled on 25 July, and after prior signature by the two principal nuclear powers, was opened to others to join. Canada signed on 8 August.[48]

As negotiations for a test-ban proceeded, the log-jam at the IAEA suddenly broke in June 1963. The Soviet Union announced its support for control and inspection, as Goldschmidt says, 'without any outward sign of embarrassment.' The IAEA acquired a significant function which would allow it to take inventories of both basic (natural uranium compounds) and fissile (enriched uranium or plutonium) nuclear materials. It took time to work out IAEA procedures, so that they were not in place in time to affect directly the Indo-Canadian negotiations over a CANDU for Rajasthan. Even the possibility of converting the IAEA back to something like its original purpose was productive of optimism, and it was in that climate that negotiations with the Indians moved towards a close in the fall of 1963.[49]

The agreements were announced in parliament in November 1963, and signed at New Delhi on 16 December. Gray, on tour, did the signing. Construction could now proceed.

By the spring of 1964 work on the project, dubbed RAPP I (RAPP stands for Rajasthan Power Projects), was well under way. In April NPPD started to hire fifty-six extra staff to cope with the Indian reactor. In May Foster reported to the board that Montreal Engineering was hard at work on the conventional civil engineering while NPPD was modifying the Douglas Point designs to accommodate Indian conditions and concerns. Tenders had been invited for condenser, turbine generator, heaters, pumps, and diesel generators. Montreal Engineering's resident engineer had arrived at the site, but although provision had been made for Indian engineers to come to Canada for training they had not, as yet, arrived.[50]

With RAPP I launched, the Indians began to agitate for RAPP II. This would be a reactor of similar size and design. Gray found on a trip to India at the beginning of 1965 that 'Dr. Bhabha is emphatic that RAPP II must go ahead immediately,' and that Montreal Engineering should manage it on the same basis as RAPP I. Bhabha was equally emphatic about safeguards. The Canadians had beaten him down over RAPP I, but they would not get their way with the next. If there were to be safeguards they should apply only to the reactor core, uranium fuel, and heavy water, and not to what were described as 'items of equipment that are normal items of commerce.'

The board may not have sympathized with Bhabha, but it agreed that he had a point. 'It would be a great pity,' it decided, 'if Canadian industry were denied the opportunity to participate in this work by reason of the application of a political decision on safeguards of doubtful merit.'[51] The board's jaundiced perception was not shared in the more idealistic reaches of External Affairs. In fact, the spring of 1965 saw the enunciation of an even stronger Canadian policy on safeguards, this time for uranium fuel. It was well known that Canada had applied a double standard to uranium supplies – one for its old partners, Britain and the United States, and another, forbidding military use, for the rest of the world. In response to French activity in the Canadian uranium market, the cabinet moved to prohibit the export of uranium to any country that refused to accept a non-military limitation on its use. 'This,' the cabinet soberly minuted, 'would remove the so-called discrimination about which France had complained.' Pearson made an appropriate statement to the House of Commons on 19 June 1965.[52]

Pearson's statement coincided with a further development in international arms-control negotiations, where the United States and the USSR had combined their diplomacy to secure an objective considered to be in both countries' interests. They wished to stop the spread of nuclear-weapons-making capability – a capability that had recently expanded to include first France and then China. A Non-Proliferation Treaty was in the offing, and although its details took years more to negotiate, the Canadian government strongly supported its purpose.

With the wind in its sails, the Canadian government took a strong line on linking not just safeguards, but *Agency* safeguards, to RAPP II. After Bhabha's death in a plane crash in January 1966, Gray visited India to further the cause of RAPP II. He took the occasion to explain to Bhabha's successors the Canadian position that RAPP I and RAPP II constituted two units of the same station, and that the safeguards applicable to one applied to the other. Attempts by the Indians to escape Canadian safeguards by procuring only the conventional parts of the reactor in Canada while making the rest in India were not acceptable. (That was not the only reason for producing components in India: 60 per cent of the total cost of RAPP I had been foreign exchange; the figure would be 40 per cent for RAPP II.)[53] Gray hoped that RAPP II was sufficiently well established in India's priorities for its sale not to be overturned by the firmness of the Canadian position.[54]

In September 1966 Gray told his executive committee that Paul

Martin, the external affairs minister, was 'adamant' in his view that the price for RAPP II should be Indian acceptance of IAEA safeguards. In response, 'Mr. Martin was made aware as to what the attitude of the Indians would be in the face of a condition of this kind.' Gray's scepticism made little or no impact on either Martin or his department; it is reasonable to conclude that External Affairs would not compromise as it had in 1963 and miss an opportunity to get the Indians to comply with international standards and purposes. For what had been true in the 1950s was still true in the 1960s: India was a bellwether in the underdeveloped world – what was coming, by this point, to be called the Third World. An Indian precedent was no small thing, and worth some trouble and disappointment to get.[55]

The Indians naturally took a different line. To them a Non-Proliferation Treaty would do one thing and one thing only: it would freeze the existing nuclear weapons oligopoly and divide the world into two classes of states: those with nuclear weapons, who could go on expanding their arsenals and threatening others directly or indirectly; and those without, who would be prohibited from countering the power of the atomic weapons club members. This was no small consideration given the state of Indian relations with China, now a nuclear power.[56]

Despite Indian reservations, a Non-Proliferation Treaty was negotiated. It was signed on 1 July 1968, and came into force in 1970. It took a great deal of time to negotiate, and it failed to attract universal support: some eight significant states, including France, refused to sign the document, although France let it be understood that it would not subvert the treaty's provisions. (Three others, Germany, Italy, and Japan, did not sign until 1975–6.)

The treaty had the advantage of a simple, comprehensible core: those states that had not already publicly acquired nuclear weapons by 1967 were to promise not to make or otherwise procure any. To enforce this provision, it gave the world a safeguards regime that included *all* nuclear facilities in signatory non-weapons states; in return, the weapons states promised to negotiate controls over the nuclear arms race. Like the original Atoms for Peace proposal fifteen years before, the Non-Proliferation Treaty promised technological benefits, including the sharing of information from peaceful nuclear explosions, to those who chose to forgo the weapons path.[57]

It is in this context that RAPP II had to work its way. India and Canada were on contradictory if not conflicting paths; both countries knew it; yet at the same time they were continuing, for reasons that were both political and economic, to co-operate in building reactors for the Indian atomic program. The political rationale was beginning to wear thin, as Indo-Canadian relations frayed in a number of areas, and the vision of the 1960s, that an intercontinental bridge of friendship could be constructed between these two formerly British countries, was becoming blurred.

The bridge lasted long enough for RAPP II to travel over it. The key was whether Indian acceptance of safeguards similar to those on RAPP I was sufficient; if that problem was disposed of, senior Canadian officials were agreed that 'Canada was all but committed to participation in RAPP II,' at least as long as the Indians were. But it was not disposed of easily. The RAPP I safeguards were bilateral, not Agency; they depended on government-to-government agreement and co-operation. Discussions with the Indians were difficult, though not fruitless. Their intricacies need not concern us; in the event, an exchange of inspections, under IAEA auspices, was agreeable. Principle was upheld, though with difficulty, and the RAPP II reactor could go ahead with Canadian export financing.[58]

RAPP I and RAPP II were completed, respectively, in 1973 and 1981, four years behind schedule for RAPP I and seven and a half for RAPP II. They were the progenitors of a family of Indian-designed and manufactured pressurized heavy-water reactors, four in number, of which two (plus the two RAPPS) are at the time of writing in operation.

If this were the whole story the history of Indo-Canadian atomic collaboration would be a happy one, satisfying to both parties. It is not, of course, the whole story; instead, what might have been the fabled bridge between east and west or north and south took a sudden jolt in 1974. It was a nuclear jolt, from the 'peaceful nuclear device' the Indians tested using plutonium that derived, ultimately, from CIRUS, the Canadian-supplied, American-moderated research reactor that had started the whole cycle of Canada's multilateral atomic diplomacy.

AECL had obviously played a major part in this story, but it is not quite the part that enthusiasts for the supremacy of a politically based and centrally directed foreign policy imagine it to be. It is clear that in its origins Canada's atomic connection to India was political; that it was aimed at influencing India's attitudes to and relations with the Western alliance, of which Canada was a self-conscious and active part. It was also convenient to AECL, which was anxious to acquire prestige for what was increasingly becoming a specialized Canadian heavy-water design. In this it was successful, since the Canadian offer displaced a British graphite model which would otherwise have been chosen. It also served to underline Lewis's belief that nuclear power offered a release from poverty and deprivation for the Third World; here was proof positive that nuclear power would work even in an underdeveloped setting.

The timing of the Canadian offer was crucial. Because it occurred before any prospect for international safeguards was apparent, and because diplomats and company officers had barely begun to think in such terms, the original offer to the Indians, for which both the Department of External Affairs and AECL were responsible, did not include any political terms, such as controls over the end-use of spent fuel. The attractiveness of the Canadian offer to the Indians was presumably enhanced by the omission. On this dubious foundation a bilateral relationship was established, in the expectation that it would later be incorporated in a stronger multilateral framework. That could always be later, when Indo-Canadian relations were stronger and better established, once Indian goodwill and trust had been won. But with the moment lost, the opportunity, if it had ever existed, was gone forever.

Subsequent attempts to manoeuvre the Indians into permitting another country, even the donor country, to inspect their nuclear inventories produced very slight concessions, but none that would seriously have inhibited the development of an Indian bomb, even a 'peaceful' one. Realistic or enforceable safeguards ran squarely into Indian national pride, and the determination of a newly emancipated people to insist on their own, untrammelled sovereign power. It was that power that Canadians were coming to conclude was the principal obstacle to halting the uncontrolled, 'horizontal' spread of nuclear weapons. But as one of the original atomic club and as a supplier of uranium and plutonium to the American and British bomb programs, Canada was ill-suited to object.

There was still politics. Canada placed considerable trust in the political reliability of the Indian government. It was possible to reason that once a nuclear reactor was in place, only the domestic constraints of the Indian political system would keep it reliably 'peaceful.' Here it would seem that the Canadian government misestimated the strength of Indian nationalism, and underestimated the external threats that the Indian government was probably going to have to confront in the 1960s. It would have seemed wildly implausible that India and China would ever go to war, or that China would in short order build its own bomb, with short-sighted Soviet aid. Both internal and external factors operated to bring the Indian government to value its own interests ahead of those of its distant Western partner.

In the 1960s these perceptions and assumptions changed. The government of Canada came to consider the government of India in the context of the international system as a whole. Co-operating with the United States and Britain, Canada shaped its foreign policy to reinforce the *system*; it accordingly devalued the bilateral co-operation between Canada and other countries, such as India, that had been the object of its special affection in the 1950s.

Unfortunately for the system, the bilateral arrangements that had been set up in the atomic field between Canada and India worked very well indeed. Both the Indian Department of Atomic Energy and AECL concluded that the bilateral connection served both their interests, AECL because it wanted to establish its reactors in an international marketplace where India for all its remoteness and poverty was nevertheless a Canadian shop window; the department because it was the Indian government's chosen instrument for the rapid industrialization of India and because it found the heavy-water pressurized system worked very well for this purpose.

Unfortunately for AECL, other parts of the Indo-Canadian relationship did not work very well, and worked worse as the 1960s dragged on. The war in Viet Nam found the two countries in basic disagreement after 1966, with Indian policy there and elsewhere veering off into a distinctly anti-Western orbit. Atomic diplomacy, in New York, Geneva, or Vienna, frequently found Canadian and Indian spokesmen on opposite sides of the fence. Instead of defusing confrontation between Western-aligned Canada and non-aligned India, atomic relationships actually sharpened it and gave it a focus that made the political arms of the Canadian government more and more unhappy as time passed. It seemed obvious to them that Canada had given a hostage to fortune in giving and selling first one, then two, and finally three nuclear reactors to the Indians; and they tended to view AECL's part in the relationship with an increasingly jaundiced eye. Ironically, it was AECL that had continued in the original spirit of the agreement while everything around it was changing.

Hinton and Temptation

AECL's first venture into the Third World was its first significant intrusion into the nuclear marketplace. Its reasons were basically political, and had more to do with the general policy of the Canadian government than with the commercial interest of the company. That would not be the case with AECL's second international expedition. Unlike the Indian adventure, it had no political dimension; unlike the Indian transaction, it was unsuccessful. This time the quarry was Canada's close ally and second-largest customer: Great Britain.

Great Britain was a logical choice for a nuclear partner. The Canadian and British atomic establishments sprang from a common root. The British Atomic Energy Authority (AEA), the umbrella organization that included Harwell, maintained a liaison office at Chalk River, and held annual conferences with AECL to exchange information and compare notes. AECL briefly reciprocated with its own liaison office in London, an initiative founded on hope but terminated in disappointment.[59]

Britain had a sophisticated engineering industry as well as a world-class scientific establishment. It was, even more than the United States, the country where AECL's top management and senior scientists felt most comfortable. Mackenzie, Hearn, and Lewis demonstrated the point when in 1954 they approached the

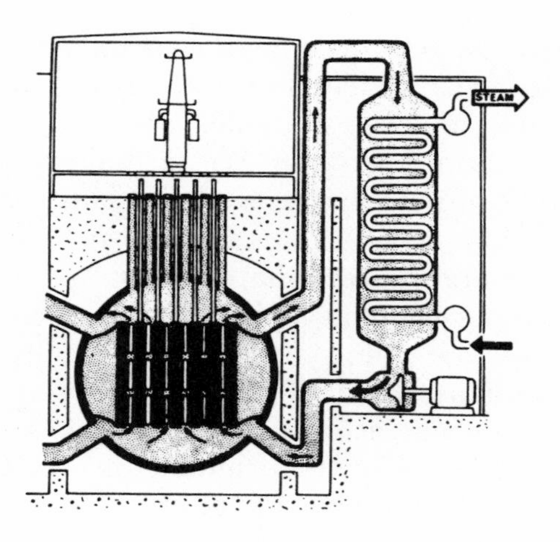

Magnox reactor. *Nuclear Engineering International*

British for assistance in design, training, and manufacture; and though Bennett recoiled from the full-scale integration that his associates proposed, he retained close and cordial relations with Sir Edwin Plowden, his British counterpart. The failure of a joint study of heavy-water design in 1955 frustrated any immediate hope of co-operation on a practical reactor; it would be followed, for a variety of quite different reasons, by an attenuation of links between Canadian industry and British suppliers, attested by the replacement of the pressure vessel of NPD, a British product, with pressure tubes and a calandria, Canadian-made. To anyone familiar with the Canadian electricity industry, this must have seemed an inversion of the natural order of things.

It is true that under the Diefenbaker government this tendency was perceived, defined as a problem, and subjected to remedial action, to counter the growing inclination even of the Canadian governnment to ignore the capabilities of British exporters. Inevitably, however, discussions took time, and solutions were caught up in larger questions of Canadian trade, and it was January 1960 before AECL was approached for its opinion. AECL protested that the problem was more complicated than simple ignorance or prejudice. Given the British origins of so many of the company staff, it might be thought that if they had a bias it would be in favour of the United Kingdom. The plain fact was that Chalk River was in the market for items such as electronic components, and these were made to an American standard, unavailable in Britain. It might be possible to alter procedures in certain respects, so as to give the British a better chance, but there was little that could be done to alter Chalk River's needs.[60]

In the larger context the two countries were drifting, perceptibly if slowly, apart, and while older connections persisted and retained much of their force, new ones were not being made. The older, essentially colonial, attitude that came naturally to Mackenzie was giving way to a more self-confident and assertive posture, best symbolized by the offer of a reactor to India. As the AEA brought its power reactor program to maturity, commercial considerations also appeared: help to Canada could be assistance to a potential rival. Access to atomic information diminished proportionately.[61]

Canadian success in pre-empting the British in supplying a heavy-water rather than a graphite-magnox reactor to India was, at the time, astonishing. The British government seemed to be eminently justified in planning for a nuclear Britain in the near future, and in announcing, in March 1957, plans for a network of magnox stations that would supply 6000 megawatts to the main British electricity grid, the Central Electricity Generating Board (CEGB).[62]

The significance of the British program to Canada at this point lay in its potential as a competitor in third countries. There was little prospect that the British-designed reactors would be adopted in Canada, and, given the apparent strengths of the AEA, little likelihood that Canadian reactors would have any chance in Great Britain. There were, however, considerable differences in the organization of the two countries' atomic programs that proved to have some relevance in altering this evidently hopeless situation.

As in Canada, atomic energy and electricity utilities in Britain were creatures of the state, although in the United Kingdom they were emanations of the same state and did not have to contend with the complexities of federal and provincial jurisdiction. Unlike Canada, the two entities were not closely linked in the development of reactors; the CEGB was much more a customer of the AEA than it was a partner. It was in the hope of co-ordinating the two elements, supplier and customer, that the British government appointed Sir Christopher Hinton, the AEA's engineering wizard, to be chairman of the CEGB in 1957.[63]

There was, however, another reason behind Hinton's appointment. Hinton and his engineering staff at Risley had the responsibility of developing reactors. They had targets and deadlines, and found the contemplative and academic atmosphere at Harwell uncongenial and inappropriate to their work. By the mid-1950s, Hinton and Cockcroft, Harwell's director, were barely on speaking terms. Plowden, faced by a schism between his two principal subordinates, solved it by moving Hinton out. 'One of them had to go, and he chose the wrong one,' a Hinton supporter later commented. Hinton did not as a result cherish great affection for Harwell.[64]

Hinton did not take long to conclude that the magnox reactor was inadequate for economical operation. An AECL delegation touring British facilities in the fall of 1960 was told that there were 'two basic problems besetting the U.K. power programme – there was no assurance that the graphite moderator in their power stations would stand up over the lifetime of the stations, and there were doubts whether the U.K. fuel could be burned up much higher than half the estimated burn-up of 3,000 MWD/t.' This was in fact pessimistic, but it was true that the Magnox reactors required uncomfortably frequent changes of fuel, a 'limitation' which, according to Walter Patterson, a prominent critic of the AEA's operations, was 'one of several factors which eventually terminated the Magnox programme and provoked a search for another approach.' At the time, it was also obvious that the magnox could produce power, but power that was more than a third more expensive than power from conventional sources.

The Canadians returned home reassured that their selection of the heavy-water, natural-uranium line had been 'correct'; soon afterwards, possibly by coincidence, they began to receive intimations that the British might after all be interested in pursuing 'the AECL approach to nuclear power.'[65]

These hints were sporadic, and gradual in their effect. Lorne Gray had begun to make regular trips to India to inspect progress there, and to Vienna for IAEA conferences, as well as more customary journeys to Britain for the AECL-AEA conferences, and en route he took the opportunity to sound out interest and opinion on Canada's heavy water system.[66] There was indeed some, enough to persuade Gray to develop a policy for the licensing of heavy water technology: the CANDU-200, the Douglas Point model. This was run past the cabinet, and approved, in May 1961. It provided for the payment of a fixed fee for the transfer of detailed design information, with subsequent payment on a sliding scale related to the amount of Canadian-manufactured equipment or fuel that the purchased reactor consumed. A proper fee, Gray thought, would be in the region of $200,000–$250,000.[67]

Encouraging news followed. Hinton wanted to send out a mission late in the summer to inspect the designs of NPD and CANDU; this visit, interestingly, was independent of the AEA and apart from the annual conference exchanges. This was significant, and provides an instructive comparison for the Canadian atomic structure. For, as the historian of British electricity reveals, Hinton had come to resent the fact that the CEGB was 'a captive buyer' from the Atomic Energy Authority and the industrial consortia it had organized to manufacture and market its products.[68]

What Hinton was thinking of was revealed in December when, in the British *Three Banks Review*, he published an article admitting the deficiencies of the magnox and speculating about the kinds of reactor that might replace it. It might be possible to follow along from existing British designs, improving the graphite reactor, or, he wrote, 'it might be necessary to abandon the use of graphite as moderator and use heavy water instead.' In all probability, the CEGB chief continued, 'the use of heavy water' would be the ticket.[69]

It would be an understatement to say that Hinton's article caused a scandal in the AEA. Britain's most important electrical executive, not to mention a genuine 'atomic knight,' honoured for his services to the country's nuclear program, had stated in public that it would probably be necessary to go outside the United Kingdom for a technological boost. What had the British taxpayer been paying for? The AEA was not charmed by the question, and Hinton's former associates and staff were even less pleased.

Hinton's was not an argument calculated to attract politicians, nor did it. Yet it was at the political level that the clash between the

AEA and the CEGB had to be resolved. Once viewed as a political problem, it became a matter of smoothing over the differences, papering over the crack, or, in the last resort, forcing the two government entities to come to terms that would be publicly acceptable. In a trial between the electricity utility and the still prestigious AEA, it was the latter that held the high ground, not to mention the low, for some of the AEA's defenders presented Hinton 'as an unreasonable turncoat.' Hinton had to muddle through with propositions about efficiency, expenditure, and the public interest; in a political context, however, such arguments can seldom carry the day against considerations of national prestige, jobs, and simple protectionism.[70]

As for the heavy-water reactor, the AEA did not reject it outright. Rather, it reasserted its claim to be the exclusive purveyor of nuclear technology to the British market, a position that the government was bound to accept. Among other things, the AEA could and did contend that it must be in a position to pass on the efficiency and safety of any nuclear installation; the heavy-water reactor must first be proven according to its own standards.[71]

This need not, on the face of it, contradict AECL's licensing program. The first signs were good. Early in 1962 The Nuclear Power Group (TNPG), one of the British 'consortia' (aggregations of industrial firms organized around a particular variety of reactor), approached AECL for a licence. The consortia, AECL was told, had recently done a study that showed that the heavy-water reactor 'was as good as any other design considered' – possibly better, because the economics of the other types were improved by a 'plutonium buy-back' allowed to enriched reactors.

Gray was pleased. There were complications, because AECL's engineering services were fully occupied. This might make it necessary to rely on those available in Britain from the AEA or CEGB, requiring a system of licences and sublicences with the AEA (or CEGB) acting in effect as AECL's British agent. An early suggestion that an arrangement be made with CGE was lost, presumably because CGE was limited to the NPD model and size, while what the British wanted was a reactor or pair of reactors in the 300 megawatt range. It was a rational arrangement but it had, of course, a problem, what Hinton called 'the danger ... that it places an organisation which may be biased against overseas systems in a position where it can weigh the evidence against your reactor.' The CEGB chairman continued to hope that any arrangement made 'would not preclude us from buying a reactor of Canadian type from a Canadian firm if this proved attractive.'[72]

Negotiations nevertheless proceeded with the AEA and its chairman, Sir Roger Makins. At first it seemed that they would be crowned with success. In July Gray reported to his minister, Gordon Churchill, that he was 'largely acting as a salesman for the Canadian nuclear technology' and that 'it seems to be starting to pay dividends.' Churchill, Gray noted, 'strongly supports' such efforts. In August Gray reported to the Advisory Panel that financial arrangements were almost complete: the AEA would pay an initial fee of $250,000 and a per-kilowatt fee thereafter.[73]

But the negotiations were not quite finished. When Gray took his executive committee to England in November for the annual exchange of visits, he and they knew that the British were dragging their heels, ostensibly because they were trying to sort out the impact that a heavy water technology would have on their consortia. In examining the hand that they would have to play, the executive committee understood that the British already had, as a result of existing exchange policy, a great deal of information on the design and construction of a heavy-water reactor; what they did not have was the 'know-how' – the detailed information on how to build one. AECL, because of its construction and ownership of NPD and Douglas Point, did. Such information would, Gray hoped, be marketable.[74]

When the executive committee arrived in London a few days later, it did not find encouragement. Instead, it concluded that the British were 'conducting a delaying action.' Although the minutes do not state it, it may have been that the British were trying to evolve their own heavy-water design without reference to the Canadian one and, of course, without payment, either monetary or psychic: the Canadians would, under these admittedly hypothetical circumstances, not even have the prestige of knowing that their design was being adopted by a major power. The executive committee decided that it would halt the flow of commercially valuable information to the British forthwith.[75]

AECL therefore evolved a new commercial policy that established a series of technological 'packages.' These might range from Douglas Point minus the know-how down to the calandria or the fuel bundles. Each sale would be on a one-time and per-country basis; thus, it could not be retransferred by the AEA to a reactor in another country. It also furnished a more comprehensive system for dealing with the transfer of information, much of which was already available to the AEA through holes in AECL's existing commercial program.

That was certainly a negative stimulus; more positive was the feeling that selling components of CANDU would be more attractive to foreign customers than a commitment to a whole new system. More immediate, however, was the belief that the British were taking longer than they should to come to terms; the new policy, Gray told the Advisory Panel, 'might jolt the AEA to activity in this regard.'[76]

The jolt, however, was mainly to AECL's traditional relationship with the British. Whatever its impact on the British, it was very upsetting to the Canadians. Mistrust lingered after the disastrous November 1962 conference; and in March 1963 Gray could only tell his minister that he was 'sorry to say that we are making very little progress' in selling CANDU to the British. 'I suspect,' he continued, 'that our relations with the UKAEA are at their lowest point in years and, although there are indications of an up-turn,' it was a real question whether the negotiations should be pursued. If they were, should AECL stand firm, 'with the possibility of further deterioration in our collaboration,' or should the company make an agreement to restore what had already been lost 'at almost any price.' In this, his last meeting with Churchill as a minister, Gray got the word to 'Try to make a deal.'[77]

This was easier said than done. AEA officials travelled to Canada in June to go over 'what AECL had to offer in experience and knowledge on heavy water moderated systems, and to make an assessment of its worth to the UKAEA.' Two sets of consequences flowed from this trip. A commercial agreement moved perceptibly closer and, early in the fall, the two sides finally reached an accord. It provided for the payment of £250,000 ($700,000) for an exchange of information. The exchange would include fuel bundles, but not the design specifications of the CANDU reactor en bloc. Although there might still be an approach from a British manufacturer, there was no longer any thought of granting a wholesale manufacturing licence to any foreign firm without first 'safeguarding the position of Canadian industry.'[78]

This may have seemed satisfactory, but there was another, and more indicative, sequel. Only a month after Gray announced the conclusion of the agreement to his board (and ten days after the cabinet ratified it), the executive committee took up some disquieting reports that had reached Canada, to the effect that 'certain UKAEA officials have been misrepresenting the value of the Company's program and hence the value of this Agreement vis-à-vis the Authority.' Gray was authorized to complain to the AEA, with what result we do not now know. What was happening, however, was plainly just what Hinton had prophesied. The AEA was now able to act as a judge in its own cause, and the verdict was never really in doubt.[79]

No British consortium ever came forward to adopt the CANDU, any more than the AEA itself ever embraced the Canadian design. Instead, the AEA went forward with advanced gas-cooled reactors, intended to overcome the low-temperature limitations of the magnox, and then with a steam-generating heavy-water reactor, the SGHWR. They were expensive and elaborate, but an SGHWR prototype was in fact built. Using enriched rather than natural uranium, it started up in 1967, just about when Douglas Point was getting underway. It operated well, but at 100 megawatts (electric) it was only half the size of its Canadian cousin and, unlike Douglas Point, it had no progeny. The British heavy-water program was to be abandoned in 1976 after the AEA had tried and apparently failed to upscale its heavy-water design. It did not subsequently choose the CANDU; it opted instead for more of the AGR, a circumstance that prolonged the 'civil war' that Walter Patterson suggests had become the norm for the British nuclear-electrical industry.[80]

Two questions suggest themselves. Could the civil war have been avoided? And what role did the CANDU play in the war? As to the first question, Hinton was later asked whether the AEA's research and development program 'was guided by the CEGB's requirements.' In reply, he said, mildly, 'their activities are guided by what they think our requirements ought to be.'

The prime evidence for his contention was the fate of the British CANDU. The AEA did not want it and would not accept it. It took information from the Canadian side and used it – or did not use it – for its own purposes. 'The Authority, by his account, had even declined to furnish the Board with the technical reports on which their decision was based.' It was, he said, 'a mistake in policy for a research organisation to launch out into expensive prototype work without taking their main potential user with them.'[81]

The 'expensive prototype work,' however, may have been the point. Having an expensive prototype, with all the research and development involved, was just what the AEA desired. It was an R&D organization, after all, and it had to develop something. To take a CANDU-200 or the later CANDU-500 and to accept the Canadian stations as prototypes, as the Indians had done, was inconceivable in terms of the state of the British program. This reluctance could be justified in terms of the AEA's own specialized knowledge of the deficiencies of the Canadian reactor, inadequacies which AEA representatives did not fail to point out, either in private, as we have seen, or in public, before British parliamentary investigating committees. One of these committees plumped for the heavy-water reactor, and claimed to be as impressed by what Canada had done as they were depressed by the cost and slow start-up of Britain's own AGR program, one of whose stations, Hunterston B, seems to have been the British counterpart to the Glace Bay heavy-water fiasco.[82]

At the same time, the British atomic civil war was the fruit of a rivalry between scientist and engineer, between designer and customer, and between two rival monopolies. The AEA flew high in the 1950s, too high in the opinion of its largest potential client, the CEGB. Failing to get what it wanted, the CEGB under Hinton established its own development and design facilities; by the 1970s, it was the CEGB that was moving into the ascendant and the AEA that was losing staff to its rival.[83]

The division of opinion and divergence of interest at the top of the British atomic electricity program was a phenomenon that Lorne Gray clearly hoped to exploit. He was finding his feet as a 'salesman,' as he told Gordon Churchill, and he was happy to practise his talents on the British. As assets he had drive and optimism, but his gifts had a negative side as well. More than one British negotiator remembered Gray's personality fondly, while adding that he was, after all, something of a pirate cruising in their placid waters. He was not quite trusted, even though he was much liked.[84]

The British-Canadian atomic collaboration bore bitter fruit. Although scientific exchanges and personal friendships were usually immune to the contortions of official nuclear diplomacy, the British example is very instructive as to the difficulties in trading nuclear technology to another nuclear power with its own well-established and technically sophisticated nuclear industry. For reasons of pride, ideology, inertia, and industrial policy, AECL's British half-sister was unable to contemplate the supersession of its own R&D agenda by the British counterpart to Ontario Hydro, the CEGB. For similar reasons, the AEA and its industrial partners, the consortia, were unable to take seriously the possibility of competition from a Canadian reactor. As one British industrialist observed to Lorne Gray, Canada was just not in the running; it followed that, for whatever reason, Canada's reactor would fail to make the grade. But it could be said, with equal justice, that had the roles been reversed it would have been very unlikely for a British reactor to be adopted in Canada.[85]

CGE

The expansion of AECL's business meant prosperity for the companies that supplied it. In Peterborough it was an optimism tinged with nervousness. As the home of CGE's Civilian Atomic Power Division, Peterborough had an interest in the progress and prosperity of the city's largest industry. Because atomics employed many of the city's highly skilled workforce, what happened to CAPD was more than usually noticeable. And what happened to CAPD depended to an unusual degree on the Canadian government and its agent, AECL.

This was because so much of CGE's business was selling to AECL: fuel bundles, for example, or refuelling machines. CGE was the builder of the OTR at Whiteshell, and one of the few Canadian companies that AECL judged competent to build the heavy water plant at Port Hawkesbury. It was a delicate matter because CGE was not merely AECL's supplier, or client; it was also its competitor, ostensibly for the export market but in fact for the favour and support of the federal government. Each company needed the other, and so, in the early 1960s, AECL shaped its policies towards the end of keeping CAPD in being and in business – up to a point.[86]

CAPD, however, wanted more than AECL could offer. It had a design and a manufacturing capacity. It was the only Canadian organization, apart from AECL, that could design and build reactors, using an upscaled version of NPD as a basis. It was believed that CGE could build a reactor up to 250 megawatts and, as long as a purchaser wanted one off the shelf, CGE could supply it.

It was different if a purchaser wanted either more or less than CGE was equipped to offer. Ontario Hydro wanted less: by breaking up the business into its components, the Ontario utility hoped to improve its own position by enforcing competition on its suppliers. That led to the exclusion of CGE and the creation of NPPD. If a purchaser wanted more – irradiation data, performance reports, and the like – CGE could not supply them. They belonged to AECL. And if a customer wanted more still, for example the creation of a complete nuclear program in another country, then only an analogous Canadian government agency could satisfy.[87]

CGE nevertheless retained a marginal advantage. Projects that were unattractive to AECL, or that were beyond the existing capacities of NPPD, could be, and were, passed along to CGE. When the NPD fuelling machines malfunctioned, CGE bid for, and got, a firm price contract to replace them with its more satisfactory variety.[88] In the early 1960s, with Douglas Point, Pickering, RAPP, and the prospect of Gentilly, NPPD was definitely overloaded. CGE, however, was hungry for business, and business came its way in the form of the Pakistan Atomic Energy Commission.

Pakistan was India's Islamic twin and rival, created in 1947 out of the old Indian Empire and staffed by Muslim counterparts to Homi Bhabha. But it was smaller and poorer than India, and drew on a smaller industrial and technical base. It took proportionately longer for Pakistan to evolve a desire for nuclear installations above the laboratory level. The rivalry between India and Pakistan found expression in a variety of ways. India was, in the 1950s and 1960s, ostentatiously neutral, though strongly democratic. Pakistan, in contrast, cleaved to the Western camp and belonged to several western-sponsored alliances. Yet Pakistan early on abandoned democratic forms in favour of an autocratic military regime; when later it briefly returned to civilian rule it was in the context of war and military defeat.

Pakistan's wars were with India. The first was in 1947–8, an unofficial conflict over the ownership of the state of Kashmir, but no less severe because of its undeclared status. The second was in 1965, effectively a draw, and the third in 1971. This last event was a disastrous military defeat for Pakistan and resulted in the independence of its eastern province of Bengal, thereafter Bangladesh. The great powers strove to remain uninvolved in the first two, although the USSR played a mediatory role in 1965; in the last India was supported at a distance by the USSR, and Pakistan by the United States, also at arm's length.

For Canada to be involved in both Pakistan and India was obviously a question of some delicacy. It was important to play a balanced role as between two Commonwealth partners, but at the same time important to keep Pakistan happy inside the western camp. Nor could the Canadian government completely ignore the drift of public opinion, either of the informed or of the mass variety, away from India's frequently censorious critique of the West and particularly the United States. That relations with India were rather better than the public perceived was not, in this period, effectively conveyed. But even confidential relations with India gradually deteriorated after the mid-1960s, though never to the same extent as those between India and the United States.[89]

Relations with India were in fairly good order when the Pakistan Atomic Energy Commission first came calling, in February 1961. The Pakistanis had followed Canadian aid to the Indians, and they were anxious not to fall behind. The Pakistani atomic chief, Dr Usmani, had had it in mind to ask for an experimental reactor similar to the NRX clone already operating in India, but he was persuaded that basic training of Pakistani staff at Chalk River would be a more logical first step.[90]

This first step coincided with a crisis in CGE's affairs. With work on NPD winding down, the CAPD was, or would shortly be, overstaffed. CGE therefore sought out further work. With the Bechtel corporation, an international engineering consulting firm, it searched for business in India, but without success. Argentina was another possibility, but it too disappointed. When, in the spring of 1962, AECL learned that Dr Usmani was now considering a small power reactor, the news came at an extraordinarily convenient time. Usmani wanted an 80 megawatt (electric) facility, just in the range of the uprated NPD model that CGE was then working on. By the end of the year the proposition had matured to a 132 megawatt plant, which offered better economies than a smaller version, and CGE had undertaken a design study. Such a reactor would probably cost in the range of $55–60 million.[91]

The next year – the year in which the RAPP agreement was made with India – was spent bringing CGE and the Pakistanis close to an agreement. The position of CAPD continued to cause concern, to the extent that AECL found it prudent to buy fuel from CGE long in advance of probable need, so as to keep CAPD going. Assuming that Ontario Hydro proceeded with Pickering, all would be well, and the stockpile would eventually be needed. It was understood that Pakistan wanted to buy a package. Like Ontario Hydro it wanted competitive pricing, an object that Gray persuaded them was not feasible; for one thing, there *was* no competition for some of the items that a CGE reactor would include. AECL would act as a friendly observer, but it would not get involved. In Ottawa, Gray did his best to firm up financing and to pacify those perturbed over safeguards.[92]

The Pakistanis emphasized that they would be responsible; the Indians, in contrast, were certainly building a bomb. CGE remained anxious to get the contract, and AECL, for its own reasons, was happy to have them get it. In April 1964 it consented to guarantee Pakistan heavy water – for a Canadian reactor – even if it had to be bought from the Americans and resold to CGE. It may seem that the problem of keeping CAPD fully employed was an excessive concern; but one of the conditions of the Pakistan contract was that it be completed as quickly as might be, on a tight schedule of fifty-two months, and for that a full and experienced team was needed. By the fall of 1964 the project had passed its last financing hurdle and secured Canadian cabinet approval. Meanwhile, Pakistani and CGE negotiators managed to satisfy one another that the estimates provided were reasonable.[93]

The export of heavy water to Pakistan therefore became a consideration in the decision to construct a second heavy-water plant in Canada. When a letter of intent arrived from Pakistan at the end of 1964, and when a contract was finally signed (a quarrel over the tax status of CGE personnel was the final hurdle), Pakistan's reactor joined AECL's other heavy-water burdens.[94]

The Pakistan project did not finally resolve the question of how CGE was to be supported. Its hopes had received a set-back in 1963, when AECL gave priority to developing the boiling light-water–cooled reactor over CGE's organic design. Lewis confirmed the preference in February 1965 when he reported in favour of the BLW variety for Hydro-Québec. Even so, Les Haywood, vice-president, engineering, urged that CGE be allowed to compete for the BLW reactor to be constructed at Gentilly.[95]

Haywood argued three main points at a meeting of the board in February 1965: first, that CAPD would be looking for another reactor sale within three years to keep itself busy; second, that competition between NPPD and another firm would be beneficial; and third, that CAPD would take less time than NPPD to develop the BLW reactor. No other firm could compete for it. He proposed to keep competition alive and lively, but Gray, Ross Strike, the Ontario Hydro chairman and AECL board member, and D.M. Stephens of Manitoba Hydro were unimpressed. CGE's services were important, even vital, but neither AECL nor Ontario Hydro would contemplate spending money on a project whose design 'rested with a single manufacturing organization.'[96]

Conditions would remain as they were, a situation which Haywood correctly diagnosed as inherently unstable. CGE was not standing still; under Ian MacKay's direction its engineers had produced a design for a 300 megawatt unit, and were preparing to work on a conceptual study for a 600 megawatt job. To organize and finance it, CGE turned to the government for help but not, this time, to AECL.

Instead it approached the Department of Trade and Commerce and the Department of Industry. Trade and Commerce was C.D. Howe's old department, and Churchill's, and it had once housed

Jack Davis and his energy economics project, but there the connection ceased. The department was no longer the same, having been split into two parts. Industry belonged to Drury and was run by his enterprising deputy, Simon Reisman. It had to encourage the development of industry, such as CGE. Trade and Commerce, which in 1965–6 fell under the direction of Bob Winters, another Howe disciple and a veteran of the St Laurent cabinet, was concerned with exporting industrial production. CGE had a project, indeed several projects, it believed it could export.

CGE had some success at the ministerial and departmental levels. As an NPPD executive enviously commented, 'they appear to maintain semi-permanent siege forces on the Departments of Trade and Commerce and Industry in Ottawa,' with the result that the bureaucracts had become 'thoroughly indoctrinated.' NPPD suggested that the Canadian Nuclear Association be cultivated as a means of diminishing CGE's domination of the private nuclear market and as a means of spreading business more equitably among Canadian private companies.[97]

CGE for its part seems to have argued that NPPD be wound up, and its responsibilities transferred to private industry. Gray, whose relations with his minister were at best uncertain, worried about Drury's reaction. 'I have been told,' Gray wrote to the minister, 'that you are sympathetic to these approaches and feel that AECL should be out of this business even at this stage.' That would be unwise, Gray continued, because it would put the Canadian utilities at the mercy of a single design-and-manufacturing concern, Ontario Hydro's old argument. Drury, to Gray's relief, 'agrees with our position,' the president wrote, and '*strongly resents* CGE position relative to our support not directly and definitely needed for AECL programmes.'[98]

That was not, however, the end. Some Trade and Commerce officials took an interest and got support. In July 1965 Trade and Commerce produced an 'Export Programme for Nuclear Reactors,' which bore traces of CGE's influence. The argument in the paper seemed to go beyond official policy, but was not basically objectionable in Gray's view.[99]

Next a Nuclear Reactor Export Promotion Committee was struck, most of whose members hailed from Trade and Commerce, with representation from the Departments of Industry and External Affairs, the Export Credits Insurance Corporation, and AECL. Gray was by this point thoroughly mistrustful, but not yet able to put his feelings into action; and so the committee proceeded. Together, the committee and Trade and Commerce got up to $2 million worth of funding for CGE's export promotion campaign; the rhythm of the committee's meetings was punctuated thereafter by requests for more.[100]

CGE was, in 1965–6, bidding for a Finnish contract. It had also devised a vertical natural-uranium reactor (Venture for short), and it wanted funding. AECL used the export promotion committee to express its lack of enthusiasm for Venture, but it did not prevail. CGE wanted a modest amount of money, and it got it.[101] It was a sign, however slight, that Gray's control over AECL's Ottawa environment was not complete.

The committee's agenda was familiar. Why not recruit 'an industrial partner' for the BLW? The NPD example showed the way, and establishing such a partnership should be a matter of 'first consideration.' Why not establish yet another committee? The resulting Nuclear Power Projects Committee spawned a series of working papers, investigated export markets, and called for a 'total' effort to overcome Canada's apparent handicaps in exports of nuclear technology. Meanwhile, the original committee generated its own report; CGE continued to bid for a reactor contract in Finland, and sought assurances that necessary supplies would be available.[102]

The draft report was ready at the end of September 1966. It had to pass scrutiny by the committee before it went on to the cabinet. It began with an elephantine statement that obscured its object behind qualifications and modifiers: 'Canadian efforts for commercial exploitation of the Canadian nuclear power generation technology show signs of being less successful than they could be.' There must be 'a much more precise definition' of 'national policy objectives,' as a prelude to action. In an age of task forces – the 1960s saw a proliferation of the breed – it called for another, to report on national objectives and 'to recommend steps for the effective application of the total Canadian resources in the nuclear power generation and associated equipment fields for the fuller exploitation of Canada's industrial development potential in this field.'[103]

It was not exactly a clarion call, but it had a real meaning. Trade and Commerce, or a faction inside it, wanted the Canadian nuclear establishment reorganized. The woolliness of its language probably reflected a belated effort to avoid an all-out confrontation with AECL, and with Gray, but in vain. It was, he told his executive committee on 21 November, 'an attack on AECL.' He had already told the Advisory Panel so; and he had expressed to Gordon Robertson, the clerk of the Privy Council, his outrage at the implication that AECL had not been properly advising the cabinet or senior officials of what it was doing, and how it was spending its time and money.[104]

The two committee reports were considered by the Advisory Panel on 10 November. Gordon Robertson, who chaired the panel, urged it to draw up the statement of national objectives by itself, and then to seek advice from 'outside specialists' on implementing them.

Gray then weighed in. AECL, he told the panel, knew of no 'significant dis-satisfaction in the industry,' certainly not of the kind that had manifested itself in the two committee reports the panel was considering. The industry was in fact healthy, and, although CGE needed foreign sales in order to maintain its existing 'turn-key' capacity, that was not by itself a measure of the viability of the Canadian nuclear sector. The amount of orders, presently $150 million a year, was enough to keep going, whether or not CGE's CAPD structure was preserved in its existing form. In reality there was no overcapacity, and the rivalry that Trade and Commerce discerned between CGE and AECL was, in terms of the national interest, the ultimate red herring.

The panel was not especially impressed. The economists, led by Reisman and Denis Harvey, an assistant deputy minister in Trade and Commerce, attacked the existing system as uncompetitive and therefore costly and inefficient. Why could not a company like CGE compete domestically? 'Full market forces' had not been allowed to operate in establishing Pickering; if they had been, 'total program cost and not just equipment production cost' would have been considered. Harvey next took up the theme.

Unless a company like CGE (or more precisely CGE, since there was no other) was allowed to sell its off-the-shelf or turn-key reactor product domestically, not enough was being done to assist. It was urgent that more be done, especially because the relatively few reactor sales anticipated for the late 1960s would turn into a flood in the 1970s and 1980s. Buyer preference should be established early if not immediately. As for national objectives, they should be easy to establish through a judicious combination of old panel minutes with the conclusions of the two contemporary committees.

The panel minutes show that Harvey and Reisman had masked Gray's position, but they had nevertheless not had their way. Robertson's response was to delay. The Privy Council Office was authorized to draw up a list of pertinent questions that would serve as a prolegomenon to consideration of a policy in the field of nuclear power. The questions were actually composed, but nothing was done with them. The Trade and Commerce offensive had petered out. Gray won time, keeping discussion of the issues squarely on an official level and away from the politicians. He did discuss the matter with his minister who would, he thought, take a reasonable view of it. The minister, Jean-Luc Pépin, one of the younger and more talented members of the Pearson cabinet, saw the point, and nothing more was heard on the issue. Gray recognized that some change had become inevitable, and at Haywood's urging he resigned himself to setting forth, in writing, AECL's commercial policy, for what it was worth.[105]

CGE chose this moment to throw in the sponge. It was not a spur-of-the-moment decision, according to Herb Smith, the company's chief executive officer. The approach to Finland was as much to keep up morale in CAPD as to get the contract. It was, in fact, a contract he hoped he would not get.

Without a domestic base, CGE could not rely on regular employment for its whole staff. Without a large staff, it could not hope to compete. But even had it succeeded in stymying AECL, there was little or no possibility of persuading Ontario Hydro to take on CGE as a supplier. Smith decided that the Pakistani reactor, now called KANUPP, would be its last full reactor project. He asked for terms from AECL, and the two companies worked out an arrangement whereby its employees could be kept together as a unit. It was agreed that those CGE employees not working on KANUPP would be set up as a division of NPPD, but located at Peterborough. Henceforth the company would take the role assigned it by AECL and Ontario Hydro – a contractor for components rather than a full-scale competitor for reactor business.[106]

Trade and Commerce was dismayed. Bob Winters, its minister, offered to keep trying for Finland, but CGE's reply was negative. It would keep building parts, and that was all. It was the end of the line for CAPD, but only the beginning for AECL. CAPD did not, of course, disappear all at once. Gray and Herb Smith continued to discuss keeping the CGE team together, and Gray even had it in mind that should Hydro-Québec 'get into nuclear power in a significant way' it should be with CGE and CGE's organic reactors. CGE's team of engineers was valuable, and Gray had done something to help out, by assigning contracts to them through NPPD.[107]

As for the Trade and Commerce–Industry study, that rested in Gray's hands, and it was Gray who put the finishing touches on it before it went, recommended, to Pépin for cabinet approval. We shall examine the results in the next chapter.[108]

As Gray a year later put it to his minister, by now Joe Greene, 'We are in effect the Canadian agency in the export field for nuclear power facilities. We are bidding on two sizes of plant for Rumania. We have a negotiation started with Czechoslovakia. We have active interests in New Zealand and Australia. Brazil and the Argentine are also good prospects.' And, Gray added in his own hand, 'Taiwan in today.' It was 17 July 1968; AECL was well and truly launched in the export business.[109]

CGE's last reactor, KANUPP, was not completed quite on schedule. The company had assistance from AECL in keeping its staff together, but by October 1969 Gray and Foster had to admit that there was little prospect of continuing to do so; the budget would simply not allow it. A small management and senior technical team, made up of former members of CAPD, would deal with KANUPP, and keep it as far as possible on schedule.[110] *Force majeure*, in the form of the Indo-Pakistani war of 1971, forced the evacuation of the Canadian engineering team as bombs and bullets whizzed by the power lines. But, allowing for the delay and damage caused by war, the delay was not excessive. The reactor began commercial operation in December 1972. It cost, on its fixed-price contract, $63 million. It is rumoured that CGE made a profit.[111]

Conclusion

The elimination of CGE as a principal player in the production of nuclear reactors terminated, finally, the opening to private enterprise that Bennett and Howe had attempted in 1955. The policy had always depended on the co-operation of the major customer, Ontario Hydro, and after 1957 Hydro made it abundantly clear that it was not prepared to co-operate. Under Gray, AECL made no attempt to get Hydro to change its mind; by the mid-1960s, AECL was institutionally committed to its own consulting engineering business and perceived itself as a competitor of CGE, despite Haywood's misgivings.[112]

Committed to a large engineering division, for essentially the same reasons that we have applied to CAPD, Gray also had to be committed to sales. AECL was a successful seller of isotopes and therapy units, which catered overwhelmingly to the export market, but there was no attempt to link Commercial Products' specialized business and sales experience to the promotion of the CANDU abroad. Instead, Gray took on the job and functioned after about 1962 as sales manager as well as chief executive officer. There was a reason: he headed a government-owned company and he could speak to other such beings around the world. But unlike Errington, he had little or no sales experience.[113]

The world was changing, and not to AECL's advantage. At the beginning of the 1960s several advanced countries maintained an interest in heavy water technology. The British, in spite of the ill feeling surrounding Hinton's war with the AEA, supported a heavy-water project that eventually turned into the SGHWR. The Americans built their own Carolina-Virginia heavy-water reactor. The Swedes took an interest, and the Swedes as well as the Americans sent representatives to NPPD. The Americans contributed to some of the costs of NPPD, and co-operated with Whiteshell on the organic reactor.

By the late 1960s most of the interest was gone. Sales to Germany, Italy, and France by Westinghouse, with enriched fuel guaranteed, were followed by Spain, Switzerland, and Japan. Sweden had gone so far as to build a heavy-water reactor, but even before it was completed Swedish interest had switched to PWRs; the Swedish heavy-water reactor, at Marviken, was never commissioned because of fears that it might prove unstable in operation.[114]

The key event was France's decision to abandon its domestic, gas-cooled model, and to opt for the Westinghouse PWR. France, like Britain, experienced an atomic civil war, with the Electricité de France on one side and the Commissariat à l'Energie Atomique (CEA) on the other. The CEA lost, and with it went the principal remaining supporter of a natural-uranium reactor outside Canada.[115]

American support also dried up. There were sporadic expressions of US interest, including a suggestion from a commission member that AECL go ahead and build a heavy-water model in the United States. But even in the United States, the supply of money to support a broad research program was not infinite. That was inherent in the decision to let Canada run heavy-water research while buying an interest in the results, if any, for the AEC. Even that limited commitment was soon to run out. Milton Shaw, who became director of Reactor Development for the AEC in 1964, discovered that the commission had been running no fewer than thirty reactor development projects. Shaw determined to concentrate his resources rather than spread them thinly across a series of rickety possibilities. His choice was a breeder reactor; in the housecleaning that followed, the co-operative program with Canada was ended.[116]

The countries of the 'First World,' the western democratic states, had by and large made their choice of reactor model by 1970. The 'Second World,' the Communist states, was naturally dominated by the Soviet Union's nuclear technology, even if some states, for reasons of their own, had other preferences. The Third World, the underdeveloped nations, remained. This had always been a field of preference for W.B. Lewis; by 1970 it perforce became one for AECL as a whole.

11

Big Science

The 1960s was a prosperous decade for Canada and Canadians. The gross national product in constant dollars almost doubled between 1961 ($39.6 billion) and 1972 ($71.7 billion). Growth was not evenly distributed, as between sectors of the economy and certainly not as between regions – though no region suffered an absolute decline. The service sector was growing, but the manufacturing sector was, after 1966, shrinking, at least in terms of jobs.

Energy consumption did not decline. Less electricity might be required for making steel or processing pulp and paper, but more was being used in homes and on farms, or for street-lighting. In Ontario, energy consumption took longer to double than had the GNP, but not by much: it doubled between 1958 and 1971, reaching 1835 trillion BTU in the latter year.[1]

Many of the other trends of the early 1960s continued. Canadian politics were preoccupied throughout the 1960s with the distribution of powers between the federal government and the provinces. Attention focused on Quebec, where a vigorous separatist movement called into question the province's continuing attachment to the rest of the country. The task of dealing with 'the Quebec question' fell to the Liberal party which, under L.B. Pearson and then Pierre Elliott Trudeau, formed the national government from 1963 until 1979.

Under Pearson, and even more under Trudeau, the government pursued the objective of making its apparatus more responsive to political priorities. The civil service was the most profoundly affected; where government departments or ministries had been remarkably stable for decades, new ones suddenly sprang up. Industry was recombined with the Department of Trade and Commerce to become the awful acronym 'IT&C' before submerging in an expanded and reconstituted Department of External Affairs. The former Mines and Technical Surveys expanded and acquired a modern definition to become Energy, Mines and Resources; when it was established in 1966, AECL migrated to become part of its minister's flock of extra-departmental responsibilities.

The first minister of Energy, Mines and Resources was Jean-Luc Pépin, from 1966 until July 1968 when he was promoted; he was succeeded by the amiable Joe Greene, who served from 1968 to 1972. It should perhaps be stressed that Energy, Mines and Resources (otherwise EM&R) had emerged from a very *junior* department, Mines and Technical Surveys, whose ministers had generally been nonentities. A rise in public concern over energy and energy supply brought the department some way up; Greene had been a candidate for the Liberal party leadership in 1968, and was recognized as a fine speaker and an accomplished politician, even if not of the first rank. This rise in departmental status appeared to be confirmed when, in 1970, Trudeau appointed Jack Austin to be its deputy minister – Austin being recognized as a man of the future, and possibly the near future, in Ottawa.

Of the three, Pépin was the most active in and around AECL, a relief for Gray, after Drury. Though Greene was a good politician, administration was not his long suit. Austin was kept at arm's length. Gray hewed to the old principle that the company reported to a minister and not a department; as far as AECL was concerned, Austin would stay strictly on the sidelines.[2]

The Trudeau government enjoyed a stable parliamentary majority in its first term, from 1968 to 1972, and Canada received at his government's hands an unexceptionable period of administration. There were government surpluses and no acute financial crisis, but there was the appearance of crisis and a tight rein on expenditure in the name of 'austerity.' A foreign exchange panic in 1968 led to a run on the dollar, and a firm promise by the Pearson government, then in its last days, that a very firm lid would be kept on expenditures. This was a ritual confidence-building exercise, but it was real enough for all that, and it meant that exceptions and exemptions would be few. It was a particular burden on those federal agencies with pretensions to a capital budget, because much if not most of the federal budget was mortgaged either to social welfare programs or to provincial budgets – transfer programs that could be touched only with the utmost difficulty and political risk. For AECL, whose capital expenditures in 1967–8 touched $12.6 million, and whose operating expenses were to rise between 1968 and 1969 by $6 million, this had a clear meaning. It was not a meaning that everyone saw.[3]

ING

ING is the abbreviation of intense neutron generator. ING was described in the 1967–68 *Annual Report* as 'a 65-megawatt proton beam, bombarding a liquid lead-bismuth target [in order to] produce neutrons by spallation reactions.' The two key components were 'a highly efficient accelerator' to generate the proton beam and 'a liquid metal target' that could withstand bombardment by the beam. There was a third component, unmentioned but crucial. ING needed $4 or $5 million to start, $150 million in capital investment, and $20 or $21 million in annual operating expenses. It was, as the saying goes, 'big science.'[4]

ING had its origins in W.B. Lewis's concern for the present status and future direction of the scientific program (and particularly nuclear physics) at Chalk River. Nuclear physics was well launched, with NRX and NRU, and with several smaller research reactors built since, such as ZED, a zero-energy facility. Chalk River was well established as an international centre for research. But its edge depended on other factors – the availability of up-to-date equipment, for example, or of abundant funds to keep it running; equally important was encouragement to the staff to keep the establishment ticking over at a respectable rate of scholarly production, and continuity of recruitment.

Organization played a part; so did personality. Chalk River was hierarchical but it permitted independence. There was not as much differentiation in research as at more lavishly funded American laboratories, or at Harwell, nor were scientists allowed to forget that Lewis was the responsible agent for the laboratory. That was oppressive to some, but almost all the staff recognized that Lewis embodied their international status. In Lewis they had star quality, and stars were forgiven their quirks and occasional outbursts. Certainly no other member of the laboratory staff had Lewis's standing, a reminder that his stature derived from much more than his title as vice-president, research.[5]

Chalk River recruited the best available between 1946 and 1960, or so its officers believed. In the 1960s they had an uneasy feeling that this was no longer so, and that some of the staff were casting their eyes and their ambitions elsewhere. Chalk River in the late 1960s was estimated to be exporting annually twenty-three man-years of professional experience to Canadian industry, and eleven to Ontario Hydro. It was well known that some sixty former AECL staff members had left for Canadian university appointments (some of those departing were very senior, witness the annual export figure in man-years: thirty-nine), and while that was a matter of pride it might be something else too.[6]

Gray and Lewis accepted that Chalk River had its limits. Harwell showed what would happen if a laboratory over-expanded, and Whiteshell had been founded on that understanding. From that point, however, their conceptions diverged. Gray was obliged to look at a wider world. He took pride in what he called 'the good reputation enjoyed by the Company at Treasury Board.' It was an asset, not to be squandered. He was aware that recruitment of

ACCELERATORS

A generic term for machines designed to impart high kinetic energy to ionized particles. Cockcroft used an accelerator at Cambridge in 1932 to produce nuclear transmutations. Accelerators of one form or another have always occupied a high position on Chalk River's shopping list and in its actual inventory. Their most recent manifestation is the Tandem Accelerator Superconducting Cyclotron, which accelerates particles to the highest energy ever achieved in Canada. W.B. Lewis's Intense Neutron Generator (ING) project of the mid-1960s was a variety of accelerator, albeit one considered at that time to be very expensive and speculative. Accelerators also travel under the names of cyclotrons, betatrons, and linear accelerators.

junior personnel was becoming more difficult, and he encouraged Whiteshell to develop strong links with western universities.[7]

Lewis was not indifferent to universities, or to university liaison, but he took a less political, and perhaps less sensitive, view of the matter than Gray. When high-energy physics attracted Lewis's attention between 1962 and 1964, it carried a price tag. A tandem accelerator, for example, would cost $5 million, and that was only the beginning. Lewis saw the need for major expenditure and large-scale growth.[8] It was a price that Gray was reluctant to pay. It would make Chalk River too large, he told his senior managers. It would cost between $60 and $100 million, and such a sum could not be obtained without the support of NRC and the universities. If it made Chalk River too large, then the obvious response was to retire some item that had become obsolescent – NRU, for example.

The question, Lewis replied, was what Chalk River would be doing ten years from 1964. Even an economic objective, narrowly construed, would require more high-energy physics, since he was certain that the price of fuel would rise and that more fissile material would be required. He assumed that Chalk River should continue to have a 'pioneering role in scientific exploration and technological progress,' and assumed also that such a role would be related to what already existed. In that area NRU, the high-flux reactor, would soon be surpassed by reactors in other countries, and it was not practical to improve NRU to keep it abreast of or superior to the competition. That left a fairly large field, in which seven specific areas were singled out for special study. They were thermonuclear fusion, direct conversion of heat to electrical energy, magnetohydrodynamics, possible developments in the long-range transmission of electrical energy, fuel cells, nuclear fission, and an intense neutron generator.[9]

It was high flux that turned attention to an intense neutron generator, called ING. The idea of generating neutrons through the bombardment by protons at an energy level of 1 GEV of an appropriate element, such as molten bismuth-lead, was not new, either in Chalk River or at other laboratories. (The lead-bismuth was to be surrounded by heavy water.) To do it would require a mile-long linear accelerator operating at 1 GEV; that was, Lewis pointed out, tiny compared to the accelerator the Russians were building, which would reach 70 GEV. But it would be important, and it was something that nobody else was then doing.[10]

We should pause here to stress one aspect of Lewis's thought: the idea that expenditure on 'the growth of advanced technology' would enrich the Canadian economy. He believed that up to 1964 the object had been attained so far as AECL was concerned. Gray in 1960 had suggested that an annual growth in atomic R&D of 8 per cent, as in Britain, was too high. Lewis differed. Scanning the figures, he concluded that it had been achieved with an annual 'expansion of 12% ... and this rate may be used as a guide.' It was this expenditure, among other things, that had established Chalk River's stature as a world-class nuclear physics laboratory.[11]

The origins of the ING proposal did not involve consultation outside AECL. It was, of course, a matter of some pride that AECL could do its own work entirely by itself; it saved time and energy, especially for a man like Lewis who liked to establish a clear view of his objective and then take the shortest possible distance to reach it. His objective, it must be stressed, was not merely ING: it was the preservation and enhancement of the morale of his staff by giving them an important, first-line project. By this token, Lewis's rationalization of his project did not, and need not, involve a detour to other scientific facilities.

During the war and after, this was entirely possible, given the decision by the politicians to accord atomic energy a high strategic and then national priority. By the 1960s, however, it was no longer so easy, because alternative research centres had reasserted themselves, or had been for the first time established elsewhere in the country. As the experience of nuclear power (and especially heavy water) showed, Canadian politics in the 1960s took regional or local desires into account; centralized structures, whether governmental or private, reacted accordingly.

That this was so had occurred to Lorne Gray, who on this and other issues acted as a lightning-rod to protect the company from the more bizarre consequences of Canada's political realities. In the spring of 1964, just before Lewis's proposal reached the Board of Directors, Gray took a tour of western Canadian universities to discover how their research programs could be co-ordinated with, or integrated into, the program at Whiteshell. At the same time A.R. Gordon, dean of graduate studies at the University of

Toronto as well as an executive committee member, approached his own university. His report, and Gray's, encouraged the executive committee in favour of greater interchange with the universities, including the welcoming of academic visitors who wished to use the company's facilities while on leave. It was a program analogous to the one AECL had instituted a decade before for engineers wishing to familiarize themselves with nuclear power.[12]

Another factor should be borne in mind. In the years after the war many aspects of Canadian cultural life, including science, had been increasingly 'nationalized.' This was obvious from the example of Chalk River alone, which served as Canada's national centre for nuclear research on the basis of annual parliamentary grants. But other centres had been encouraged too, with grants either from their provincial governments or from Ottawa. Realism suggested that there was only so much money that Canadian governments would give to research; yet science was unendingly expensive. Nor were science departments across the country less inclined than the staff at Chalk River to want new and expensive equipment. There was a sense that money spent on Chalk River was not simply money spent on science; rather, it was money *not* spent in one university or another. There were many universities, and there was only one Chalk River. The universities had learned to make a national cause of their separate interests, where money and grants were concerned.

Research and Development, R&D, was one of the primary issues of the day, and the Pearson government, with its ear attuned to considerations of what was modern, was eager to respond. A Science Council was in the offing, to advise the government on quirks and quarks, just like the Economic Council of Canada on graphs and curves. In the white heat of modern technology, to use a contemporary term, Chalk River's aloofness, its pre-eminence, could be turned to negative account. ING offered the very best opportunity.

The board approached Lewis's project with some trepidation. Lewis had great authority and prestige. He was AECL's scientific figurehead, respected for his prophetic talents. It seemed little enough to allow him to redirect some of Chalk River's energies towards ING, especially if, in its first stages, ING cost relatively little.

Later on, it was true, Lewis predicted costs of between $4 and $10 million a year, and such sums would weigh heavily on the budget. When that happened it would be necessary to seek cabinet sanction.[13]

From its inception ING caused controversy. At first it was within AECL. Lewis saw ING as part of his forest realm at Chalk River; Gray preferred to place it elsewhere, possibly as part of his co-operative university program at Whiteshell. Both Gray and the board saw difficulties in asking the government, through the Treasury Board, for more money, a consideration that stimulated Lewis to address a general paper on the financing of science to them. It is of interest as an illustration of Lewis's views on the subject, and an indication where a commitment to science, once made, may lead.

Lewis took as his starting-point the existence of Chalk River and Whiteshell, 'a teenager' and 'a baby' respectively. They could not be abandoned or starved any more than a family could budget its children out of existence. 'The Canadian government has brought them into being and is committed to their nurture.' But they were not just part of the Canadian government family; they were the Canadian entry in the atomic energy international league, 'a big league business in which we are still very small.' The government could easily, and should, commit an extra 10 per cent or 20 per cent to the 0.8 per cent of its budget consumed by AECL. Lewis's enthusiasm for a bigger atomic budget derailed his syntax: 'that would not break the country, so even a large increase should not be rejected out of hand but rather our proposed estimates should be considered on their merits.'

Those merits included, as the board already knew, outclassing the new American and Russian high-flux reactors. ING would therefore be able to lower 'the cost of production of transuranic isotopes that have their chief application in space technology, as well as accelerated tests on materials and more refined studies of the vibrations and other characteristics of solids and liquids.' That was, perhaps, an ultimate goal, but en route Canadian industry would develop a unique competence in advanced technology. Lewis did not confine himself to the ING project; all the divisions at Chalk River had worthy projects, and he commended them all. Their total real worth he estimated at $1 million more for the next fiscal year than Gray had projected; and more thereafter.[14]

The board was impressed enough to recommend an increase in the projected estimates, to $55.5 million. This was $1 million more than Gray's figure, and $2.5 million more than the executive committee thought the Treasury Board would accept. (The *Annual Report* for 1965–6 showed expenditures on research and capital programs of $60 million and total income of about $14 million; I am open to suggestions as to how these figures may be reconciled.)[15]

Gray had the duty of finding a way of carrying his revised estimate forward. To Drury he argued that both the boiling-water project and ING would suffer if the estimates were reduced below $55.5 million. They were, he wrote, 'the type of programme that keeps an organization such as ours alive.' Without them morale would droop. Drury apparently acquiesced; there were, in a year when heavy-water plants were first on the agenda, more expensive things to worry about.[16]

ING survived. It moved on to the next stage, a symposium on the subject, held at Chalk River in 1965. The symposium was the first public sign of what was in Lewis's mind, and it produced two different sets of results. To Lewis it meant that ING was being ratified by scientific opinion in open congress; to rival scientists and engineers it meant that AECL would be ravaging the national treasury of funds that might be spent elsewhere and, by definition, better.[17]

Lewis proceeded. Gray suggested involving industry at an early stage; if industry could see what kind of contracts would be available, support for ING would grow. Lewis established an ING Study Advisory Committee to work up interest in the universities. It would meet seven times between November 1965 and February 1967; it was thought in official circles to have had a beneficial influence, though one that was possibly not entirely to Lewis's taste. Its members suggested, among other things, that Chalk River was not the place for such a facility – too far, too remote, and, for francophones, too unilingually English.

The location was left unspecified, as were other considerations such as organization and accessibility. The result was unsettling as far as the scientific and engineering communities were concerned, since it seemed that this was just another sham consultative body of a kind that university experience had presumably sensitized them to. On the positive side there was consideration given to making ING a true national laboratory of the kind that had taken root and flourished in the United States – Brookhaven, for instance, which though a 'national' laboratory was really run by a consortium of universities.[18]

1965 passed into 1966, and Lewis carefully tended ING. Its next hurdle came in the summer of 1966, when it was sufficiently mature to be presented to the board and executive committee. Maturity meant that its costs had been worked out in greater detail, and inevitably these costs were higher than the first estimates. ING would cost $150 million in capital investment, plus $20 million per annum in operating expenses. It would not all have to be spent at once, but no matter how the subject was presented it was evident that it would cost double the company's annual budget, and then some.

The executive committee and the president were in a dilemma. They believed, so they told Lewis, that 'ING would be a valuable facility.' That was 'common ground.' But how valuable would it be when it was compared to other projects? That required further thought and, inevitably, a presentation from their scientific vice-president. The committee had met in July; it would make ING 'virtually the sole subject of consideration at the next meeting.'[19]

That was not quite true. The next meeting, on 12 August, had to deal with heavy water, as every meeting did, but once that was out of the way it gave its attention to Lewis. Gray, and possibly the other committee members, knew that ING was already under assault from the Treasury Board, whose secretary had written Gray on 8 August to say that he could not 'see how the Government can agree to starting such a multi-million dollar project in 1967–68 and therefore think that this project should be removed from your estimates.' AECL's budget was already growing by quite enough. Gray would do his best to defend ING, but his response to the Treasury Board, sent on 19 August, showed that he would prefer to defend Whiteshell and research on isotopes before he launched himself into the lists for ING.[20]

The executive committee had, therefore, other pressing concerns than ING when it met to hear Lewis. The committee had grown from the three-man body of earlier days: it had Gray, David Golden, a former deputy minister under Howe and Drury now working in the private sector, Dick Hearn, Harry Thode, a veteran of the wartime atomic project and latterly president of McMaster University, and Brigadier F.C. Wallace, another Howe-vintage executive. Engineers predominated, but engineers whose personal experience had illuminated the value of even far-fetched science to Canada's industrial and intellectual capacity. The two were closely linked, they knew, but the linkage had a price. Could Lewis make the connection?

Lewis started the day with a two-and-a-half-hour presentation by himself and his senior staff. They gave the highlights, emphasizing the positive side, and ended by suggesting that the project could be carried for two more years inside the company before the next stage of decision was necessary. The executive committee drew its breath. In the minutes it recorded its appreciation of 'an exciting and imaginative project.' But it was evident that Lewis's imagination had overreached itself. It was very doubtful, in the executive committee's view, that the federal government would pay for ING, and it would be improper to continue such a project without first seeking the government's consent.

The minister already knew about ING; so did his department. His department thus far was not especially impressed by the 'industrial fall-out' – an unfortunate term – that would result from ING. The civil servants' lack of enthusiasm was matched only by alarm and resentment in the universities, which felt that they had as strong a claim on basic research funding as AECL. Stronger, in fact, some of them argued, because they had graduate students to train; those students might well leave the country unless graduate schools got some of the funding that AECL was proposing to have granted to itself. If the government was going to take the blame for ING, it wanted to know very clearly what it was getting into. Drury wanted a full-dress presentation to as many ministers and senior civil servants as could be assembled for the occasion; once again Lewis was to do the presenting.[21]

This new presentation took place before the end of August, and by the time the executive committee met again, on 12 September, the critical notices were in. It had been a disaster, what the minutes termed 'disappointing. Some of the speakers tended to be too long and the overall presentation [too] highly technical.' R.B. Bryce, the longest-serving bureaucrat in the room and himself an engineer with an interest in atomic energy, finally told Lewis that he had better cut short his speech, and translate it into layman's language. The two cultures, scientific and bureaucratic, had met, and discovered that they could not understand one another.

Although Gray hoped to arrange a return engagement, the effect of Lewis's performance determined the clerk of the Privy Council, Gordon Robertson, to move him to another theatre. The Pearson government had created a Science Council to advise on just such subjects; let Lewis go before it and see whether he could do better. Gray told the executive committee that ING might yet survive, since despite Lewis there was still a great deal of sympathy for 'big science' projects among the senior civil service. If it did it should not be in the form of an attachment to Chalk River. It should be an independent institute that AECL should encourage, help design, and even build. It should not, however, belong to AECL. That, in Gray's opinion, was the maximum that scientific opinion outside AECL could now be brought to accept.[22]

Variations were still possible. The Americans were taking an interest in ING, the board was told in October, and would consider a joint project. The suggestion attracted very little support, with both Lewis and Haywood opposed to losing control of a national enterprise. It was left to Gray, one more time, to tell Lewis and the board that there might never be enough money to fund this particular example of 'big science.' But ING went merrily on.[23]

For Lewis, life now consisted of one meeting after another, gatherings essentially political in nature, in which he had to beg the support of his audience. November's meeting was with the deans of engineering of Canada's universities. Gray sent Donald Watson, the company secretary, to ride herd on Lewis. Although Lewis felt that the meeting 'had gone well,' Watson differed. There was still 'a hard core of opposition' – all but four deans. The four had to dissociate themselves from the condemnation of ING that issued from their colleagues.

Gray's view of ING was perceptibly hardening. He would not offend Lewis and was prepared to utter the project's praises in due form, but his qualifications were increasingly severe. It would be difficult to build and while it might be 'state of the art,' who knew really what that was? Nor should ING be built at Chalk River. It would add 1000 people to the workforce and 3000 to Deep River, and that was more than the existing plant could bear. ING should, instead, go to a university, preferably one with a major airport close by and, of course, it should be an independent institute and not part of AECL.[24]

The Science Council, when it had considered the matter, did little more than repeat Gray's concerns. ING got a benevolent wave, subject as always to budgetary priorities – whatever those might be. It noted in passing that AECL had already spent $10 million on ING; Lewis disputed the figure. As Gray suggested, its approval was subject to ING's practicality. ING should have an independent institute, and it should be close to a university. The similarity to Gray's view may have been entirely coincidental, but it may have owed something to the simultaneous presence of Harry Thode on both the Science Council and the AECL Board of Directors.[25]

The Science Council pronounced in March 1967. ING returned to the political agenda with every sign of remaining there for the very indefinite future. Lewis grumbled that the Science Council's exclusion of Chalk River was a mistake – that the whole idea had been to revitalize AECL's science staff because 'the glamour that had been associated in the early days with atomic energy had been dissipated.' Now the company was prepared to squander its research future elsewhere. By the time he spoke, Chalk River had been reorganized out from under him; a management firm had recommended to Gray, who had accepted the recommendation, that AECL should be parcelled out in more formal divisions, with a vice-president responsible for each. In the allocation, Les Haywood, who had been vice-president engineering, was appointed to be vice-president in charge of Chalk River; Lewis retained a general responsibility as vice-president, science.[26]

Lewis's concern for ING did not abate. At the end of December 1967 he wrote to Harry Thode of the executive committee that he was certain that the climate of opinion would change in the next few months. Surely the opposition would not deny that Chalk River was the only place in the country that could actually manage a project like ING. Even the hostility of the Engineering Institute of Canada would surely be seen as absurd, amounting as it did, in his view, to the implication 'that Canadians should merely buy scientific journals and leave the actual work to other countries.'[27]

It took almost another year, a new government, and a federal election finally to resolve the ING question. The project was kept alive with small injections of capital: $270,000 in January 1968 alone. While keeping hope alive, Gray steadfastly refused to do more. If the government in its wisdom decided it had money to spend on ING, then AECL would do what was necessary. And when, in September 1968, the cabinet celebrated Canada's latest austerity program by finally cancelling the project, Gray did not protest; instead he tried to salvage something – 'general accelerator research' – from the wreckage. Lewis, who wanted nothing less than the fullest support, tendered his resignation, which was refused.[28]

There the matter ended. Opinions still differ on every aspect of ING. Was it feasible? Was it original? Was it worthwhile? Was it, above all, worth expending so much of Chalk River's political capital only to discover, in the event, that the laboratory's support among the scientific community was insufficient? To Lorne Gray the answer to some of these questions was plainly no; to others, the answer to all of them was negative. ING nonetheless remains a great romantic concept, a great project whose scale evokes admiration for the grandeur of its conception. W.B. Lewis wanted two great monuments in his own lifetime, Chalk River and ING; in the event he got only one.

Atomic Supermarket

ING was not AECL's only worry. While Lewis and his scientists contemplated Chalk River's future, Gray had to be concerned about the increasingly depressing supply of heavy water. This was a question that was growing more, not less, acute, because of the decision of Ontario Hydro to commit to two further CANDU units for Pickering in 1966, with all the supply problems that that posed.

Having fought off Trade and Commerce's incursion into reactor development and sales, he and AECL still had to deal with the question of the rhythm of Canadian development, given the necessity of keeping together a viable Canadian team while juggling their availability for both the domestic and the export market. At the same time, Gray and the board had to come to terms with the fact that Douglas Point, their one full-scale power reactor, was simply not functioning well.

It is possible to establish what Gray and his advisers thought through a memorandum they drew up for their minister, Jean-Luc Pépin, in May 1968. It was founded on an assessment of the competitive nuclear environment. The heavy-water technology needed access to the international market, lest its American light-water competition become too strongly entrenched. Unfortunately, and through no particular fault of its own, CGE's attempt to become Canada's international standard-bearer had failed because it had no domestic base.

AECL had such a base, but that did not mean that it could afford to pass up foreign sales. It had a strong engineering division in NPPD, which was essential for the design of domestic reactors; should AECL disband its staff, Ontario Hydro would set up its own, or so it said. To keep NPPD in being at a proper level of staffing efficiency, it was necessary in turn to counteract the cycles of domestic reactor design. But these cycles were outside AECL's control. A five-year period offered the best planning interval for staffing and investment. 'The only way, then,' Gray concluded, 'that a full workload could be provided over the five-year period and beyond would be if Canada were to gain orders for nuclear power stations abroad at reasonably regular intervals.'

Gray listed twelve countries, from Argentina to Turkey, where there was hope for sales. He admitted that 'in a number of cases' – Mexico, for example – they would require financing. This could be done by guaranteed export credits. It could be anticipated that a heavy-water customer would also demand an assurance of heavy-water supply at a fixed price. The continuing unavailability of Canadian-made heavy water made this guarantee an onerous one, since the price had to be fixed below what the Americans were willing to charge for the substance. Essentially, Canada had to offer terms comparable to those of other suppliers, bearing in mind that the competition was very well financed. Gray wanted authority to carry on 'a vigorous nuclear power plant export sales effort,' and, as a consequence, to establish his own small sales force.[29]

Gray had always been his own principal salesman. He had sales efforts ongoing with France, Taiwan (then recognized by Canada as the Republic of China), Czechoslovakia, and Romania; at the same time Canatom, a nuclear consulting conglomerate with which CGE was associated, was bidding to supply a reactor to Turkey. This latter initiative Gray discouraged, on the grounds that if any of the other contracts came through there would be too few Canadian nuclear engineers to meet the demand. As usual, there was friction between AECL and the Department of External Affairs, with AECL charging the diplomats with foot-dragging and the diplomats reciprocating with accusations of irresponsible haste. But as Gray travelled around, he had more and more to deal with another problem: Douglas Point. For the moment, Douglas Point was where the international and domestic concerns of the company converged, and the news from Douglas Point was mostly bad.[30]

The bad news reverberated throughout AECL, and beyond. By January 1968 the president was consulting the board on ways and means of countering bad publicity. It was a matter of concern, and debate, whether Douglas Point's record of malfunctions, outages, and losses could account for the apparent lack of success of the heavy-water design elsewhere in the world. AECL officials on tour abroad found that quite a bit was known about CANDU. Reporting on a European visit in the fall of 1967, Gib James reported that his interlocutors 'all expressed sympathy at the difficulties being experienced but generally they were not at all impressed by the complexities of the Candu system. They left no doubt in my mind that they felt that Candu thus far had not contributed much to the confidence in D_2O reactors.' The Europeans did, however, express the hope that 'things would drastically improve with time.' This is what Lorne McConnell of Ontario Hydro summed up as over-design. Or, as John Foster delicately expressed it, Douglas Point was marvellous, when it worked.[31]

In most places where reactors were being built, they were made to American designs and touted by American salesmen from the two great US electrical conglomerates, GE and Westinghouse. Once the Americans had swept by, there was very little left for the Canadians; nor was morale improved by the abandonment of official American interest in heavy water and the decision by the Swedes to go for a light-water design instead of the previously favoured heavy-water model. To top it all off, Ontario Hydro was becoming distinctly jittery. The knowledge that AECL had guaranteed heavy water to Pakistan when it was certain that there would be a serious shortage at home shook the faith of some of the Ontario utility's higher management. It was small consolation that AECL was now – only now, in 1968–9 – setting the wheels in motion for a third heavy-water plant at Douglas Point and even considering doubling its capacity before it was built. The heavy-water plant, after all, was premised on the construction of a 3000 megawatt (electric) station alongside the existing Douglas Point reactor, a painful reminder that that reactor was itself not the best advertisement for its genre.[32]

Douglas Point had had problems of another kind. AECL as its owner and promoter had an interest in seeing the plant establish the best conceivable record – the maximum time on, at full power, that could be managed, with the minimum number of incidents or slippages. Ontario Hydro was naturally all in favour of these objectives, but its enthusiasm was tempered by the necessity to train crews for the soon-to-be-commissioned Pickering reactors. Thus, as far as manning policy was concerned, AECL's interest was on the side of stability and continuity, Hydro's on that of variety and exchange. AECL had the added incentive that every hour that the reactor was down was money lost to the company – $1.3 million between April and August 1968 alone. By May 1969, after listening to another worried report from Foster, Gray directed that Hydro must operate the plant to meet the concerns of its owner, by keeping it on-line.[33]

This was a minor but not insignificant irritant in relations between AECL and Ontario Hydro. It should not be exaggerated; Harold Smith, as Hydro's chief engineer, and his staff continued fully to support Ontario's nuclear option. But George Gathercole, the Hydro chairman, could not be quite so committed. He had Pickering already far advanced: four units of 500 megawatt each

(Pickering A), with the first due for commissioning in November 1970. The next project, four 750 megawatt reactors at Douglas Point (renamed Bruce, after the county), had been authorized but was still at a very early stage. He warned the AECL board in February 1969 that he was having second thoughts about Bruce and even about what he called 'the viability of Ontario Hydro's nuclear power program'; this, he said, was because of 'the shortage of heavy water.'[34]

It was in this context that Foster was asked to appear before the board to defend not merely Douglas Point but also the validity of the CANDU. He remained convinced, he told the directors, that the heavy-water-moderated reactor remained the right choice for AECL and for Canada – an indication that the directors wished to be reassured on the subject. That said, he admitted to the several manifest shortcomings of Douglas Point. There had been problems with the fuelling machine, the most recent being its seizing up while attached to a fuel channel. No one, fortunately, had been over-exposed to radiation as a result. Rotor blades in the turbine had cracked, and the turbine would require a major overhaul in the relatively near future. Gib James, from Chalk River, criticized the lack of 'tightness' at Douglas Point, and praised the greater care taken in Europe with conserving heavy water. Above all, however, there was a shortage of trained personnel, not just on the operating side but among specialist consultants; the trouble was that these consultants had been shifted to Pickering and more recently to the Bruce project, with the result that Douglas Point could only be properly serviced by the constant shuffling of crews.

If that was all, then the matter could surely be remedied, although doing so, in the board's view, had become a matter of 'the highest priority in the Company's program.' It was recognized, presumably by others than just Gathercole, that if it were not, 'Ontario Hydro's program would be in serious jeopardy.' Gray, however, remained 'optimistic,' or so the minutes stated.[35]

At the next board meeting the directors pondered the other half of Ontario Hydro's problem – the supply of heavy water for Pickering. That, they resolved, would have to be met somehow, even if it meant shutting down Whiteshell, NRX, NRU, and Gentilly.[36]

The heavy-water controversy dragged on. The realization that Pickering units three and four could not be commissioned even if every ounce of deuterium in Canada were shipped to the station gave these concerns some urgency. Understandably, Hydro's concerns did not diminish. Inflation, costs, and interest rates were slowly rising, increasing the uncertainties of finance and driving Hydro's projections higher than the utility found comfortable.[37]

The result was a lingering uncertainty, a doubt that CANDU as presently embodied in Douglas Point or Pickering would be or should be AECL's last word in designing reactors. The focus of attention in the Power Reactor Development Program Evaluation Committee (PRDPEC) was more and more the BLW design for Gentilly, although the organic reactor continued to have its fans. Ara Mooradian, from Whiteshell, was active for the defence. His ultimate argument, however, was that 'if enrichment were permitted, a different plant would be designed and the result would be much better.' His observation could have been extrapolated to other varieties of heavy-water reactor, and it was. His was not the last word on the subject; two years later, in 1970, Gray mentioned to his executive committee that his senior staff were anxiously debating whether some enrichment of the reactor fuel might not after all be advisable 'under Canadian electric utility conditions.'[38]

At Ontario Hydro the situation became a grim joke. 'We don't have to retire,' the nuclear engineers told one another, 'we'll all be fired if this thing doesn't work.' To defend against it, they tried to incorporate more of the lessons of NPD than of Douglas Point, as we have seen; to the outside world they said little or nothing and hoped that Hydro's ultimate management would take their assurance on faith, that the lessons had already been well and truly learned.[39]

Nerves frayed, but Gray was successful in scouring the globe for any stray deuterium. Meanwhile debate raged over whether it would be better to scrap Glace Bay entirely, or to attempt to rebuild it. Port Hawkesbury, when completed, proved to have its own teething troubles ('maturation,' according to AECL's official handouts); research continued on whether the existing process could be improved upon, and whether such an improvement would be in time to help with Bruce.[40]

There were, moreover, problems with the Bruce heavy-water plant but, compared to Glace Bay and even Port Hawkesbury, they were slight. AECL gave the contract to Lummus, an American-owned firm that had built Port Hawkesbury for CGE. Using the Port Hawkesbury experience as a guide, and relying on information supplied by the AEC's contractor, du Pont, as well as hands-on experience at Savannah River, Lummus got Bruce heavy-water completed in 1972. Steam was diverted from the Douglas Point reactor to the heavy-water plant. Its capacity of 800 tons of heavy water a year got Ontario Hydro over the hump and supplied an abundance of deuterium for any conceivable Canadian reactor, probably for the balance of the century if not longer. (That, plus the eventual capacity of the two Nova Scotia factories, would give Canada about 1600 tons a year – if they ever worked simultaneously.)

Among all the public and governmental references to the subject, Bud Drury, by then president of the Treasury Board, was alone in predicting that there would soon be a surplus of heavy water rather than a shortage; the only problem was how 'soon' was defined. Fortunately, Bruce did work, and because it worked it relieved pressure on supply, definitively. Though built at AECL's expense and owned by the federal company, it would be sold to Ontario Hydro in 1973.[41]

That, however, is to anticipate. What AECL and Ontario Hydro knew that they had, as of the winter of 1969–70, was bills, an interminable problem with heavy-water shortages, and a decline in their standing in public opinion. The bills were getting heavier. Units one and two at Pickering had been estimated in 1966 at $285,650 million; in 1968, that rose to $320 million; and as of October 1969 the estimate was $383 million. Units three and four had been rising at a similar rate. This was not necessarily a strange experience, as every reactor from NRX forward had demonstrated to some degree; but the magnitude of the sums gave pause.[42]

Heavy water stimulated the creation of yet another official committee, this time between senior officials of AECL and Ontario Hydro. But while heavy water was the occasion, it would not be the only item to be discussed by the committee. By the time it was established, Pickering One was on the verge of becoming operational, and Hydro's fit of nerves had passed.[43]

Public criticism was the remaining problem. It was especially frequent that winter. One prominent theme was that the Americans had done, and were doing, better, a phenomenon that was ascribed by some to good old American know-how as embodied, for example, in CGE. To Harold Smith, it was the old story of a Canadian inferiority complex vis-à-vis the richer and more successful southern neighbour. In a letter to a critic in February 1970 Smith suggested that 'if the current trend of criticism and opposition in our country continues and reaches into many more high places, we shall, as my son would say, "blow it." If we do,' he added, 'history will record that technics and economics were not the cause, but rather the Canadians' readiness to undertake a major scientific and engineering development without recognizing and dealing with their own shortcomings, and to their lack of guts to see it through when the going got tough which, by definition, was bound to happen.'[44]

It was not, in Smith's opinion, a matter of Canadian inferiority, but a question of scale. The Americans were ten times bigger and had a five-year head-start on the Canadians in the power reactor business. There were reasonable explanations for Douglas Point's difficulties, explanations that had to do with Canada's relative size but not with Canada's immutable inferiority. It would remain immutable, he argued, only if the Canadians refused even to try and insisted on placing their faith in American infallibility.

It was certainly true that Americans, and other foreigners too, were more succesful when it came to overseas sales. Canada lacked a consistent national sales policy, not enough resources had been devoted to a sales effort, and Canadians themselves were skilled at proclaiming their lack of confidence in their own reactor system. 'We seem to have a facility,' Smith wrote, 'of ensuring that critical and doubting statements by influential political persons, financial commentators and editors, who have no competence whatever in the subject, are reported to our potential customers overseas just prior to a sales meeting. Compare this with the difficulty of finding out anything about major blunders by our competitors even two or three years after the fact and you have a picture of the difference in approach.'

Eventually, Smith hoped, there would be external sales. If the sales could not be reactors, because of sales difficulties, then they might as well be of power. Build it in Canada, operate it in Canada, and 'export the energy product to the south of us where electric energy costs are high.' The CANDU, with its low fuel costs, was made to order for that kind of export market, and in return, 'we would be getting the right kind of dollars.' It was, in the final analysis, a matter of faith rather than ability. The ability was there, that Smith knew. But how to recapture public faith?

The answer was provided in February 1971 when Ray Burge of the AECL staff sent out a press release telling the world that Pickering One had started up.

Starting up was a relative term. On the first day of its operation, the reactor was functioning at 0.01 per cent of full power. Over the next few days it was raised to 0.1 per cent (8 March), while adjuster rods were slid in and out and reactor physics work proceeded. Everything, inevitably, did not go smoothly. Experience prepared operators for heavy-water leaks, and leaks there were. But finally, on 16 March, the first steam was fed to the turbine and then, on 4 April, the first electricity was fed into the Ontario Hydro grid. By the end of April electrical output had reached 250 megawatts (electric), and a month later, for the first time, Pickering reached full power.[45]

To the general relief, if not surprise, Pickering did not operate like Douglas Point. On the contrary, once operating experience had accumulated, Douglas Point might even be operating like Pickering. To his board, Gray reported in June 1971 that Douglas Point was running very well, at 208 megawatts (electric); that Pickering One was operating at full power, or near it: 470 megawatts (electric); and Gentilly, while shut down, was at any rate working on schedule. Even Port Hawkesbury sent relatively good news: it was operating at 32 per cent of capacity, and Haywood was confident that production there would exceed current estimates.[46]

It was at least a breathing space. Everyone recognized that AECL and its reactor program were hardly out of the woods yet. They would not be until adequate supplies of heavy water were finally assured, until experience with Pickering was seen consistently to surpass that with Douglas Point, and until some solution was found to the eternal problem of exports. For the moment it was sufficient that public enterprise – and not public caution – had prevailed.

Marketing and Enterprise

Gray had less reason to be concerned about Commercial Products than about other areas of the company. He preferred a 'hands-off' style of management, punctuated by meetings of the Senior Management Committee to keep himself informed. As the president's attention was directed more and more towards markets and marketing, the occasion for intervention in Commercial Products' affairs was slight. But, as with other parts of the company, Gray was also determined to keep his inheritance intact. This became a point of some interest in 1958.

Commercial Products did not make a profit until 1956, but when it did the result was gratifying. Most of its sales were exports, because economy of scale required a high volume of production. They were achieved under internationally competitive conditions, through marketing arrangements in more than forty countries; over time the size of individual shipments tended to increase. More strikingly, the larger part of Commercial Products' revenue tended to come from sales of beam units: 65 per cent in 1956. AECL reports underlined that it was only because of foreign sales that Canada could have an isotope industry at all; the domestic market was far too small (14 per cent of AECL's total isotope sales went to Canada) to sustain such an operation. To encourage sales, the board approved sales on the instalment plan in July 1957.[47]

The expansion of Commercial Products' business led management to give its requirements second priority (after lucrative plutonium sales) in the use of NRU. Experimental work came third; nevertheless the balance sheet continued to show a loss in commercial operations, even if the magnitude of the operations went steadily up from $1.5 million in income in 1955–6 to $2.3 million in 1958–9.[48]

Commercial Products attracted a certain amount of outside attention. As we have seen, in 1958 Canadian Curtiss-Wright, which had just bought a related firm, Isotope Products Limited, approached Gray with an eye to buying Commercial Products too. The board recoiled, but though it rejected the offer it gave thought to the implications of operating a profit-making venture in an area where private companies were active. A government operation was always open to criticism on the grounds that its ownership gave it special privileges, and this was especially the case where business was potentially profitable. But selling off an entity like Commercial Products also raised the important questions of how, and to whom. In particular, it raised the question of whether Commercial Products should be sold to an American-owned firm like Curtiss-Wright. Directors also noted the high proportion of CP's income devoted to research and development – 20 per cent of gross revenue, as opposed to 5 per cent in a 'normal' commercial establishment. It ought to be reduced, or some other source of funding should be found. It was clear that if the rest of the company had contributed to that item as it had once done, there would have been a profit and not a loss in CP operations. It was true that competition with private firms must be considered; fortunately, in the board's opinion, the only possible competitor was in a substantially different line of work.

The result of the board's inquiries was wholly favourable to CP and its continuing relations with the rest of AECL. CP was a credit to the company and there was no good reason, either in business or in consideration of broader public policy, why it should be disposed of. On the contrary its links with the rest of the company, and especially Chalk River, ought to be strengthened. A joint technical assistance committee was therefore established for the study and approval of development proposals. As a small token, the executive committee was instructed to find some money for research. It found $215,000. At the same time, the research budget should not be diminished.[49]

It was expected that with this assistance Commercial Products should break even in fiscal 1959–60. It did. Instead of the $285,000 loss it had shown in 1959, CP had a $525,000 profit in 1960. Errington was pleased, but not, he told the board, entirely optimistic. Other companies were entering the business and demanding tariff protection in their respective countries. To forestall them, CP had concluded licensing agreements for its beam units in Germany and would soon do the same in Italy. In France, they were resorting to heavy discounts. There was renewed competitive danger from the United States, where the Americans had convinced themselves that they should lower the price of cobalt-

60 since there was surplus capacity in the industry; and they did. Nevertheless Commercial Products racked up another profitable year in 1960–1: $239,000 to the good.[50]

By this point, CP was making radiation units for both human and industrial use; the Theratron and the Eldorado, and the Gammacell, respectively. It wanted to branch out with an agricultural product. Irradiating food would preserve it, and CP received permission from the Department of National Health and Welfare to try out its theories on potatoes, so as to prevent sprouting, withering, and blackening. The agricultural unit was placed in a large truck and carted around from place to place to familiarize farmers and processors with its use and, especially, its availability.[51] Commercial success nevertheless eluded this particular line of products, and twenty-five years later AECL was still trying to secure its adoption by significant sectors of the food industry.

Despite the slow growth of its food processor line, CP's business expanded. The early 1960s were good for business around the world, and CP got its share. Beam units remained the foundation of the division's prosperity, but industrial machines had an increasing appeal. For customers with specialized needs, a consulting service was offered. By 1962–3 business had reached the $4 million level, of which 16 per cent was spent on research and development. The next year, sales were up another 10 per cent, and the company entered the business of making machines for sterilizing medical dressings and sutures. The US market had been especially good, the directors were told, so much so that the AEC was becoming embarrassed. It might be as well to establish an AECL subsidiary to deal with the trade directly. That, however, would raise once more the question whether the Canadian government should be in such a highly profitable and competitive business.[52]

With $5 million in orders in prospect for 1964–5, Errington had to contemplate moving his shops, this time to a building outside Ottawa's city limits; and with some some shuffling a move was approved. It took CP out to Kanata, in the summer of 1965. It was costly but business was very good; besides, there would soon be more cobalt-60 from the new CANDU reactors and if so, there would also be customers to sell it to. Indeed, if cobalt could not also be irradiated in Ontario Hydro reactors an automatic ceiling would be placed on the growth of CP. Given the cost of the relocation, there was an obvious reason to select a strategy for growth, and with some caution Gray did so. Against objections from one of the senior directors, Gray decided that he would accept the assurance of his staff that burn-up (and hence the cost of fuel) in the reactors would not be adversely affected. There would be more cobalt-60 production, in tune with CP's projected needs. In the summer of 1965 AECL and Ontario Hydro agreed that CP would put its cobalt in the future Pickering reactor; the cost, Errington believed, would be tantalizingly low.[53]

Prosperity brought another re-examination of Commercial Products' current standing and likely future. 'Senior government officials,' Gray told his directors, were worried at CP's success, so reminiscent of Polymer, the government's artificial rubber and chemical company at Sarnia, Ontario – like AECL itself, a Howe legacy. It had to be borne in mind that CP depended on the export market for 90 per cent of its business. As before, directors worried that CP had benefited from its Canadian government ownership through disguised subsidies – R&D grants, or low interest loans. Would not the US government be tempted to intervene and slap on a countervailing duty or some other form of prohibition?

The directors repudiated any idea of a subsidy, hidden or otherwise. As a subsequent meeting of the executive committee indicated, they had some misgivings about the R&D money they were pumping in; though some of it could be justified as part of AECL's program of free or low-cost public service to industry and to universities, there were, or ought to be, limits. And, since CP was such a roaring success, they were not in the least inclined to put it on the market. They knew that under the Pearson government, even more than under Diefenbaker, there were serious political reservations about the transfer of a Canadian company to (possibly) American hands. As a matter of national policy as well as national pride, CP would not be easily sold. Nor was it; an offer from a small Canadian company to purchase it was rejected in August 1965.[54]

Strangely, the rise of CP in the company's ledgers was the occasion for friction rather than harmony. Errington, it appeared, wanted more. He wanted more cobalt, meaning an assurance that it would be irradiated in Douglas Point as soon as that reactor was ready. And he wanted more money for research: $856,000 instead of the $700,000 the executive committee considered prudent. The former question was not solved until the following May, after appeals to the authority of W.B. Lewis. Foster, as head of NPPD, gave the necessary assurance that four channels of Douglas Point could be used without reducing its power output. Gray accepted his assurance, and there were no further objections.[55]

Still, the size and potential impact on the rest of the company's programs of CP's expansion gave pause. Errington gave consideration to a separate reactor, solely for isotope production, a reactor that Gray thought might possibly be built by CGE, but the idea apparently came to naught.[56] In another scheme, AECL's headquarters, which was just then expanding, would move out to Tunney's Pasture to replace the departing CP staff. To focus executive attention, Errington drew up a master plan, which Gray reviewed with some of his deputy ministerial dining club; it would be as well, the president decided, if some senior and respected economist could pass on Errington's plans before AECL was totally committed to them. He might be able to say whether, as CP predicted, competition would stabilize, and whether the food irradiation business would become as big as Errington thought it would. Because it was intended to be commercially viable, the CP program would be financed out of loans and investments and not from government appropriations.[57]

Although, because of a shift in overall government economic priorities, Errington was asked to postpone his construction plans for a year, Gray did give him something in return: $2 million in research and development funds rather than the $700,000 that the directors had been contemplating only a year before. Head office would stay in downtown Ottawa; given Gray's complicated network of bureaucratic friendships it is hard to know how he would have been able to operate out of Tunney's Pasture; and Tunney's Pasture continued to house CP's laboratories.[58]

Sales figures remained very good. Income reached $9,300,000 in 1967–8, up 12 per cent from the previous year. By the spring of 1968 there were 600 Theratron and Eldorado beam units installed in no fewer than forty-eight countries; some hospitals had even come back for a second unit. But while sales were up, competition was fierce. A survey in 1968 indicated, as salesmen already knew, that linear accelerators from Varian Associates in the United States were pre-empting more and more of the market. Unless AECL reacted, it would be pushed more and more to the fringes of the market by the new product. With some misgivings – which at least one member of the Board of Directors shared – CP turned to Chalk River for help. But the help was slow, and in 1974 Gray and Grinyer helped negotiate an arrangement with CP's existing marketing agent in Paris, CGR-MEV, for a pooling of technology. CGR would supply the accelerator components, and CP the mechanical side.[59]

In spite of the prospect of an expensive adaptation, Errington projected a $400,000 profit. But profits decreased, to $90,000, which caused Gray to pause and reflect. The next year, 1968–9, income was up by over $1 million, but profits were only $107,000. The explanation was the very high cost of research and development, which reached $1,121,000 in 1968–9, up $200,000 from the previous year. Income went down in 1969–70, and so did profits: $35,735 in all. At least the government would not have to face the charge that it was mining a lucrative business; yet, all in all, it was not discreditable to run a consistent profit alongside a very costly research and development program for more than a decade.[60]

Errington, however, was worried. His solution was to expand, this time into new areas of instrumentation, and to increase funding for research in the applications of radio-activity and radio-isotopes. He won, in February 1969, the partial consent of the board; but he had to accept a detailed study of the proposed market for his products. He had, as well, to accept an investigation of CP's financial procedures, which gave rise to some suggestions for improvement.[61]

The marginal profitability of CP led the board back to considering, for the third time in just over a decade, whether AECL would not be justified in selling off the division. This time the topic was raised by Gordon Shrum, the chairman of BC Hydro and, like his Ontario, Manitoba, and Quebec counterparts, a member of the board. Shrum's institutional memory did not carry him back to the previous occasions, but Brigadier Wallace, who had helped get Errington his job in 1946, was willing to supply the lack.

Commercial Products, he told his colleagues, was 'a good thing for Canada,' and there was a real question if, once sold, it could continue to serve the Canadian public interest as it had been doing. Moreover, in purely business terms, they would not make as much as they would like if they sold it; the big expansion that Errington had implemented was still working itself out and, in conjunction with the marginal profits, would adversely affect any sale.[62]

And so, with renewed expressions of affection, CP was reconfirmed in AECL's corporate family. It had not, after all, done badly either for itself or for the company. From AECL it received ample R&D funding, not to mention budgetary implants in the lean years during the 1950s as the division established itself in the international market. AECL basked in the prestige that the Eldorado and Theratron beam units gave it: the clearest and most tangible demonstration of the wholly beneficent aspects of atomic energy. During the 1960s, CP's profitability reflected well on its corporate parent and if, from time to time, AECL had different kinds of misgivings about its offspring, it consoled itself with the thought CP was a credit both to itself and to the nation.

CP's accomplishment was, like its counterparts' elsewhere in the company, the work of a single generation. That was not unusual in Canada in this period; indeed it may be more common than not. It gave an unusual stability to the direction and purpose of CP, the more so because its basic products – the several beam units and the isotopes they carried – were designed and successfully produced at the beginning of Errington's stewardship. They acted as a foundation for twenty years of sales for CP; and without them the division would have been unable to carry on its extensive R&D program that made it such an unusual commercial company in Canada in the 1960s.

But if Errington shaped and led CP to commercial success, he was also instrumental in expanding it. From the expansion that he directed flowed a different organization, one that he hoped would be adapted to the 1970s and 1980s.

12

Epilogue; or the Price of Success

Nuclear power in Canada meets the classic definition of 'industrial policy': a sustained effort by government 'to promote growth, productivity and the competitiveness of Canadian industries.' Nothing in the definition says that the effort is bound to succeed and, despite major elements of success, Canada's nuclear policies are still a subject of debate and uncertainty as the country approaches the 1990s.[1]

During the 1970s Atomic Energy of Canada faced considerable changes in its environment. The first change seemed promising. Canada, like most of the developed world, began to apprehend an energy shortage in the early 1970s. The government responded with expert studies, including an examination of atomic power. One study, in the Finance Department, noted a trend to thermal generating stations, accounting for 25 per cent of Canadian electricity production in 1972 (up from 6 per cent in 1950). The future, it suggested, was nuclear.[2]

The study was received and filed on 5 October 1973. The next day, 6 October, war broke out in the Middle East. Canada was confronted with drastically rising petroleum prices imposed by an oil producers' cartel. The Canadian federal government reacted with a flurry of activity. The maritime provinces were exposed to the rise in prices and the shortage of oil, even though Canada at the time was producing enough oil to meet its own requirements. The government froze oil prices within Canada, in the expectation that it could then allow them slowly to rise; it temporarily subsidized the lower Canadian price through an export tax on oil. But its hopes were blasted as the cartel price continued to rise; what had been temporary turned out to be permanent, especially after the Iranian revolution of 1979 set off another large increase in oil prices.

To secure greater control of Canadian energy supply, and in the belief that petroleum prices would continue to soar, the Canadian government instituted a National Energy Policy. The NEP was concerned with petroleum; but in the considerations of government nuclear energy occupied a place, and continued to do so even when energy prices fell and the NEP was dismantled. It was not only the federal government that was concerned. New Brunswick solicited proposals for nuclear reactors from Westinghouse and General Electric, as well as from AECL; soon a New Brunswick reactor was committed, from AECL.[3]

Insecurity of external energy supply remained a powerful argument for domestic control of energy sources, even at some cost. As task forces and consultants' reports proliferated, it was difficult to escape from the conviction that the national interest demanded that Canada retain its nuclear option even if the final payoff in most of the country awaited the exhaustion of other, cheaper, forms of energy supply.

The alteration of Canada's energy supply situation for the worse was outside AECL's control. If the oil crisis improved AECL's prospects, other factors diminished them. The most damaging was an accident at the Three Mile Island station in Pennsylvania of the Metropolitan Edison company on 28 March 1979. A combination of mechanical failure and human error turned the core of Reactor Number Two into what an observing scientist called 'a pile of rubble' – a radioactive pile of rubble, needless to say, with a bubble of hydrogen gas hovering above it. Fortunately, there was no explosion, but one was feared, and women and children were evacuated from the immediate area of the plant, all in the glare of publicity and instant television replays.

Ironically, the 'TMI' incident, as it was called, occurred a few weeks after the release of a sensational movie, *The China Syndrome*, about a fictitious fuel meltdown in a reactor: the China in the title referred to the eventual destination of the burning fuel on the other side of the Earth. The nuclear industry's attempts at rebuttal made little impression. Public opinion, which had been veering away from trust in experts, diverged some more. A decade that had started with Viet Nam (popularly believed to be an experts' war caused by the Best and the Brightest) ended with the explosion of another dream of the 1950s – the nuclear dream. That it had always been only a dream, at least in the version subscribed to in public, made little difference. As the American atomic historian Spencer Weart observed, atomic energy to most people had been a metaphysical rather than a physical fact, not 'something that might be handled with much the same commonsense methods as facts about oil reserves or chemical poisons,' but an elemental force to be approached with 'feelings of awe and terror, little different from what we might feel if confronted with a mad scientist's monster or a divine apocalypse.'[4]

AECL, founded on faith in science and sustained by a national commitment to research and development, was ill-prepared to meet the changing climate of opinion. It was, of course, surprised; unfortunately, it was also disarmed.

Gray and the Search for Markets

Pickering, Gray believed, was not enough. Nor would Gentilly-1, when it came on stream, suffice. CANDU and its boiling-water cousin were a wonderful technical achievement, or would be when Gentilly was proved, but their future would only be secure when other countries adopted heavy water technology.

The signs were not unpromising. India was, as far as Gray knew, a success story, a major and notable example of technological transfer between the 'First World' and the 'Third.' In 1971 Canada and India concluded an agreement applying IAEA safeguards to the Rajasthan project, a good sign. India's prime minister, Mrs Gandhi, daughter of the peace-loving Nehru, had assured Canada's Prime Minister Trudeau that she would keep Canada's gift of technology peaceful. Many Canadians wanted to believe her, and took pride in the success of this particular foreign aid program.

Taiwan was another success. AECL had bid for a fixed-price contract, $30 million, and came in on target and with a small profit.

But Taiwan was a research reactor, not a power system; moreover, Canada was in the process of de-recognizing the Taiwanese government in favour of the Chinese Communist government in Beijing. As far as the government of Canada was concerned, Taiwan no longer existed, and it was impossible to sell such a sensitive item as a reactor to a government that no longer had any legitimacy in official Canadian eyes. That the Taiwanese wanted another reactor and would pay for it in cash made no difference to the political facts.

Finally, KANUPP, even though it was the swan-song of CGE's export program, seemed like a triumph. But outside Canada, KANUPP was regarded as the product of foreign aid, not the result of a truly commercial sale. Special circumstances, it could be argued, made KANUPP unrepeatable.

Gray tried to compensate with bids to Australia and Mexico. Elaborate and expensive sales programs were mounted. But hopes rose only to fall again. Neither country would buy CANDU. Nor would Italy, where AECL had entered a partnership with Italimpianti, a very large local engineering firm, with an eye to an Italian CANDU. Not everything about the project went sour. There were long-term hopes, and Italimpianti appeared impressed with what AECL had to offer. It was just that AECL was not able to sell it.

Why not? A few years later, Gray commented that one mistake was to have used some of his senior scientists and engineers as fuglemen for CANDU. It was, he explained, 'impressive ... [but] they did not produce contracts.' Yet Gray and his associates did not trust Canadian industry to do the job either. A visitor to Chalk River in 1973 discovered the view that 'Canadian industry [was] unwilling to accept risks and, in the long run [was] unable to think in terms of a nuclear industry.' The visitor was doing more than compile casual comments; as a representative of the federal Department of Finance, he was pleased to report that these Chalk River opinions 'coincide with ours.'[5] And while Canadian ambassadors and trade representatives abroad plugged the CANDU as a matter of course and duty, they seldom played much more than an auxiliary role.

Ross Campbell, the ambassador to Japan, went further than most. He was 'extremely active in promoting CANDU,' Gray told his directors in April 1974. The Japanese were, therefore, considering a licensing arrangement, which would bring CANDU prestige if not much financial benefit. External Affairs 'had written the Ambassador advising that he be less vigorous in his promotional effort.'[6]

Gray reflected on the experience of trying to sell CANDU and the CANDU concept, as any intelligent executive might. He was struck when a New Yorker came to call in 1968. 'He felt,' Gray remembered, 'that we had a Sleeping Beauty in this reactor system but we did not know what to do with it, we did not know how to market it.'[7] Italimpianti had the same advice. They had got to know AECL pretty well, Gray testified in 1977. 'Although they were more than satisfied with AECL's technical competence they said our marketing competence left something to be desired.' As AECL searched for markets beyond the North Atlantic and European world where its scientists and engineers felt most at home, marketing competence assumed a new importance.[8]

AECL was looking outside what Gray described as 'the cream' markets, and its attention became concentrated on two: Argentina and Korea. Argentina is often thought to be a 'Third World' country and perhaps, by some very broad geographical definition, it is. Realistically considered, however, Argentina has much in common with Canada, a broad agricultural base, a large immigrant population, and close cultural ties with Europe. Early in the twentieth century Argentina had a higher standard of living than Canada. But its advantages were frittered away, some by uncontrollable circumstances, some by misgovernment.

After years of demagogic rule by the popular dictator Juan Domingo Perón (whose wife Evita had been a candidate for the first Eldorado-model therapy machine), the Argentine military had rebelled and expelled Perón. There followed a dismal succession of coups and counter-coups, the return in the 1970s of the moribund Perón, followed by the brief reign of his improbable successor, his second wife Isabelita. Politics, Argentine-style, featured an extremist tinge whether of the right or the left. In the mid-1970s the Peronist government was faced with a guerrilla war, opened by leftist terrorists; it responded brutally. In the opinion of the military it was not sufficiently forceful, and in 1976 a military government was imposed. The military eventually succeeded in putting down the insurgents; they also murdered a large number of other opponents, a fact that caused an international scandal.

As for Korea, it had only recently emerged from colonial tutelage as a Japanese territory. Two Koreas appeared on the scene, the North being Communist and dominated by a perpetual dictator, Kim Il-Sung; the South following the path of free enterprise and fettered politics. It was freer, certainly, than the North, and its political system made possible, or at any rate did not interfere with, an economic explosion almost unparallelled in twentieth-century history. In South Korea, as in Argentina, the military was never very far in the background; unlike Argentina, South Korea was a country in the economically ascendant.

Gray would have preferred other markets, but these were the ones available. In Argentina he entered a partnership with Italimpianti. Italimpianti would supply the conventional side of the reactor, AECL the nuclear components. The Italians would also direct the marketing campaign in Buenos Aires. The cost would be $5 million, of which AECL would contribute half; the marketing was to be placed in the hands of an agent. Gray and the Italimpianti president agreed that the Canadians would ask no questions; even the identity of the agent was not disclosed.

For South Korea he hired Shaul Eisenberg, an Israeli businessman, whose United Developments International was headquartered in Tel Aviv. Eisenberg told Gray that his fee would be 5 per cent of the contract; the 5 per cent was customarily added to the contract price. Gray asked other companies whether the arrangement was fair, and was told that agents' fees were a fact of life in the Far East; they ranged between 4 and 9 per cent and, by those standards, the price quoted AECL was not unreasonable. The arrangement with Eisenberg had the advantage that if he failed he would not be paid, no matter what his expenses in the meantime.

Gray confirmed his arrangements with Eisenberg in a letter on 28 November 1972. Eisenberg's fee was not specified; the details were written down only two years later. In the interim, Gray told his board that he had employed Eisenberg as agent, and praised Eisenberg's skill and familiarity with the Korean market.[9] Then, on 11 June 1973, he brought the matter before a meeting of the cabinet committee on operations.

AECL was the subject under consideration that day. Should it have an export program? What should its capital budget be? The ministers agreed that the company should remain in the exporting business; if that were so, Gray told them, they should expect AECL to hire agents to sell the product, perhaps for considerable fees. This observation was incidental to the more important matter of export policy, and it was the latter, apparently, that was recorded in the minutes.

Three and a half years later a dispute arose as to whether the subject had in fact been discussed. In testimony before the Public Accounts Committee Alastair Gillespie, minister of industry, trade and commerce in 1973, said that he could not recall it. He was contradicted by Gray, and also by Donald Macdonald, then minister for energy, mines and resources. Gray had indeed discussed the agents with him prior to the relevant cabinet committee meeting, Macdonald told his fellow MPs, and had certainly raised it there. What was done, he later said, was done with his knowledge and his authority.[10]

During 1973 and 1974 AECL had other business to concern it: the development of Pickering and Bruce, problems at Gentilly-1, and the continuing curse of heavy water. How much would be needed? How much would it cost? The answers were unknown and possibly unknowable. The Canadian capacity for heavy-water production turned on export sales and on the British heavy-water reactor, the SGHWR; with Gray flying around the world in quest of markets each trip, each passing month held a tantalizing if distant promise of fulfilment. AECL therefore lived, and built, in hope.

The bad news, when it came, was unexpected. On 18 May 1974 the Indian government exploded an atomic device – possibly a bomb – in the Thar desert. It was, the Indians told the world, a 'peaceful' explosion in keeping with their peaceful diplomatic traditions.[11]

The Indian explanation was not generally believed. India was not a signatory to the Non-Proliferation Treaty. It was simultaneously working on rocket development. It was worried about China and Pakistan. India had made a bomb, most thought, and another power had joined the club of 'nuclear weapons states' – NWS in the jargon of arms control. As for how the Indians had done it, they had used the Canadian-supplied research reactor, American heavy water, and foreign (but not, so they said, Canadian) uranium. In Washington there was consternation. 'Spending much time blaming the Canadians, Secretary of State Henry Kissinger chose to deny that the United States had contributed any nuclear material for the bomb.' He would, however, do nothing constructive in the circumstances. Perhaps there was nothing he could do.[12]

In Canada, a general election campaign was underway. There was a rush to blame, with External Affairs officials discreetly letting it be known that the trade-happy salesmen at AECL had brought the trouble on themselves. Prime Minister Trudeau was shocked at the Indian action, particularly since he had obtained assurances from Indian Prime Minister Gandhi that a bomb was the farthest thing from her mind.[13]

Mitchell Sharp, the minister of external affairs, condemned the Indian action and derided the idea that a 'peaceful' explosion was somehow different from other kinds. It was, he said, 'a severe setback' to any hopes for controlling nuclear weapons proliferation. On 22 May Sharp announced the suspension of all nuclear shipments to India, the termination of co-operation with India, and the recall of AECL's Indian representative. In total, $12 million worth of heavy water, a $6 million turbo generator, $1 million in spare parts, and credits and loans were frozen, pending discussions with the Indians.

The discussions took place in Ottawa at the end of July. There was no progress made. The Indians had conceded that CIRUS was the source of supply for their bomb program. In subsequent talks in New Delhi, they attempted to patch up the damage, but nothing they offered was sufficient for the Canadian government. The Canadians refused to believe them. There was no progress, and the suspension of nuclear relations was made permanent.[14]

The special atomic relationship was dead. There is no doubt that India's atomic program was retarded and even damaged by Canada's action. There was a price for joining the nuclear weapons club. The Indians resented Canada's action, and publicly condemned it as hypocritical. Canada for its part decided to dispel any possibility of ambiguity from its nuclear export policy. Allan MacEachen, Sharp's successor as external affairs minister, announced the new shape of Canadian policy at the United Nations in September 1974. 'Canada,' he told the General Assembly, 'intends to satisfy itself that any country using Canadian supplied nuclear technology or materials will be subject to binding obligations that the technology or material will not be used in the fabrication of nuclear explosive devices for whatever purpose.'[15]

MacEachen did not intend to terminate the nuclear export program, but the Trudeau cabinet definitely intended to limit it to states that would accept stringent bilateral Canadian safeguards. Trudeau, speaking to the Canadian Nuclear Association in June 1975, argued that nuclear power was a tremendous weapon in a worldwide war on poverty. 'It would be irresponsible,' he said, 'to withhold the advantages of the nuclear age – of power reactors, agricultural isotopes, cobalt beam-therapy units.' Other Canadian representatives made it clear that the benefits of nuclear technology transfer would be confined to states that rejected proliferation.[16]

The ramifications of this policy took time to become manifest, but there were a number of coincidences. The year 1974 was W.B. Lewis's last with the company, as it was Lorne Gray's. Both 1972 and 1973 were not entirely happy years for Lewis. His interest in the organic reactor had never died, and he objected to the cancellation of the organic reactor development program. He invoked his right to appear before the board, granted when Gray became president. As with his presentation of ING in 1967, it failed in its object. Lewis argued that Whiteshell's morale would suffer unless the organic reactor were revived; Gray disagreed. Lewis attempted to analyse Ontario Hydro's rate structure, and was sharply contradicted. He next suggested that the proceeds of an export tax on oil be applied to the development of organics. That idea, a board member snorted, 'had no chance of acceptance,' and was irrelevant to whether an organic reactor was worth proceeding with. The board was unimpressed, so much so that it declined even to appoint Lewis as a consultant to the company to attend international meetings.[17]

In the fall of 1973 Lewis again prepared a position paper on the subject of organics; Gray, however, refused to give it official status or even to distribute it. When Lewis protested, both the executive committee and the Board of Directors backed the president. There was simply no interest in the organic reactor, Gray explained, from any utility, and no prospect of funding from the federal government. (Even in Whiteshell's home province, Manitoba, AECL's principal activity was building long-distance transmission lines from a hydro site in the north to Winnipeg.) However good the organic reactor might have been, it was stillborn and the company must face the fact.[18]

In June 1973 Lewis had turned sixty-five; he waited a year, and then took his retirement. Lorne Gray had also been thinking of retirement, and in 1974 he decided that it would be at the end of the year. When a parliamentary committee asked why a vigorous sixty-one-year-old would quit, Gray replied merely that he had served his time, qualified for his pension, and 'done an awful lot of travelling. ... I did not,' he concluded, 'know my wife and son.'[19] These were cogent reasons, but Gray told others, including Sir John Hill, the head of the AEA, that the Indian explosion was a factor. Long afterwards Hill contended that the Indian explosion broke Gray's spirit and decided him to leave. Probably the true explanation is a combination of the personal and the political.[20]

Gray had his successor picked: John Foster, head of power projects and a twenty-year veteran in the nuclear business. Foster was moved up to Ottawa to be shown the ropes; in October Gray announced to the board that he would be leaving at the end of the year. It was a year of departures, for in addition to Gray, Haywood and Errington also left.

Just before leaving, on 30 December, Gray wrote and sent two letters to Shaul Eisenberg, confirming the details of Eisenberg's Korean agency: 5 per cent if and when he secured a reactor sale, and a $3 million service contract thereafter. John Foster would find them on his desk when he reported for work in the new year.

The Valley of Humiliation

John Foster was the logical successor to Gray. There were other possibilities – Haywood, for example – but Gray preferred Foster. He was the head of the company's largest division, he had international experience, and next to Errington he was the longest-serving senior officer. He had a well-deserved reputation as a gentleman. He was identified with Gray's policy of growth, and his connections with Ontario Hydro, AECL's largest customer and partner, were excellent.

Donald Macdonald, the minister, approved. He had met Foster while he was still vice-president: 'a great gentleman and a fine engineer,' entirely appropriate to head an engineering company. If Foster had a flaw it was in marketing, where AECL as a whole was admittedly weak. But Foster promised to work hard; some of his subordinates credited him with the belief that no problem could not be mastered, if only one worked hard enough, intelligently enough.

Good fortune at first attended him. The Argentinian contract was won, and then, after complicated negotiations, the Korean. There were the usual worries, mostly to do with heavy water in Cape Breton, and with a plant at LaPrade, Quebec, to service Hydro-Québec and provide for expected export sales. There were problems with Commercial Products' medical line: one product, the Therasim, had to be taken out of service pending urgent repairs. There were concerns about Gentilly, but these could be balanced by progress in Ontario. In any case, Hydro-Québec remained enthusiastic; its president, Roland Giroux, assured Foster, Gillespie, and Jean Chrétien that Gentilly-3 would be committed in 1977.[21]

Macdonald worried nevertheless. Gray's 'system' was still in good order in the early 1970s. AECL proposals arrived before cabinet committees and were supported. There were no surprises for the minister, even though there were continuing difficulties over heavy water, with consequent infusions of cash; these were offset by the operating performance of Pickering One, which surprised and pleased. But Gray's generation was passing. R.B. Bryce left for a job in Washington, other colleagues were approaching retirement, and their subordinates were not especially fond of the way that Gray had circumvented departmental influences. Foster reaped the unintended consequences of Gray's system; paradoxically, the more effective Gray had been, the more resentment Foster had to overcome.

There were concerns about the Argentinian contract: the possibility, for example, that it had not provided a sufficiently large margin to hedge against Argentinian inflation. If that were so, AECL was facing not merely a loss but a very large loss on its South American business. There was the weakness of the sales force to consider: perhaps, on balance, a change was indicated.

There were also larger concerns. Macdonald was a supporter of Canada's nuclear program, believing that CANDU represented the best in Canadian technology. It was an outward sign of the country's industrial and technological maturity, 'the most significant Canadian engineering achievement since the Second World War.'[22] Atomic energy was undeniably becoming more controversial, as well as much more expensive as Ontario, Quebec, and then New Brunswick turned to nuclear technology; in June 1973 the Ontario government approved Ontario Hydro's 'Generation and Development Plan' which provided for another four-unit station at Pickering ('B'), another at Bruce ('B'), and a new station at Darlington, on Lake Ontario near Bowmanville; besides these, Ontario Hydro ran its own heavy-water plant at Bruce.

Nuclear energy was by this point becoming more directly political. In the Privy Council Office a study of AECL was carried out in the early 1970s in the context of federal social policy programs (other subjects studied included federal involvement in Cape Breton coal mining and federal research incentives).[23] Public pressure groups developed to oppose the siting of nuclear plants in California, New York, and elsewhere. It has been suggested that their energies had been directed elsewhere during the Viet Nam war; with the war over they felt free to turn their energies to another cause.[24] Their example was noted in Canada, which was developing a vigorous environmental movement of its own. The environment had become a lively issue; in 1970 Canada got a minister of the environment, a post that would have made C.D. Howe snort in astonishment. Health, waste disposal, regulation, all these created concern.

No minister had intervened directly in AECL's administrative affairs since Howe had caused the company to be created in 1952 and had stocked it with directors in his own image: engineers and utility executives. Directors came and went, according to the minister's pleasure and sometimes his political preference, but the same basic pattern was always maintained. Macdonald changed that: off the board went the utility presidents who had made the utilities as much partners as clients. Arm's-length relationships gradually superseded the old arrangements. Within six months of Gray's retirement Macdonald approached Ross Campbell, Canadian ambassador to Japan, to take on the post of chairman of the AECL Board of Directors. As Macdonald remembered it, Campbell was to be 'Mr Outside,' representing the company in its dealings with the outer world; Foster 'Mr Inside.' Campbell hesitated; as he chewed the matter over there was a cabinet shuffle and a new minister of energy, mines and resources, Alastair Gillespie, repeated the offer. Campbell accepted, resigned from the foreign service, and returned to Ottawa at the end of 1975.

Foster had to make the best of the situation. This was no small task, for Campbell did not arrive alone. Foster was told that two deputy ministers were to be added to the board, T.K. (Tommy) Shoyama, former deputy minister of energy, mines and resources and current deputy minister of finance, and Gordon MacNabb, his successor at energy. The idea appears to have been Macdonald's, though it was implemented by Gillespie; as Macdonald later explained, it was difficult to master the intricacies of all the special agencies, boards, and crown companies that reported to him, and almost impossible to get them to follow a single path. The solution, therefore, was to place his deputy minister on them. The advantage, Macdonald later explained, was that the minister had a better source of information and a means of co-ordination; the disadvantage was that the deputy became committed to the policies that the relevant board adopted.[25]

The reinforced board soon had more on its plate than it could reasonably handle. The mid-1970s were a disquieting period in Canadian politics. Part of the malaise blew north from Washington, where the aftermath of Viet Nam and the Watergate scandal weakened confidence in American institutions. American business was not exempt from criticism: it was a nine days' wonder when the Lockheed company was revealed to have bribed Japanese politicians to secure a lucrative contract. It was a signal to investigate whether other firms engaged in such unwholesome practices, and suspicion did not stop at the border. It was in this atmosphere that representatives of the auditor-general of Canada took up the problem of the financial management of AECL.

The auditor-general of Canada is an officer of parliament, to which he annually reports. His function is to ensure that government financial records are properly kept and that public funds are properly spent. If anomalies exist, the auditor-general reports them; his report has become a major annual event as the media focus on oddities and follies in the government's spending practices. It is always the anomalies that attract attention and that stimulate investigation via the standing committee on Public Accounts of the House of Commons. Unlike other committees of the House, the Public Accounts Committee is always chaired by a member of the opposition, so as to escape the tyranny of the majority party in the House.

AECL had never had any particular trouble with its audit. It received money from parliament and spent it on research and development; almost all its expenditures were within Canada, and the sums were not large by comparison with the whole federal budget. The company made provision for its financial task by establishing an Audit Committee chaired by an 'outside director' in 1975. The committee had to discuss with representatives of the auditor-general the possibility that the Argentinian contract would lose them money; but it did not take very long for the auditors to zero in on the agency agreements for Argentina and South Korea.

With the conclusion of the Korean contract in January 1976 the agent's fee became due to Shaul Eisenberg. Campbell, scenting trouble, renegotiated the fee downwards and urged Eisenberg to produce receipts for expenses incurred in negotiating the contract; but the auditor-general's office did not find the documentation adequate, and so reported in November 1976. The Public Accounts Committee summoned AECL's officers to appear before it, and the winter of 1976–7 saw a parade of witnesses appear to justify or condemn AECL's sales agency arrangements.

What had the money gone for? Bribes to corrupt officials? It could not be proven, but neither could it be definitively disproven. Gray, who testified at the end of January, defended his conduct and his company's reputation. It might be true that AECL was light in the accounting department, and that was regrettable; but nothing dishonest had been done. If it had been, the RCMP, who had taken up the case, would presumably uncover it. As for Gray's authority, he contended that he had properly reported both to his political masters and to his Board of Directors. Donald Macdonald, who had been the minister responsible, agreed with Gray. And on that point the investigation eventually ground to a halt.

The damage to AECL's reputation, and to that of the CANDU, was very great. *Maclean's* put it succinctly: 'If we have to loan people money at subsidized interest rates to buy CANDU at prices below cost and then bribe them to do it, how great is the accomplishment?'[26] That other companies or governments also used substantial inducements was generally unmentioned; nor did critics respond adequately to Gray's assertion that in employing agents he was merely following the custom of trade.

What was clear was that parliamentarians intended to hold AECL, and other crown companies, to a higher standard of conduct than would be expected or accepted from private firms. AECL was a public company, and its practices could not be justified merely because others did the same thing. (It is also possible that Canada, though a trading nation, has never completely mastered some of the sales arts necessary for market penetration abroad; seen from this standpoint, AECL's sales tactics would be unusual even in the private sector.) The dilemma of a government enterprise operating in the world of private commerce, where neither it nor the Canadian government made the rules, was highlighted, but not solved.

The auditor-general's office, for its part, used AECL as an awful example, not merely of bad accounting practices but of bad business judgement. Poor decision-making had once been accounted a political or administrative matter, to be fought out between a crown company president and the appropriate minister. After 1977, under the banner of the 'comprehensive audit,' the auditor-general proposed to judge whether company officers, or the Board of Directors, had been sufficiently economical or efficient.[27]

The reporting of a loss of $130 million, including $22 million for heavy-water for Argentina, $30 million on heavy water sales, and $10 million on AECL's engineering contract for Gentilly-2, was a further blow. It made a change of regime inevitable, as the board recognized when it met at the end of June 1977.

The wreckage at AECL was already considerable. Three senior company officers left at the end of 1976; more new directors were appointed (and old ones shuffled off, in the middle of their terms) to bring the company into line and, so it was hoped, to strengthen its commercial capabilities. As Ross Campbell told the Public Accounts Committee, 'We needed people accustomed to doing business and making a profit at it thereby relieving undue burden on the taxpayers ... [and] a better equilibrium between government and business in the affairs of the Corporation.' The equilibrium was secured by appointing three deputy ministers.[28]

The Privy Council Office, which was expanding its influence as a co-ordinating agency for the bureaucracy, took a profound interest in the remoulding of AECL. It was working on a reorganization of crown companies, and AECL was proof positive of how much the companies needed a new and firm hand. Michael Pitfield, clerk of the Privy Council, was prepared to supply it. AECL was not merely gathered into a departmental fold, but into the larger lines of governmental authority that Pitfield was sketching out. What Arnold Heeney could not accomplish in the age of C.D. Howe, Pitfield easily managed in the age of Trudeau. Against the new men, operating in a new era, the veterans of the age of Howe could not prevail.

It came as no surprise that in July John Foster was fired. He went quietly and without complaint. His friends were indignant, and it is safe to say that Foster's reputation among his fellow engineers was in no way damaged. His departure, abrupt, public, and painful, was symbolic of the government's determination to manage things differently. How differently, only time would tell.

New Management and Old Problems

Atomic Energy of Canada Limited was to a very large extent the creation of a single generation, men (and, as we have seen, very few women) who were reaching retirement in the late 1970s. They came to maturity just before and during the Second World War, when their talents found direction and focus.

Of the figures who had shaped the company, some had already passed from the scene. C.D. Howe died in 1960. C.J. Mackenzie, his student and friend, died in Ottawa in 1984; almost to the end he drove in to the office in the National Research Council building that he had occupied first in 1939. Sir John Cockcroft became master of Churchill College, Cambridge; he died there in 1967.

W.B. Lewis, who more than any other person was responsible for AECL's power program, visited for a time at Queen's University, in Kingston, Ontario, but he remained Deep River's most prominent and eminent citizen. After some years of ill health he died there in 1987. Lorne Gray was one of the pall-bearers at his funeral; a month later Gray died in his sleep. George Laurence, who started the cycle of nuclear piles in NRC in 1940, died in November 1987. Bill Bennett went on to become president of the Iron Ore Company of Canada; he now lives in Montreal. Roy Errington, a lifelong smoker, faced a bout with cancer after his retirement; recovered, he now lives in Picton, Ontario. Dick Hearn outlived his entire generation, working as a private consulting engineer until shortly before his death in 1987. Harold Smith, his protégé, served as Ontario Hydro's chief engineer, and took early retirement. John Foster became an engineering consultant, and then president of the World Energy Conference.

It is too soon to give a historical assessment of AECL's most recent decade. Certain salient facts can nevertheless be noted. The new regime began with the appointment in February 1978 of a new president, James Donnelly, a forty-seven-year-old Scotsman who had moved to Canada in 1974. Donnelly was a product of the British electrical industry, with extensive experience in nuclear engineering, overseas projects, and commercial operations. He worked for a time with Ross Campbell, but Campbell left the chairmanship in 1979 to preside over AECL's international sales effort. After a disagreement with the government, he resigned that post and became a private consultant in Ottawa.

It was up to Donnelly to salvage what he could of the past, something that Foster could never do. Reflecting on the situation as he had found it, Donnelly later commented that things were both better and worse than he had expected. On the positive side, the company had preserved its 'technical integrity' – its ability to produce good, imaginative designs. If that capacity could be maintained and enhanced, Donnelly reasoned, AECL had a future. In any case, if one looked into the long term, what chance did Canada have of energy security unless nuclear power were employed? The facts had not changed since the 1950s, when Jack

Davis sketched out Ontario's energy alternatives: despite agitation for a 'small-is-better' approach to power generation, utilizing small rapids and waterfalls, or alternative technologies, the large quantities of power demanded in Ontario could only be provided by imports or by nuclear stations.

But prediction, even by the sophisticated statisticians of Ontario Hydro, was fallible and known to be fallible. The long term might well be absolutely certain, but the medium term was not, or not entirely. The decline of the expert's standing in society thus took its toll even in the energy field.

Donnelly also concluded that it did no good for AECL to produce superb drawings if their products could not be shipped on time, or matched with reliable cost estimates. AECL's performance in this department was not good. Nor was the central administration up to much, as anyone who had followed the Public Accounts Committee knew. It was too central: too much depended on an overworked head office staff, too little was left to the discretion of the divisions. A management study in 1976 pin-pointed some of the problems; unfortunately there was nobody around who could absorb the recommendations, far less implement them.

Donnelly decided to devolve authority within AECL, creating a research company, an engineering company, a radio-chemical company, and an international company, all of them reporting back to a corporate office, to be sure, but each enjoying substantial autonomy. As he pointed out, NPPD alone had a 1600-person engineering staff, comparable to the largest engineering companies in Canada, and it was simply impossible adequately to administer the group from a distant head office. Establishing a new engineering company was also a way of bringing on new talent, and he was convinced that there was plenty of that. The corporate office made its contribution through revenues received from Ontario Hydro (Pickering and Douglas Point). Commercial Products (Radiochemical) was also expected to make a positive cash contribution. Combined with the corporate cash flow, this would fund the engineering company. As for research, it was expected that it would be paid for directly from government appropriations. The reorganized company was therefore very different in structure from the AECL described in these pages.

There was plenty to do. New Brunswick started work on a

nuclear power station at Point Lepreau in 1975. It would supply 30 per cent of the province's power needs. Korea was under way, and the Argentinian project was going forward. There was nuclear construction in Ontario too, but less business than that fact by itself might have suggested. In 1958 AECL and Ontario Hydro had opted for in-house capacity and joint projects which, until Pickering A, were also jointly paid for.

Hydro, as a result of its activities, had started to accumulate its own nuclear expertise. Its in-house activities were no longer shared with AECL, and increasingly it relied on its own staff rather than on those of its federal partner. The replacement of the utility presidents on the AECL board did not help, and even though combined board meetings were scheduled once a year they did not stimulate the same sense of common purpose that had informed the first twenty years of Ontario's nuclear power program. Disputes over money and its distribution occurred from time to time. Though hardly cataclysmic, they consumed time and effort better spent elsewhere.

Ontario in any case remained the chief advertisement for the CANDU. Its reactor system remained one of the best and most reliable in the world (the four operating reactors at Pickering averaged 93.4 per cent capacity in 1976), and one of the cheapest, cheaper, according to company estimates, than its American light-water rival.[29]

There was a loss when pressure tubes sagged in the Ontario reactors, more or less as predicted, in the early 1980s. Expected the event may have been; welcome it was not. The reactors had to be taken out of service and, with no power being produced, no revenue was generated for their investors. Shutting down the reactors, not to mention replacing the pressure tubes, was an expensive proposition for Ontario Hydro. Firm in its belief in nuclear power, however, Hydro proceeded with the refitting. AECL's cash flow suffered, and some financial scrabbling followed; nevertheless, the shut-down was not so serious as to impair AECL's and Ontario Hydro's sense that they needed one another, or to prevent speculation about further future co-operation.

Quebec, so hopeful in the mid-1970s, modified its nuclear power policy shortly after the advent of the Parti Québécois government in November 1976. Negotiations were prolonged, complicated as they were by considerations of heavy-water supply to Quebec, by the costly LaPrade project, which the board wished to discontinue, and by concern over the political effect in Quebec of the cancellation of so large a federal project. There would be no Gentilly-3 in the immediate future; however, Quebec did not rule out another CANDU station sometime in the 1980s, either of 600 or 900 megawatt power. In 1978 Gentilly-1 was closed, primarily for safety reasons. The Quebec government in the spring of that year issued a White Paper on Energy, forecasting diminished electrical demand over the next decade. For development, it placed its hopes in hydro power. This made any further nuclear reactors beyond Gentilly-2 most unlikely.

The federal government continued to drag its heels well into 1979 on LaPrade, because of its political implications. The delay involved considerable loss, a loss which would probably not be recovered, since heavy water was by 1978 abounding. (Port Hawkesbury and Bruce were performing better than expected.) Quebec sought compensation; as the two governments bickered, AECL stood by, an essentially helpless bystander. Given that LaPrade's estimated cost would have been $1 billion (and climbing) and given that the expected markets for the heavy water product were unlikely to materialize, its eventual cancellation was more a relief than a disappointment. In return, the federal government forgave the debt incurred on the Nova Scotia and Quebec heavy-water plants.

Commercial Products, renamed the Radiochemical Company, had a crisis of its own. It had made its reputation on medical therapy units, the by-product of AECL's abundant isotope production, enhanced in the 1970s by isotope production in Ontario Hydro's power reactors. When it was faced with a rival line of therapy machines, accelerators that were electronic rather than chemical or radioactive, it responded by attempting to produce (in partnership with a French company) its own line of accelerators, the Theracs.

The competitive strategy did not work. Rival products dominated the market and continued to do so. Early in the 1980s Radiochemical abandoned the struggle. Its accelerator line was no more. It would, in a minor way, continue to make radioactive therapy machines, but it would never again dominate the market. Instead, it concentrated on medical isotopes. As with therapy machines, it sold most of its product abroad.

The existence of a powerful engineering group at Sheridan Park was, as Donnelly recognized, a source of strength. Their prime purpose was to design and consult on reactors, but in the world around there were very few reactors being built and sold. As the Korean and Argentinian contracts were completed it became apparent that the Mississauga operation was too large to be comfortably supported. The Mississauga staff therefore began to shrink.

This left the company, and the government, in a quandary. If the AECL engineering company was allowed to dwindle with its market, its talents would be dissipated and eventually lost. There were very few similar engineering companies in Canada and, despite the growth of Ontario Hydro's nuclear engineering force, the disappearance of the AECL engineering company would mean a substantial diminution in Canada's advanced technical capacities. Governments, especially in Canada, have traditionally worried about such things. Every government, and virtually every political party, in Canada has argued for a larger and stronger high-technology sector as a means of employing Canadian talent and increasing Canadian trade. It follows that no government would lightly abandon 'high tech.'

In 1985 the government did agree to abandon one high-tech experiment that it regarded as definitively failed – the heavy-water plants in Cape Breton. At the very least they were making a product for which there was no conceivable market. With the constriction of heavy-water experiments around the world, and the completion of AECL's current construction projects, there was a glut, not a shortage, of the heavy isotope of hydrogen. Even Ontario Hydro's Bruce heavy-water facility was moth-balled, and although the closing of two factories in employment-scarce Cape Breton was a serious matter, it could no longer be avoided.

The company put a premium on innovative research with an eye to diversification. The two laboratories, Chalk River and Whiteshell, had always paid part of their own way. With budgets shrinking in the mid-1980s, income from inventions and services became more important. Even so, shrinkage was indicated as the government reordered its priorities to place its research money in different channels.

Though afflicted by a shrunken international market, the Canadian nuclear industry had prospects: the construction of four 881 megawatt units at Darlington, the possibility of another CANDU at Point Lepreau, hopes for Romania, where economic conditions put the nuclear program on hold, and, perhaps, contracts in Japan and the United States, whenever that country's nuclear possibilities began once again to unfold.

1986 was the year of the Chernobyl nuclear disaster in the Soviet Union, another blow to nuclear power. Despite Chernobyl, however, construction proceeded at Darlington under a new provincial government whose attitude to extension of the nuclear industry had been uncertain. Ontario Hydro found it opportune to volunteer its services to the Soviet Union to help deal with the aftermath, a sign of life and optimism in the face of adversity. Not long afterwards, its chairman urged the provincial government to start thinking about the next generation of power generating stations, after Darlington.[30]

Remarkably, such was the strength of the CANDU's domestic reputation that surveys in 1987 showed that Canadians had not translated their perceptions of the Chernobyl disaster to their own country's nuclear program. People might have reservations about nuclear wastes, or about the costs of nuclear power stations, but they did not demand that existing ones be shut down or that stations under construction be terminated. It was quite the opposite: CANDU, where Canadians were concerned, was its own recommendation.[31]

Conclusion

The completion of Pickering One followed by Two in December 1971, and Three and Four, both actually ahead of schedule, in 1972 and 1973, and then the rest of the Pickering and Bruce atomic complexes, gave Ontario over 10,000 megawatts of base-load nuclear power. Operating at a lifetime capacity of 89.3 per cent, Pickering

Seven led the world; year after year, Canadian-designed heavy-water reactors were at or near the top of the world ranking of capacity factors. Taken together, they marked the culmination of a thirty-year cycle that stretched back to George Laurence's modest experiments in the basement of the NRC building in Ottawa.

The Canadian nuclear program had benefited, more than it had suffered, from a fair degree of coincidence. It was the coincidence of war that gave nuclear research both urgency and finance, and that brought a critical number of scientific and engineering talents to Canada. Yet it did not bring quite enough, for neither in money nor in the magnitude of its staff could the Canadian project match its American cousin, with which it was originally supposed to be integrated. It became a cinderella enterprise, dependent on the kindness of strangers. Its wartime director, Hans von Halban, was confident that a heavy-water, natural-uranium reactor was the ticket. Although he was not able to create a viable laboratory, his successor John Cockcroft did. The kindness of a stranger, the American general Leslie Groves, eventually guaranteed that its reactor would be completed, and that it would be the heavy-water species that the Americans themselves were not for the moment pursuing. But it was not on time for the war. This fortuitous circumstance made what was intended to be a prototype for a bomb factory into a research reactor with a very high neutron flux, the famous NRX.

The Canadian government, as such, had no idea what to do with its reactor. But, because of the prestige that being an 'atomic power' brought to Canada, because the responsible minister, C.D. Howe, believed however vaguely in an eventual connection between pure science and economic progress, and because the cost of maintaining NRX and its supporting laboratory, Chalk River, was not exorbitant in terms of the federal budget of the late 1940s, NRX survived and even flourished. It attracted to itself an exciting group of young and well-trained men of a kind that characterized Canadian society in the period.

Although the Chalk River laboratory was notoriously badly managed, that did not matter very much. It was able to secure necessary equipment, to procure its budget more or less when it should have, while all the time doing basic research and learning about the nature of the reactor that the war had bestowed upon it. Because the reactor was unique it was much sought after, and its facilities were rented out to the Americans for their experimental work on nuclear fuel that might power submarines. The reactors they were developing were, of course, power reactors, and the American experiments, when extrapolated, were a stimulant to Canadian ideas in the field.

It took a strong personality to stir the scientific pot and keep it boiling on top of the NRX 'stove.' The talents required of a laboratory director, especially of a large laboratory like Chalk River, are unusual. W.B. Lewis, a late choice for the job and not the man whom the government had sought, proved to have the necessary talents. He had experience and intelligence. He had to an unusual degree the arrogance of his convictions, an arrogance that he was able to make prevail in battle with some of the senior sabre-tooths in his newfound Canadian lab.

Lewis was inspired by the idea that nuclear power would be the salvation of industrial society, if not actually a staff of life and light for all mankind. He took this conviction and applied to it the simple but basic principle of neutron economy. Using that principle, he worked over the economics of low-cost power generation, to the point where he was able to convince a Board of Directors that the hour for nuclear-generated electricity had struck. The board was predisposed to believe him, although not so predisposed as to ignore completely such alternatives as were circulating around the laboratories of Britain and the United States.

It was fortuitous, and a highly unusual circumstance, that C.D. Howe was still the minister responsible for atomic energy. Howe, using the executive talents of his former aide Bill Bennett, reorganized Chalk River into a crown company with the object of converting neutrons into (eventually) steam. Howe and Bennett were responsible for putting representatives of Canada's electrical utilities onto the board of their new company, AECL, thereby establishing a connection that was both unique and highly beneficial to the development of Canada's domestic nuclear power industry. Unlike the British and the Americans, the Canadians sought out a single line of reactors and concentrated on them, ensuring that Canada's limited resources in science, engineering, and money would not be squandered by being spread too thin.

The experience of war, very recent then, and the availability of a highly motivated and cohesive workforce, with a spirit and work ethic derived from depression and war, made it both conceivable and possible to go forward. This was the circumstance that underlay much of Canada's economic growth and political initiatives in the fifteen years following the Second World War.

Again circumstances were kind. The St Lawrence–Great Lakes basin was running out of hydroelectric power. The country was enjoying its most prolonged economic boom, one that in the 1950s was expanding Canadian industry in precisely the area that would very soon need more electric power. Howe believed in energy self-sufficiency for Canada, and so did Ontario Hydro; Howe's agency therefore formed a close alliance with the provincial utility. The alliance took shape in the form of the Nuclear Power Group at Chalk River, composed of engineers drawn from both AECL and the utilities. The Nuclear Power Group was dominated by the unusual personality of Harold Smith; it developed plans for not just one but three reactors.

It was Bennett's intention, conformable to Howe's overall policy of minimizing government risk *if possible*, to develop a workable power reactor and then turn the result over to private (Canadian) enterprise. Further expenditure would be left to the utilities and the private sector. This would have turned Canada's nuclear industry into something resembling the then existing American and British models. The chosen instrument for this policy was CGE from 1955 to 1958; the product of the policy was NPD 1 and 2 at Rolphton. It may have been that had the Liberals, and Howe, stayed in office, this plan would have been followed. But they did not, and – also coincidentally – Bennett departed.

Because of frictions and early disappointments in relations with CGE, AECL formed, before Bennett's departure, the nucleus of its own engineering group, with Smith very much in command. Smith proceeded to apply what can only be called Ontario Hydro's 'corporate culture': he demanded the greatest possible freedom of action for his utility, meaning that Hydro should not allow itself to become dependent on a single source for any item the size of a power reactor. This fitted Hydro's existing capabilities, which had been expanded for hydro and thermal power development (not to mention the great 60-cycle conversion) in the 1940s and 1950s.

Lorne Gray, the new president of AECL, agreed. Gray's reasons included his on-the-spot experience with CGE, to which might be added his belief that to have a nuclear reactor done right it had to be under direct control and not in the hands of others. In terms of the expenditure still required to develop a viable nuclear power station, Smith and Gray may well have been right. The money that would be spent might well have been spent anyway, and with less direction and control. At the same time it was admittedly easier to plot AECL's longer-term future, including that of Chalk River, if an important and identifiable public function was being served by the company. An engineering department was easier to understand than an accelerator, especially by the bureaucrats and politicians from whom the company's growing budget flowed. Gray assigned considerable importance to the engineering effort; the scientific side he left to Lewis, whose talents he attempted to supplement and, if possible, delimit by assigning a succession of engineers to run the practical side of Chalk River, as he himself had successfully done between 1948 or 1949 and 1958. As in the past this experiment had mixed results. The neat categorization of the Howe-Bennett period was therefore abandoned.

The result was NPPD, which was moved to Toronto as if to emphasize symbolically its distinction from the old Chalk River enterprise. NPPD took with it the plan for a 200 megawatt (electric) reactor, dubbed the CANDU, which it proceeded to build on the shores of Lake Huron at Douglas Point. To undertake this project before NPD (in its second manifestation) was ready to go was probably rash. The timing was dictated not so much by domestic need (Ontario Hydro did not need the power so fast) as by a sense that the psychological moment had come and might be passing: the spirit of innovation and competition that followed on the 1955 United Nations atomic energy conference at Geneva. Besides, the engineers were available and ready to go.

NPPD developed as it had to, taking advantage of the favourable climate furnished by the Diefenbaker government, which had got its fingers burned by the Arrow affair of 1958–9. In a sense that we must not exaggerate but which was nevertheless real, CANDU was the surrogate Arrow, the next logical repository of Canadians' attraction to high technology and skilled industry as a means to making their society 'complete.' NPPD was placed under some strain by reason of a Canadian commitment to help India in nuclear technology, and it began to expand to meet its obligations. At the same time, CGE experienced less success than it had hoped, facing a refusal by Ontario Hydro to buy its off-the-shelf up-scaled NPD; it was also unable to make its mark internationally. AECL and CGE maintained an uneasy truce in international markets. It is probably fair to say that AECL viewed CGE as at best a supplement to its own efforts in the international marketplace and that it did not co-operate over-enthusiastically with its private erstwhile partner. It became in the short term a pensioner of AECL in order to keep its entire team in being; in the medium term it became a rival, invoking, as it had to, bureaucratic and political support to get a slice of the Canadian nuclear pie.

The *size* of an engineering enterprise in the nuclear field was the next important internal consideration for AECL's development. It was accepted that an engineering team must be kept in being. It followed that it must be kept employed, primarily by designing reactors one after another. The design for Pickering was set on foot long before Douglas Point was even remotely ready, and even before very many results had been obtained from NPD; Bruce, in turn, began some years before Pickering One was ready. This observation will not be surprising to any student of bureaucratic behaviour; fortunately the need was there because of a boom in the Ontario economy during the 1960s.

The increasing size of Canada's nuclear sector and its spread from Ontario to Quebec after 1963 encouraged AECL to complete the national nuclear policy and manufacture heavy water inside Canada. It made economic sense to do so, at any rate as long as it could be made for less than what the Americans, up to that point Canada's sole supplier, were charging. It probably did not make economic sense to combine this desirable goal with another, the Pearson government's regional economic development policy, intended to direct economic growth, preferably of the 'modern' variety, to Canada's have-not regions.

No area was more 'have-not' than Cape Breton Island which, as it happened, had a politically skilful representative in the federal cabinet. It also had an active provincial government which did not hesitate to pump in money to finance what it hoped would be an industrial renaissance in a decaying mining and smelting region. It may be argued, however, that Cape Breton had not decayed far enough, because enough of the region's previous economic experience (essentially one of confrontation, or, as some would put it, of people's struggle) was still present to poison the work atmosphere in the new industries that were being set up.

In AECL's favour it may be said that it did not want to go to Cape Breton. Once it did, it became a captive market for a subsidized industry that was, for reasons somewhat outside Nova Scotia's control, being built on the 'wizard' principle. The wizard in this case resided in New York, the best part of two days by airplane from Cape Breton, and his managers dwelt in Halifax, a couple of hours away. The money wasted on this enterprise was considerable, but since it was provincial money it did not directly concern AECL. The waste of time did, and it contributed to a situation in which Canada – federal Canada this time – paid for not one but three heavy-water plants, built too late to avoid paying premium prices, probably more than the Americans would have charged in the first place, for the precious substance.

This circumstance, combined with the over-designed and frequently malfunctioning Douglas Point reactor, helped to undermine AECL's position both in Ottawa and in the larger sphere of public opinion in the period 1967–71. Fortunately for the company it had an exceptionally skilful corporate president in Lorne Gray; Gray's technique, which was more reminiscent of a 1930s Evelyn Waugh novel than of staid Ottawa under the confused but respectable Pearson Liberals, preserved AECL from most kinds of harm and steadily augmented its budget.

He had at the same time to cope with an exhalation of scientific steam from Chalk River, where the other resident wizard of this story, W.B. Lewis, was busily thinking up new and ingenious ways for his laboratory to stay abreast of its competitors in modern physics. As with NPPD, his motivation is entirely understandable in terms of bureaucratic theory. It may even be comprehensible in terms of the scientific agenda of the period (or even today), but it was grandiose, on a scale that probably dwarfed even the expenditure on the original NRX when eventual operating costs were figured in. It would probably have doubled the size of Chalk River. It ran inevitably into heavy weather in Ottawa, where the bureaucrats and their political masters simply did not understand it, or why it was necessary. And it ran athwart the designs of the engineering faculties of Canada's universities, who saw it as a diversion of funds that would be better spent on them. Considerations of national excellence did not impress them, since the excellence would not be under their control.

ING perished, though Lewis did not go down with that particular ship. It is fair to suggest that this was his last 'big' project, for though he did not allow himself to rest or, as far as we know, brood, his remaining time with the company was spent pursuing other forms of reactor design, such as the 'value-breeder.' The Chalk River laboratories continued to do basic research, but it may be fair to suggest that the ING episode damaged their overall reputation, at least in Ottawa and possibly elsewhere; the scientists there acquired a reputation for being over-fond of expensive scientific toys, and their future requests were inevitably scrutinized through the prism of ING.

This result was not especially Gray's fault. But since his presidency was partly founded on a compromise with Lewis, on whose scientific and evangelical skills he depended, he was not able to abort Lewis's designs until it was too late. He had better luck with his other responsibility, the development of a viable and economic power reactor. For Pickering, as we observed at the beginning of this section, was a success.

Pickering succeeded in terms of *scale*. It was no mean feat to organize the design, development, and construction of not one but four unprecedentedly large Canadian reactors. It succeeded in terms of *design*. It was not perfection, but it was a lot closer to it than NPD or Douglas Point had been; and Bruce, its still larger successor, was a success as well. They were a success in *operation*, because they provided reliable lower-cost power to Ontario, from an essentially domestic source. They were also a success by comparison with the efforts of other would-be nuclear powers, whose reactors did not equal the on-line performance of the Ontario power stations. It may, however, be noted that the Canadian nuclear industry is not completely self-contained: zircaloy, the crucial cladding for pressure tubes, still comes from the United States.

AECL was therefore a success in terms of its public purpose. Was it also a success in organizational terms? Plainly in large part it was. An organization was built on a scale and with talents that were unprecedented in Canadian terms and that made Canada at least a competitor in world markets. But there was a cost, a cost that had been recognized by Howe and Bennett, even if, under the pressure of circumstances, Gray was unable to avoid incurring it. To establish capacity meant establishing commitments – commitments to the maintenance of a large engineering department whose life's blood was the construction of reactors on a regular basis. This conundrum is not unique to AECL; obviously it could not be, since so much explicit consideration was given to the problem back in the 1950s.

Harold Smith had argued in 1970 that Canadians found it difficult to come to terms with their own success. He might better have said that they find it hard to understand, and to pay, the price of success.

Notes

CHAPTER 1: EARLY DAYS

1 S.R. Weart and G.W. Szilard, eds., *Leo Szilard: His Version of the Facts: Selected Recollections and Correspondence* (Cambridge, Mass 1978), 16
2 Harold Nicolson, *Public Faces* (Toronto nd; original edition 1932)
3 Bertrand Goldschmidt, 'Hans Halban (1908–1964),' *Nuclear Physics* 79 (1966) 1–11
4 Spencer Weart, *Scientists in Power* (Cambridge, Mass 1979), 65–6; Jules Guéron, 'Lew Kowarski et le développement de l'énergie nucléaire,' in J.B. Adams, ed., *Lew Kowarski 1907–1979* (np, nd)
5 Weart, *Scientists*, 97–103
6 Ibid., 83
7 Ibid., 137
8 Ronald W. Clark, *The Birth of the Bomb* (London 1961), 99–101
9 Halban diary, 3 July 1940. I am grateful to Bertrand Goldschmidt for letting me see his copy of the diary.
10 Margaret Gowing, *Britain and Atomic Energy, 1939–1945* (London 1964), 67
11 Ibid., 53–4
12 Public Record Office, Kew (PRO), AB1/30, J. Cockcroft, 'Report on a discussion with Dr. G.C. Laurence and Dr. Stedman, 22-11-40'; R.H. Fowler, Central Scientific Office, to C.J. Mackenzie, NRC, 28 Jan. 1941, saying that Cockcroft has suggested someone named Cook [Les Cook] and inquiring as to who he might be.
13 AB1/157, Cockcroft to R.H. Fowler, NRC, 28 Dec. 1940

14 Gowing, *Britain and Atomic Energy*, 72
15 Ibid., 75–6
16 The MAUD committee report is reproduced in ibid., 394–436.
17 Ibid., 96
18 PRO, CAB 90/8, SAC (DP)41, minutes of 10th, 11th, and 13th meetings, 3, 16, 19 Sept. 1941
19 Quoted in Gowing, *Britain and Atomic Energy*, 105
20 AB1/105, Halban, 'Possibilities of the Development of the Boiler in UK and USA,' 16 March 1942
21 AB1/34(a), Halban to Akers, 27 April 1942
22 Ibid., Akers to Halban, 5 May 1942
23 Ibid., Halban to Akers, 13 and 14 May 1942
24 See Robert Bothwell, *Eldorado: Canada's National Uranium Company* (Toronto 1984), 119–21.
25 Akers was still reluctant. Perrin's view is contained in AB1/244, Perrin to Akers, 8 and 13 June 1942.
26 AB1/207, Anderson to Malcolm MacDonald, 6 Aug. 1942; Public Archives of Canada (PAC), C.J. Mackenzie Papers, vol. 1, Mackenzie diary, 9 June 1942, giving the first indication that 's-1' (atomic research) might move to Canada.
27 Betty Lee, 'The Atom Secrets,' *Globe Magazine*, 28 Oct. 1961, quoted in John Porter, *The Vertical Mosaic* (Toronto 1965), 432
28 Mackenzie diary, 2 Sept. 1942
29 See Robert F. Legget, *Chalmers Jack Mackenzie*, a memoir for the Royal Society (London), 28 Feb. 1985, 17n.
30 On the Treasury Board and for Ilsley's speech, see R.M. Dawson, *The Government of Canada*, 4th ed. (Toronto 1963), 393–4.
31 C.J. Mackenzie, 'Introduction' to Mel Thistle, ed., *The Mackenzie-McNaughton Wartime Letters* (Toronto 1975), 4–5
32 C.J. Mackenzie interview, Rideau Club, 29 March 1976
33 Quoted in Legget, *Mackenzie*, 8
34 Ibid., 23
35 Ibid., 36
36 Quoted in ibid., 431–2
37 C.J. Mackenzie interview, NRC, 17 Aug. 1977
38 AB1/137, R.E. Newell to Wallace Akers, 28 Oct. 1943
39 S.F. Borins, 'World War II Crown Corporations: Their Functions and Their Fate,' in J.R.S. Pritchard, ed., *Crown Corporations in Canada* (Toronto 1983), 447; see also Robert Bothwell and W. Kilbourn, *C.D. Howe: A Biography* (Toronto 1979), 139–40, from which the quotation is taken.
40 Mackenzie diary, 17 Aug., 2 Sept. 1942; J.F. Hilliker, ed., *Documents on*

Canadian External Relations (DCER), IX (Ottawa 1980), 453–4: Malcolm MacDonald, 'Aide-memoire in connection with proposed transfer of "Team 94" from United Kingdom to Canada,' 2 Sept. 1942; Mackenzie's proposed organization, dated 'September,' is in AB1/123/80521.

41 AB1/123/80521; PAC, C.D. Howe Papers (MG27 III B10), vol. 14, file S-8-2 (32), record of a meeting held in the Lord President's room, 12 Oct. 1942

42 That Mackenzie harboured such sentiments need not surprise us; few Canadians ran into real 'foreigners,' except as immigrants. The result, today, sometimes makes unpleasant reading. On 10 December 1942 the NRC chief met his 'league of nations' for the first time. Placzek, Mackenzie wrote, looked like a Jew, and 'undoubtedly Halban, Bauer and Goldschmidt all have Jewish blood but Auger is definitely French.' Mackenzie diary, 10 Dec. 1942.

43 AB1/123, Halban to Akers, 25 Sept. 1942

44 Weart, *Scientists*, 191–2; AB6/674, J.D. Cockcroft to James Chadwick, 5 Jan. 1949; AB1/123, Akers to Perrin, Nov. 1942, indicates that since Auger had been hired, 'it is possible to move in the Kowarski matter'; AB1/50, Perrin to Akers, 2 Dec. 1942; Akers to Perrin, 7 Dec. 1942, suggested that Kowarski's exclusion was not personal and proposed that if Kowarski came anyway his family should not accompany him; he would probably choose not to stay!

45 The relevant correspondence, some of it reflecting little credit on its writers, is in AB1/50.

46 AB1/123, Jackson to Sir E. Appleton, 30 Nov. 1942

47 AB1/123, Akers to Perrin, 5 Nov. 1942

48 Goldschmidt interview, Paris, 10 June 1985

49 Chalk River Nuclear Laboratories (CRNL) Records, 1500 Montreal Labs – Administration, vol. 1, Vallée to S.P. Eagleson, 25 Nov. 1942. The rental was $25 per square foot.

50 Colin Amphlett interview, Harwell, 29 May 1985; Fred Fenning interview, Reading, 6 June 1985

51 AB1/124, Freundlich to Auger, 15 April 1943; Halban, 'Memorandum on organization of project administered under the title of "Special Committee on Radiological Research,"' 24 June 1943. Under NRC regulations all purchases costing more than $100 had to be put out for tender. Worse still, the tender system occasionally produced the wrong result, and the whole thing had to be gone through again. It usually took six to ten weeks for an order to be placed.

52 PAC, NRC Records (RG 77), vol. 283, file Personnel CR, Mackenzie to Gordon Shrum, 13 Nov. 1942; Shrum to Mackenzie, 24 Nov. 1942

53 B.W. Sargent interview, Kingston, 26 Oct 1985; Harry Thode interview, Hamilton, 18 Oct. 1985; PAC, RG 77, vol. 283, file Personnel CR, Thode to Otto Maass, 25 Feb. 1943; James B. Conant to Howe, 20 Feb. 1943; W.J. Bennett to Mackenzie, 10 March 1943; Mackenzie to Bennett, 13 March 1943
54 Bertrand Goldschmidt, 'Les premiers milligrammes de plutonium,' *La Recherche*, #131 (mars 1982): 369; B. Goldschmidt and Jules Guéron interviews, Paris, 10 June 1985; Leo Yaffe interview, Montreal, 17 July 1985
55 AB1/50, Perrin to Akers, 2 Dec. 1942; AB1/123, Akers to Perrin, 9 Dec. 1942
56 AB1/123, Jackson to Sir Edward Appleton, DSIR, 30 Nov. 1942
57 PAC, Munitions and Supply Records (RG 28A), vol. 173, file 1, Mackenzie to Conant, 12 Dec. 1942
58 CRNL Records, 5000 Montreal Labs, Reactor Concepts, vol. 1, Halban, Report on meeting in Chicago with General Groves, and representatives of du Pont and MIT
59 Richard Hewlett and Oscar Anderson, *The New World: A History of the United States Atomic Energy Commission*, vol. 1 (reprinted, Washington 1972), 186–8
60 Ibid., 200–2
61 The text of the letter is in DCER, IX, 464–5.
62 Mackenzie diary, 2, 3, and 7 Jan. 1943
63 Ibid., 15 and 18 Jan. 1943
64 See Hewlett and Anderson, *New World*, 268–9, for Akers's attempt to succeed where Mackenzie had failed; Mackenzie diary, 20 Jan. 1943.
65 Hewlett and Anderson, *New World*, 269; AB1/128, Akers to Gordon Munro, 30 Jan. 1943
66 Bertrand Goldschmidt, 'Les premiers milligrammes de plutonium,' *La Recherche* 131 (mars 1982): 369–70
67 RG 28A, vol. 173, file 1A, Conant to Mackenzie, 27 Jan. 1943
68 Ibid., memorandum by Lesslie Thomson, 2 Feb. 1943
69 See Bothwell, *Eldorado*, 127–44, for the uranium crisis of 1943. Halban in fact wanted considerably more uranium oxide over the longer term than Eldorado could possibly have supplied: his figure of 400 tons, which dates from May 1943, was two full years' production.
70 Mackenzie diary, 1 May 1943. Goldschmidt reported on Eldorado's sales program to Akers on 4 May 1943: AB1/80.
71 Mackenzie diary, 11 May 1943; see Brian Villa, 'Alliance Politics and Atomic Collaboration,' in S. Aster, ed., *The Second World War as a National Experience* (Ottawa 1981), 155.
72 Fenning interview; Amphlett interview

73 Mackenzie diary, 22 Feb. 1943; Halban diary, 22 Feb. 1943
74 Mackenzie diary, 25 Feb. 1943
75 AB1/105, 'Memorandum of Conference between Prof. E. Fermi and Prof. H.C. Urey on March 6, 7 and 8, 1943'; Conant to Mackenzie, 13 March 1943; Halban, 'Remarks,' 20 March 1943; Kowarski to Akers, 11 June 1943
76 Sargent interview
77 B. Goldschmidt, *Pionniers de l'atome* (Paris 1987), 244: my translation
78 Goldschmidt interview; Guéron interview; Amphlett interview
79 Goldschmidt interview; CRNL Records, file 5000 Montreal Labs – Reactor Concepts, vol. 1: Minutes of meetings, 8–10 June 1943; AB1/124, TA Technical Committee, minutes of meeting, 12 Aug. 1943. This senior advisory committee included Chadwick, Peierls, Kowarski, and Perrin.
80 Fenning and Amphlett interviews
81 Mackenzie diary, 25 Feb. 1943; Mackenzie was not especially enlightening on the subjects discussed en masse; his notes concern Halban's conference with Urey and Fermi, and personnel questions.
82 AB1/129, Notes of a talk by Halban to senior UK staff in Montreal, 1 June 1943
83 Gowing, *Britain and Atomic Energy*, 162
84 AB1/80, MacDonald to Anderson, 30 June 1943
85 K.D. Nichols, *The Road to Trinity: A Personal Account of How America's Nuclear Policies Were Made* (New York 1987), 97–8
86 See J.W. Pickersgill, ed., *The Mackenzie King Record*, vol. 1 (Toronto 1961), 503, 531–2, diary entries for 19 May and 8 August 1943.
87 See Gowing, *Britain and Atomic Energy*, 164–77, and Vincent C. Jones, *Manhattan: The Army and the Atomic Bomb* (Washington 1985), 241.
88 Text of the Quebec agreement is in R.C. Williams and Philip L. Cantelon, eds., *The American Atom, 1939–1984* (Philadelphia 1984), 40–2. The suggestion of Howe was apparently Churchill's: Pickersgill, *King Record*, vol. 1, 536, King diary for 10 Aug. 1943.
89 PAC, RG 77, vol. 283, file 'R.R. General,' James Chadwick, 'Tube Alloy Project' (diary), 8 Sept. 1943. Jones, *Manhattan*, 242, says 'Howe had not yet arrived from Canada.' Since he had not been invited to attend, this is not surprising.
90 See Hewlett and Anderson, *New World*, 193–7, 200–3. The matter was complicated by conflicts between the Chicago scientists and the contractor, du Pont.
91 RG 77, vol. 283, Chadwick, 'Tube Alloy Project' (diary), 10 Sept. 1943.
92 Mackenzie diary, 9 Aug. 1943
93 Ibid., 27 and 29 Aug. 1943; on Chadwick, Lord Sherfield interview, London, 15 April 1987

94 Mackenzie diary, 27 July 1943
95 US National Archives (USNA), RG 77, box 60, entry 5, MED Decimal File, Arthur Compton to Leslie Groves, 'Exchange of information with British at Montreal, September 18, 1943,' 9 Oct. 1943; Mackenzie diary, 19 Sept. 1943
96 See AB1/137, Newell to Akers, 28 Oct. 1943, in which Newell described his attempt to persuade Mackenzie of the possibility of building a plant in Canada, at a price somewhere between £1 million and £10 million. It was in this conversation that Mackenzie revealed his ignorance of the military purpose of such a plant. For the conversation with Chadwick, Mackenzie diary, 8 Dec. 1943; for Halban's visit to Chicago, CRNL Records, 5000 Montreal Labs, Reactor Concepts, vol. 1, 'Report on the Visit to Chicago,' 8 Jan. 1944.
97 At a meeting in Montreal on 19 Dec. 1943 Mackenzie made it plain that he thought Canadian money would depend on 'what co-operative arrangements could be worked out' (diary). On Cockcroft see Wilfrid Eggleston, *Canada's Nuclear Story* (Toronto 1965), 94–5; G. Hartcup and T. Allibone, *Cockcroft and the Atom* (Bristol 1984), 126.
98 CRNL Records, 1500 Montreal Labs, file Admin, vol. 2, minutes of 5th meeting of the TCRR, 25 Jan. 1944; this meeting led directly to Laurence's letter: Laurence to Mackenzie, 4 Feb. 1944; Fenning interview; Amphlett interview; Goldschmidt interview.
99 Mackenzie diary, 17 Feb. 1944
100 AB1/197, Chadwick diary, 23 Feb. 1944
101 Ibid., 24, 27, and 28 March 1944; Gowing, *Britain and Atomic Energy*, 272–3; Jones, *Manhattan*, 246
102 CAB 126/13, Gorell Barnes to Anderson, 13 Jan. 1944; subsequently Malcolm MacDonald was instructed to tell Howe and Mackenzie that everything would be wound up, and to go down to Montreal to tell the scientists: AB1/113, Gorell Barnes to MacDonald, 22 March 1944; MacDonald to Anderson, 28 March 1944.
103 Mackenzie diary, 14 and 15 April 1944
104 Amphlett interview; Fenning interview; Hartcup and Allibone, *Cockcroft*, 127
105 Goldschmidt interview
106 Halban diary, 19 Sept. 1944
107 Ibid., 22, 24 Oct. 1944
108 Weart, *Scientists*, 205–6
109 Ibid.
110 Mackenzie diary, 20, 22, 23, 28 Dec. 1944. Howe had just come from an unpleasant experience with the British at an aviation conference in Chicago:

see R. Bothwell and J.L. Granatstein, 'Canada and the Wartime Negotiations over Civil Aviation,' *International History Review*, 2 (1980): 585–601; on Chadwick, Goldschmidt interview.

111 CAB 126/31, MacDonald to Anderson, 4 Feb. 1945; AB1/193, Chadwick to Cockcroft, 26 Jan. 1945, told the director that Howe was insisting on more authority over the project, though he thought it unlikely that he would get 'sole control.' The control would include approving all visits to England.

112 CRNL Records, Cockcroft to division heads, 8 March 1945

113 Goldschmidt, 'Halban', 5–7; see also R. Peierls, *Bird of Passage: Recollections of a Physicist* (Princeton 1985), 159–60: 'Halban was ... impatient with obstacles and delays, and, some said, [put] speed before accuracy.'

114 See Laurence, *Early Years*, 5–6.

115 Maurice Pryce interview, Toronto, 6 Dec. 1985; Weart, *Scientists*, 206: in April 1945 Joliot was told that none of the Montreal team would work in the same project as Halban ever again.

CHAPTER 2: ERRAND IN THE WILDERNESS

1 On Rutherford at McGill, see David Wilson, *Rutherford: Simple Genius* (Cambridge, Mass. 1983).

2 See Hartcup and Allibone, *Cockcroft*, 49–54; John Hendry, ed., *Cambridge Physics in the Thirties* (Bristol 1984), 103–22.

3 Hartcup and Allibone, *Cockcroft*, 87–8; Roger Makins at the British embassy in Washington enjoyed the spectacle of the thin and lugubrious Chadwick with the round and cheery Cockcroft: Sherfield interview.

4 Ibid., 101, relying on a letter from Mackenzie dated 1980

5 Ibid., 116

6 Virtually all interviews with personnel of the Cockcroft era concur on these points; see also Dr Kemmer's reference to 'the "little notebook" technique,' in Hartcup and Allibone, *Cockcroft*, 127.

7 Gowing, *Britain and Atomic Energy*, 278ff

8 CRNL Records, 5003, Plant Site Selection, vol. 1, Greville Smith (DIL) to Mackenzie, 15 April 1944; Mackenzie to Groves, 19 April 1944

9 Correspondence on the subject is very extensive: ibid., G.S. Anderson to Cockcroft et al., 12, 20, 21 June, and 5 July 1944.

10 Ibid., minutes of meeting held at CIL House in Montreal, 12 July 1944. In attendance were Mackenzie, Jackson, Newell, Steacie, and Laurence; they decided on Chalk River, subject to water tests; memorandum from Boulton, 'Experimental Plant for the Production of NRX,' 8 Aug. 1944; this memo is scaled down from one dated 3 August. Howe objected at a

meeting held on 9 August, in his office, with Cockcroft, Greville Smith, and Mackenzie and some of their staff; a refutation was drafted on 17 August; and Howe capitulated after reading the minutes of a meeting dated 18 August.

11 CRNL Records, 2410-2 Plant Construction, vol. 1, Report No. 1 of the DIL Special Projects Department, 9 Sept. 1944. The key man at DIL was Greville Smith: David Kirkbride interview, Oakville, 9 June 1986.

12 A copy of the contract, dated 7 July 1944, is in CRNL, W.B Lewis files.

13 CRNL Records, 2410-2 Plant Construction, vol. 1, DIL Report No 1, 9 Sept. 1944

14 Ibid.; Kirkbride interview

15 AB1/197, Chadwick to Cockcroft, 5 Aug. 1944. Chadwick also alerted Cockcroft to tell him of any financial problems, for Anderson was prepared to find the money if the Canadians would not (11 Aug. 1944).

16 Hartcup and Allibone, *Cockcroft*, 129, argue that Cockcroft realized that with construction resources scarce in Canada, NRX would take time.

17 Eggleston, *Canada's Nuclear Story*, 162, quoting Cockcroft

18 Halban diary, 25, 27 July, 7 Aug. 1944; Weart, *Scientists*, puts Kowarski's arrival in August. See also CRNL Records, 6000 Reactors – ZEEP, vol. 1, Notes on meeting to discuss the construction of a Zero Energy Pile, 27 July 1944.

19 It should be stressed that, while Kowarski was chef de projet, the credit for ZEEP belongs to his team as a whole: Pryce interview, Toronto, 6 Dec. 1985; see also Guéron, 'Lew Kowarski.'

20 CRNL Records, SDDO, CCC 17, Kowarski, 'First Report on the Proposed Zero-Power Polymer Pile,' 23 Aug. 1944; Bauer, May, and Pontecorvo assisted in preparing this report.

21 AB1/193, Groves to Chadwick, 6 Nov. 1944: two memoranda, same date

22 CRNL Records, 6000 NRX-Overall, vol. 3, Newell to G.B. McEwen, 2 March 1945; Newell to Cockcroft, 29 June 1945; on period jokes, see transcript from CBC (Ottawa) Radio Noon, 4–5 Sept. 1983: Cropley interview; see also AB 1/193, Cockcroft to Chadwick, 25 May 1945, stating that the only way to speed construction was to get more labour, which was unavailable: 'We are raising wages to see if this has any effect.'

23 Hartcup and Allibone, *Cockcroft*, 131. Groves unilaterally revealed many of the secrets of the atomic bomb when he authorized the official publication of the 'Smyth Report' in mid-August. What was not revealed there then became the 'secrets' of the future.

24 AB1/205, F.W. Fenning, 'Reports on the assembly operations,' 19 Sept. 1945. This document takes the form of brief weekly reports.

25 See Paul Boyer, *By the Bomb's Early Light: American Thought and Culture at the Dawn of the Atomic Age* (New York 1985), 22.

26 Quoted in Hartcup and Allibone, *Cockcroft*, 131
27 The standard account of Mackenzie is Leggett, *Mackenzie*; also, Porter, *Vertical Mosaic*, 431, where Mackenzie is described as 'the architect of modern science in Canada'; or John Spinks, *Two Blades of Grass: An Autobiography* (Saskatoon 1980), 61: 'one of the few scientific statesmen of our time.' For a more critical view, see F. Ronald Hayes, *The Chaining of Prometheus: Evolution of a Power Structure for Canadian Science* (Toronto 1973), 47–8.
28 Pryce interview; Roy Errington, Venice, Fla, 31 Jan. 1987; Mackenzie interview, Ottawa, 17 Aug. 1977
29 See Bothwell and Kilbourn, *Howe*, 186.
30 Wilfrid Eggleston, *National Research in Canada: The NRC*, 1916–1966 (Toronto 1978), 261, 269; Thistle, ed., *Mackenzie-McNaughton Wartime Letters*, 137–8; Bothwell and Kilbourn, *Howe*, 186
31 AB1/278, Cockcroft to Appleton, DSIR, 10 Oct. 1944
32 King diary, 12 April 1944, reporting discussion in the War Committee of the cabinet
33 Mackenzie diary, 1 Sept. 1944
34 Cockcroft and Steacie, 'The Future of the NRX Project in Canada,' 27 Sept. 1944, enclosed in ibid. Cockcroft and Steacie also suggested that the Chalk River site be used for experiments in metallurgy for the sheathing of uranium rods, aluminum being inefficient, and for research into 'piles to multiply stocks of the fissile elements,' what would later be called breeder reactors.
35 Gowing, *Britain and Atomic Energy*, 288
36 *Toronto Star*, 7 Aug. 1945
37 Hartcup and Allibone, *Cockcroft and the Atom*, 131–2
38 Ibid. The date is approximately August 1945.
39 Pryce interview
40 Margaret Gowing, *Independence and Deterrence*, vol. 1 (London 1974), 24–5
41 Howe Papers, vol. 13, file S-8-2(30), Bateman to Howe, 10 Aug. 1945, and Howe to Bateman, 17 Aug. 1945
42 Gowing, *Independence and Deterrence*, vol. 1, 133
43 Quoted in ibid., 133–4
44 Heeney left behind a set of memoirs, *The Things That Are Caesar's: The Memoirs of a Canadian Public Servant* (Toronto 1972), to be used cautiously because of their frequent inaccuracies.
45 PAC, RG 77, vol. 284, file RR V.2, Hume Wrong to Mackenzie, 22 Oct. 1945, and Mackenzie to Wrong, 26 Oct. 1945
46 CAB 126/13, D. Rickett to Stephen Holmes, 7 Nov. 1945
47 Mackenzie diary, 13 Nov. 1945
48 See Bothwell, *Eldorado*, 162, 176–7. Howe's objections to the Trust arrange-

ments were eventually dispelled when it became clear that he had misunderstood the original basis for the Trust. As for the tripartite declaration, it had to be sent on to Ottawa for Mackenzie King, who had already left the conference, to sign: Sherfield interview.

49 The British wanted information on gaseous diffusion enrichment plants and on the Hanford plutonium piles, which Truman's 'basic' qualification would prevent; it should be noted that neither side expected that the wartime alliance would be formally perpetuated and adjusted their behaviour accordingly: see John Simpson, *The Independent Nuclear State: The United States, Britain and the Military Atom*, 2nd ed. (London 1986), 38–9.
50 Goldschmidt, *Pionniers*, 339, my translation
51 Howe Papers, vol. 13, file s-8-2(30), Howe to Bateman, 18 Oct. 1945
52 Privy Council Records (RG 2), file w-46-A(2), Wrong to Norman Robertson, DEA, 22 Oct. 1945; House of Commons *Debates*, 5 Dec. 1945, 2959
53 PAC, Privy Council Records, file R-100-A, vol. 1, Heeney to Howe, 20 March 1946, Howe to Heeney, 23 March 1946, and N. Robertson to King, 26 March 1946
54 George Laurence interview, Deep River, 24 July 1985
55 The question of recognizing McNaughton was much on King's mind in the spring of 1946: see J.W. Pickersgill and D.F. Forster, *The Mackenzie King Record*, vol. 3 (Toronto 1970), 253–6. See also George Ignatieff, *The Making of a Peacemonger* (Toronto 1985), 89–90.
56 See Gordon Sims, *A History of the Atomic Energy Control Board* (Ottawa 1981), 24–5; Bothwell, *Eldorado*, 171–2.
57 See Bothwell, *Eldorado*, 172–3; Sims, *Atomic Energy Control Board*, 40–1.
58 Mackenzie diary, 14 Feb. 1946, reporting Howe's views; as late as December 1946 Howe mentioned to Pearson his hope that the British might soon approach Canada 'with a view to sharing in the Chalk River enterprise.'
59 Bertrand Goldschmidt, *The Atomic Complex: A Worldwide Political History of Nuclear Energy* (La Grange Park, Ill 1982), 64–5; CRNL Records 1500 Montreal Labs – Admin., vol. 5, Chadwick to Cockcroft, 27 Dec. 1945. See also CAB 126/31.
60 See Andrew Boyle, *The Fourth Man* (New York 1979), 287–8; Norman Moss, *Klaus Fuchs: The Man Who Stole the Atom Bomb* (London 1987), 89–90; Robert Bothwell and J.L. Granatstein, eds., *The Gouzenko Transcripts* (Ottawa 1982), 344–5. The Soviets were certainly intrigued by the Canadian atomic project. One garbled version of events in 1944, citing Steacie as the ultimate (and unwitting) source, correctly identified the upper St Maurice region near Grand'-Mère as a site for a 'pilot plant' to produce uranium for bombs: ibid., 121.

61 See the introduction to *The Gouzenko Transcripts* for the context of the affair, and for King's reactions.
62 Mackenzie Papers, vol. 5, file Atomic Energy vol. 2, Mackenzie to Zinn, 17 April 1946; Mackenzie diary, 13 April, 6 May 1946; Les Cook interview, Summit, NJ, April 1985
63 RG 77, vol. 283, file Personnel CR, Howe to Mackenzie, 11 June 1946
64 AB16/268, Cockcroft to Appleton, 7 Feb. 1946
65 AB1/193, Chadwick, notes of meeting with MacDonald, Howe and Mackenzie, 23 March 1946; Chadwick to O.S. Franks, Ministry of Supply, 27 March 1946; AB16/271, MacDonald to Secretary of State for Dominion Affairs, 27 March 1946; Chadwick to Sir John Anderson, 8 April 1946
66 AB16/268, F.C. How to permanent secretary, 4 March 1946 (erroneously dated 4 Feb.), enclosing a draft letter from Cripps to Howe, dated 1 March
67 Ibid., acting high commissioner to Sir R. Makins, 17 May 1946; Dominions Office to acting high commissioner, 23 May 1946; Cockcroft (then at Harwell) to undersecretary, 7 July 1946
68 *Regina Leader-Post*, 6 Aug. 1945; *Toronto Star*, 7 Aug. 1945
69 Volkoff and Pryce interview, Toronto, 6 Dec. 1985
70 Ibid.
71 Richard Rhodes, *The Making of the Atomic Bomb* (New York 1986), 486–7. The reference was to Al Capp's 'Li'l Abner' comic strip, whose characters dwelt in a hillbilly slum called Dogpatch.
72 Joan Finnegan, *Some of the Stories I Told You Were True* (Ottawa 1981), 2–3; *Financial Post Survey of Markets and Business Year Book* (Toronto 1944)
73 Paul Martin, *A Very Public Life*, vol. 1 (Ottawa 1983), 4; in 1960 Pembrokers earned 4 per cent less than the national average. However, the city enjoyed strong population growth, 16 per cent during the 1950s: *Financial Post Survey of Markets and Business Year Book*, 1944–60.
74 CRNL Records, 2410-2 Plant Construction, vol. 2, Record of meeting held at Montreal, 18 Dec. 1944, with Mackenzie, Cockcroft, Boulton, Steacie, McEwen, Greville Smith, and Newell in attendance
75 Ibid., memorandum of meeting at DIL in Montreal, 15 Aug. 1945. Beds would be needed for 2650 men, 200 more than currently available.
76 AB1/197, Cockcroft to Chadwick, 5 Aug. 1944, enclosing preliminary estimates for the Chalk River and Deep River sites; CRNL Records, 2410-2, Plant Construction, Record of meeting held at DIL, 1 and 12 Sept. 1944; Les Cook, 'The Birthpangs of CANDU: The Chalk River Story,' unpublished ms, 91–2; Hartcup and Allibone, *Cockcroft and the Atom*, 132–4
77 Cook, 'Birthpangs,' 92; Yaffe interview, 16 July 1985; Sargent interview

78 CRNL Records, 2410-2 Plant Construction, vol. 2, Record of meeting, 25 Feb. 1946
79 Goldschmidt interview; Fred and Eileen Fenning interview. The same problem occurred at Los Alamos: R.P. Feynman, *Surely You're Joking, Mr. Feynman* (London 1985), 113.
80 Goldschmidt, *Pionniers*, 340; Alan H. Armstrong, 'Deep River: New Canadian Town,' *Canadian Welfare*, 15 Oct. 1948, 23–9; Freda Kinsey, 'Life at Chalk River,' *Atomic Scientists Journal* 3(1) (Sept. 1953): 18–23
81 Yaffe interview, Montreal, 25 March 1986; Cook, 'Birthpangs,' 100; Cockcroft described the situation to his mother at about the same time: 'They have a shopping bus once a week and the laundry, eggs, milk and bread are delivered. Later, there will be a village store.' Quoted in Hartcup and Allibone, *Cockcroft and the Atom*, 132
82 Cook, 'Birthpangs,' 93–4
83 Ibid.
84 Mackenzie Papers, vol. 5, file Atomic Energy vol. 2, Mackenzie to W.J. Bennett, 18 April 1946; Mackenzie to V.W. Scully, 27 May 1946
85 Armstrong, 'Deep River'
86 Cook, 'Birthpangs,' 94–9
87 Yaffe interview, 25 March 1986
88 Anderson, 'Deep River,' passim; Pierre Berton, 'The Mayor of Atom Village,' *Maclean's Magazine*, 1 Dec. 1947, 9, 71–3
89 See Kinsey, 'Life at Chalk River.'
90 Ibid., 20–1
91 Peter C. Newman, 'Deep River – almost the perfect place to live,' *Maclean's Magazine*, 13 Sept. 1958, 24–5
92 Rhodes, *The Making of the Atomic Bomb*, 394–7
93 Howe Papers, vol. 7, file s-8-2(10), na, 'Discussion on Atomic Problems,' 13 Dec. 1946, held at the Canadian embassy in Washington; Howe was speaking to Lord Inverchapel and Field Marshal 'Jumbo' Wilson. CRNL Records, 5003 Plant Site Selection, vol. 1, Groves to Cockcroft, 26 Sept. 1944; Jones, *Manhattan*, 246–7
94 See Eggleston, *Canada's Nuclear Story*, 175; USNA, RG 77, MED Decimal File, 334 British, box 60, entry 5, Major H. Benbow, 'Combined Meeting of the Montreal and American Group on 12 June 1944, at Montreal, Canada,' 15 June 1944
95 Cook,'Birthpangs,' 23–4
96 Ibid., 26–7
97 NA, RG 77, MED files, 334 British, box 60, Capt. J.H. McKinley to Patent Advisor, 13 Nov. 1944; Hewlett and Anderson, *New World*, 305–8; Rhodes, *Making of the Atomic Bomb*, 559–60

98 NA, RG 77, MED Decimal File, 680.2 (Montreal), box 81, entry 5, Eugene Wigner to W.W. Watson, 25 Oct. 1944
99 AB1/193, Groves to Chadwick, 6 Nov. 1944. Groves recommended that ZEEP should be scrapped, because it involved Montreal 'splitting its effort' to no very good purpose, since ZEEP could not possibly be ready in time to help the NRX design.
100 Eggleston, *Canada's Nuclear Story*, 136–7. Water flowing through NRX becomes radioactive, and is held in storage until the contamination has diminished to a safe level.
101 AB1/197, Cockcroft to Chadwick, 5 Aug. 1944; CRNL Records, 5003 Plant Site Selection, memorandum by B.K. Boulton, 8 Aug. 1944
102 AB1/353, Chadwick to Portal, 29 April 1946; Britain had decided for the time being to concentrate on plutonium rather than enriched uranium: Simpson, *Nuclear State*, 45.
103 Cook, 'Birthpangs,' 27–8, 39–40; Goldschmidt interview; Eggleston, *Canada's Nuclear Story*, 199
104 Mackenzie Papers, diary, 3 Dec. 1946, 9 and 15 Jan. 1947; vol. 5, Mackenzie to Lewis, 9 Dec. 1946; for the chemistry section, see AB16/271, na, 'A brief description of Project N.R.X.,' nd but probably late 1946; see also Margaret Gowing, *Independence and Deterrence*, vol. 2 (London 1974), 404–9.
105 Cook, 'Birthpangs,' 43–6
106 Eggleston, *Canada's Nuclear Story*, 205
107 Ibid., 2410-2 Plant Construction, vol. 1, NRX Plant estimate, 19 Jan. 1945
108 See CRNL Records 6000 NRX-Overall, vol. 2, A.B. McEwen to Cockcroft, 24 Jan. 1945; on draftsmen, 2410-2, Plant Construction, vol. 2, Report of Meeting in Montreal, 21 Jan. 1946 (shortage for the 23 Plant), Record of meeting in Montreal, 29 May 1946, where DIL reported that drafting staff were leaving at a steady rate with their drawings unfinished.
109 Gib James interview, Deep River, 27 Aug. 1985
110 C.H. Jackson, 'The Chalk River Atomic Energy Project,' *Engineering Journal* (June 1947): 265–6
111 CRNL 2410-2 Plant Construction vol. 2, Record of Meeting held in Montreal on the 20/9/45
112 Mackenzie Papers, vol. 5, Mackenzie to Cockcroft, 22 Jan. 1947
113 Cook, 'Birthpangs,' 36
114 Mackenzie diary, 21, 23 Oct. 1946
115 James interview
116 Eggleston, *Canada's Nuclear Story*, 186–7; Mackenzie diary, 2, 17, 22 July 1947

CHAPTER 3: ATOMIC ENERGY?

1 Thomas S. Kuhn, *The Structure of Scientific Revolutions*, 2nd ed. (Chicago 1970), 164

2 NRC *Review, 1951* (Ottawa 1951), 142: the visitors were thirty members of the Parliamentary Press Gallery.

3 W.B. Lewis, 'The Development of Electrical Counting Methods in the Cavendish,' in Hendry, ed., *Cambridge Physics in the Thirties*, 133; other information on Lewis is derived from his entry in *The Canadian Who's Who*, x, 1964–66 (Toronto 1966).

4 See the British House of Commons Select Committee on Energy, *First Report*, 3 vols (London 1981): Bowden stated that Canadian power stations 'are the best in the world today': vol. 2, 145–53.

5 Lorne Gray interview, Toronto, Feb. 1987

6 PAC, RG 77, Acc. 85-86/178, Boyer to Lewis, 26 March 1947; Simpson, *Independent Nuclear State*, 84

7 Mackenzie Papers, vol. 5, file Atomic Energy vol. 2, Mackenzie to L.G. Cook, 26 Jan. 1946, Mackenzie to Lewis, 9 Dec. 1946; Yaffe interview, 16 July 1985

8 See Walter A. MacDougall, ... *the Heavens and the Earth* (New York 1985), 47–50.

9 Nichols, *Road to Trinity*, 271: 'We were trying to get the British to agree to concentrate their fissionable material production program in Canada in exchange for renewal of full cooperation on building a gaseous diffusion plant in Canada and joint weapons development at Los Alamos.'

10 RG 2/18, vol. 100, file R-100-A vol. 2, Howe to Heeney, 8 Sept. 1947

12 *Foreign Relations of the United States* (*FRUS*), 1947, vol. 1 (Washington 1973), 848: F.H. Osborn to Dean Rusk, 29 Oct. 1947; 786: Dean Acheson, Memorandum of Conversation with Roger Makins, 1 Feb. 1947

13 Howe Papers, vol. 12, file s-8-2(25), Bateman to Howe, 4 Feb. 1947; Howe to Bateman, 7 Feb. 1947: 'I am inclined to let sleeping dogs lie.' That was Groves's advice too: Bateman to Howe, 17 Feb. 1947. David Lilienthal, the first chairman of the USAEC, reminded senators that 'we had a heavy investment' in Chalk River and that the Canadians were, after all, only looking for the 'most advantageous use' for it: *FRUS*, 1947, vol. 1, 876: Minutes of meeting of the US members of the CPC with senators Hickenlooper and Vandenberg, 26 Nov. 1947.

14 "Canada Doubts Atomic Power in General Use,' *Financial Post*, 14 Sept. 1946

15 Mackenzie diary, 30 Sept. 1947, reflecting a phone conversation with Bennett; Howe Papers, vol. 12, file s-8-2(26), Bateman to Howe, 16 Feb.

1948, Howe to Mackenzie, 19 Feb. 1948, Howe to McNaughton, 25 Feb. 1948, Mackenzie to Howe, 25 Feb. 1948

16 See, for example, AB6/362, Cockcroft to D. Peirson, 20 Feb. 1947. Cockcroft shrewdly commented that 'Mackenzie and Howe ... do not show their hands unless you know them very well.'

17 Mackenzie Papers, vol. 5, file Atomic Energy vol. 2, Mackenzie to Cockcroft, 18 March 1947; Howe Papers, vol. 12, file S-8-2(26), Howe to Mackenzie, 19 Feb. 1948

18 According to the *Financial Post*, 18 March 1944, DIL's operations had cost the public $600 million; fees accounted for 0.2 per cent of that sum: $1.2 million. According to an estimate in March 1945 DIL had produced $766 million worth of goods to date (*FP*, 17 March 1945). The parallel between DIL-CIL and their parent company, du Pont, is striking. According to the historians of du Pont, that company's wartime involvement in atomic energy was wound up because it believed that electric power would be next, a step 'quite removed from the company's present line of business.' Du Pont terminated its operation of Hanford in November 1946. See G.D. Taylor and P. Sudnik, *DuPont and the International Chemical Industry* (Boston 1984), 156–7. They note that in 1950 du Pont was persuaded to re-enter the field with the construction of the Savannah River reactors, with which Canada also had a connection.

19 CRNL Records 2410-2 Plant Construction, vol. 3, A.N. Budden to C.D. Howe, 23 Sept. 1946; Howe Papers, vol. 12, file S-8-2(25), Howe to V.W. Scully, 11 Nov. 1946: he had just received Budden's pessimistic report and was enquiring whether things were as black as painted; PAC, RG 60, vol. 13, 'Rough notes of first meeting of AECB,' 16 Oct. 1946; Mackenzie Papers, vol. 5, file Atomic Energy vol. 2, Mackenzie to Dr Duncan Graham, 28 Oct. 1946; Mackenzie to F.C. Wallace, 28 Jan. 1947.

20 Mackenzie Papers, vol. 5, Mackenzie to Brigadier F.C. Wallace, 28 Jan. 1947; Mackenzie to Chadwick, 12 Feb. 1947

21 Mackenzie diary, 28 Oct. 1946

22 Ibid., 30, 31 Oct., 4, 6, 7, 9 Nov., 12, 14, 16, 17, 20, 27 Dec. 1946, 3, 4, 8, 16, 17 Jan. 1947; Mackenzie Papers, vol. 5, file Atomic Energy vol. 2, Mackenzie to F.C. Wallace, 28 Jan. 1947

23 Mackenzie diary, 14, 27 Dec. 1946, 3, 8, 17 Jan. 1947. Keys asked, and apparently got, $10,000 a year and his moving expenses; Mackenzie Papers, vol. 5, file Atomic Energy vol. 2, Mackenzie to Keys, 10 Jan. 1947

24 Mackenzie diary, 17 Dec. 1946, 3 Feb. 1947, claims that CIL offered to take anyone who didn't wish to stay. David Kirkbride casts some doubt on this (interview). The five who returned to CIL in 1946–7 were those who had worked there before the war. (One CIL veteran chose to stay at Chalk River.)

25 AECB Minutes, 15 March 1947; Yaffe interview
26 On organization, see Howe Papers, vol. 10, file S-8-2(20), Mackenzie to Howe, 19 Nov. 1951.
27 Mackenzie diary, 12 Feb., 13 June 1947, 12 April 1948
28 AB2/888, J.D. Cockcroft, 'The Development of Nuclear Physics in the Universities of Canada,' 11 June 1946
29 Ibid.; Geoffrey Hanna interview, Deep River, 28 Aug. 1985
30 RG 60, vol. 13, AECB minutes, 27 Sept. 1947
31 Ibid., file 1-3-2, Minutes of AECB, 6 Jan. 1947; Mackenzie diary, 19, 21 Nov. 1946
32 CRNL Records, 6000/5 Reactor Incidents – 1947 NRX Accident vol. 1, 'Report of the Fact Finding Board Investigating the Accident at Chalk River'; 6000-NRU-Overall, vol. 1, Tupper to Lewis, 18 Nov. 1948; Gordon Hatfield interview, Rosemere, 26 March 1986
33 CRNL Records, SDDO, Keys Reports, vol. 1: April and May 1948 (8 June 1948) and April and May 1949 (4 July 1949)
34 AB16/271, Cockcroft, 'Canadian-U.K. Collaboration in Atomic Energy,' 2 Nov. 1948; for Hinton's visits, Mackenzie diary, 1 Oct. 1947, 21 Feb. 1949; on Pontecorvo, ibid., 20 Jan. 1948
35 These examples are taken from *The National Research Council Review*, 1949 (Ottawa 1949).
36 AB6/513, J.D. Cockcroft, 'Visit to Chalk River, October 1948, Part VI'
37 AB16/271, W.A. MacFarlane, UK Scientific Mission, Washington, to Cockcroft, 21 Jan. 1949; Mackenzie was pleased by a proposal from Cockcroft dated 17 Jan. 1949 that envisaged using Chalk River to study tactical nuclear weapons as well as means of defence against atomic weapons.
38 RG 60, vol. 13, AECB minutes, 27 Sept. 1947
39 W.B. Lewis office files, Lewis, 'Possible Future Projects,' 14 Oct. 1946
40 CRNL Records, W.B. Lewis, *Director's Report* (DR-1), 'World Possibilities for the Development and Use of Atomic Power,' March 1947. Of equal interest is a paper by George Laurence, produced in response to Lewis's DR-2, 'The Future Atomic Energy Pile' and dated 11 Sept. 1947: CRNL Records, 6000-NRU-Overall, vol. 1. In this paper Laurence stressed the importance of a unit capable of replenishing its fuel supply, but disputed the necessity of the 'multiplication' of fuel under Canadian conditions.
41 Ibid., Lewis to Mackenzie, 11 Sept. 1947
42 CRNL Records, 1615/F1 Future Systems Committee, vol. 1, Paper for the Future Systems Committee Drafting Subcommittee vol. 1. There is no date on the paper, but it is clearly one of the first if not the first in the series.
43 Eggleston, *Canada's Nuclear Story*, 253–6
44 Ibid.; on changing standards, see Gordon Stewart interview, Deep River, 28

Sept. 1985; CRNL Records, 1735 AECB vol. 1, Cipriani to Lewis, 11 Dec. 1947.

45 See Bothwell, *Eldorado*, 3–9.

46 Quoted in Rhodes, *Making of the Atomic Bomb*, 238

47 See Szilard to Strauss, 25 Jan. 1939, in Weart and Szilard, eds., *Leo Szilard*, 62.

48 Quoted in Eggleston, *Canada's Nuclear Story*, 257–8

49 See Richard Hewlett and Francis Duncan, *A History of the United States Atomic Energy Commission: Atomic Shield, 1947–1952* (Washington 1972), 252–3.

50 See D.G. Hurst, *The Development of* NRX *as a Research Reactor* (AECL-473: Chalk River, August 1957), 3–4, 6.

51 *NRC Review, 1949*, 54

52 Bennett to author, 18 Feb. 1987

53 Errington interview

54 See Bothwell, *Eldorado*, 225–6; Errington interview.

55 *NRC Review, 1948*, 61–2, 1949, 65, 1951, 168–9; Commercial Products, *Annual Report, 1952–3* (mimeo), 1

56 Eldorado Papers, file 1-8-15, Bennett to Mackenzie, 22 Oct. 1951; Eldorado board minutes, 25 Oct. 1951; Bennett to Mackenzie, 11 June 1952; Bennett to author, 18 Feb. 1987

57 Errington interview

58 Commercial Products, *Annual Report* (mimeo), 1951–2, 8

59 Ibid., 1949–50, 8

60 Bennett to author, April 1987

61 RG 60, vol. 13, draft minutes of AECB, 31 March 1948; Mackenzie diary, 11 Nov. 1948, 21 Feb. 1949

62 Mackenzie diary, 16 Feb. 1949

63 Howe explained his point of view in a later letter to Douglas Abbott, the finance minister: Howe Papers, vol. 11, file S-8-2(21), Howe to Abbott, 16 Sept. 1950.

64 External Affairs Records, file 50219-A-40 (partly illegible), minutes of the Advisory Panel, 23 Dec. 1948. The minutes are badly torn and words in brackets are conjectural.

65 Ibid., minutes of the Advisory Panel, 4 April 1949. Mackenzie once again referred to the presumed lifespan of NRX, which he estimated at five years; Nichols, *Road to Trinity*, 270–1.

66 CPC minutes are in FRUS, 1949, vol. 1, 529–50; Mackenzie diary, 29 Sept. 1949; see Gowing, *Independence and Deterrence*, vol. 1, 284–9.

67 Department of External Affairs [DEA] Records, file 201-F(S), minutes of meeting of the Advisory Panel, 10 May 1949; ibid., Robertson to Heeney, 8 Nov. 1949

68 Howe Papers, vol. 11, file s-8-2(21), Wrong to Robertson, 29 Nov. 1949
69 Ibid., Sumner Pike (USAEC) to Mackenzie, 24 Aug. 1950; Mackenzie diary, 28 Aug. 1950; RG 60, vol. 13, draft minutes of AECB meeting, 17 Feb. 1950
70 Howe Papers, vol. 11, file s-8-2(11), Mackenzie to Howe, 29 Aug. 1950, Howe to Mackenzie, 29 Aug. 1950; Mackenzie diary, 29 Aug. 1950
71 RG 60, vol. 13, draft minutes of AECB meeting, 5 Sept. 1950: present were Mackenzie, Paul Gagnon, Omond Solandt, and V.W. Scully, the deputy minister of national revenue and former deputy minister of reconstruction.
72 Mackenzie diary, 12 Sept., 3 Nov. 1950; External Affairs records, file 201-F(s) or 50219-A-40, minutes of Advisory Panel, 10 Nov. 1950
73 RG 2/16, cabinet conclusions, 13 Dec. 1950
74 Mackenzie diary, 17, 18 Jan., 21 Aug. 1951. Mackenzie had some leverage, since the Americans were anxious to get as much information as they could from Chalk River for their next generation of plutonium factories along the Savannah River. The AEC commissioners, Mackenzie said, were acting 'like Russians,' more inclined to facilitate deliveries of American heavy water to Canada than to contemplate Canada opening its own plant.
75 Bennett to author, April 1987
76 Howe Papers, vol. 10, file s-8-2(20), Mackenzie to Howe, 19 Nov. 1951; Howe to Mackenzie, 19 Nov. 1951
77 Ibid., Mackenzie to Howe, 5 Dec. 1951; Howe to Mackenzie, 8 Dec. 1951; Ken Palmer to Howe, 11 Dec. 1951; Howe to Saint Laurent, 15 Jan. 1952, enclosing a list of directors; Mackenzie to Howe, 12 March 1952. Howe also recommended that Steacie succeed Mackenzie at NRC, and that Mackenzie in his new post receive a salary of $25,000 a year.
78 Errington interview, 31 Jan. 1987

CHAPTER 4: A CANADIAN REACTOR

1 Norman Polmar and Thomas B. Allen, *Rickover: Controversy and Genius* (New York 1982), 124–5
2 Richard G. Hewlett and Francis Duncan, *Nuclear Navy, 1946–62* (Chicago 1974), 63–5, 79–80; Peter Pringle and James Spigelman, *The Nuclear Barons* (New York 1981), 156
3 Hewlett and Duncan, *Nuclear Navy*, 98
4 Ibid., 59
5 Ibid., passim
6 Gowing, *Independence and Deterrence*, vol. 2 (London 1974), 298–9
7 Ibid., 408–9
8 Mackenzie expressed himself remarkably freely on the subject, telling a

British scientific envoy in November 1951 that 'he attached the utmost importance to free interchange between Canadian and British scientists not merely of results but of plans for future work.' He was confident that American objections would not cripple the Anglo-Canadian relationship: FO 371/93204, Angus McFarlane to Sir Christopher Steele, 30 Nov. 1951; see also AB16/751, Cockcroft, 'Canadian–United Kingdom Collaboration in Atomic Energy,' 5 April 1952; Roger Makins, memorandum of conversation between Norman Robertson, C.J. Mackenzie, John Cockcroft, and himself, 6 Nov. 1952.

9 Mackenzie claimed as much in his November 1951 conversation with McFarlane, above, note 8; Hewlett and Duncan, *Atomic Shield*, 379, 401, 424.

10 George Mercer interview, Washington, 18 April 1985; Gib James interview. One sticking point was getting the Americans to take the financial consequences if one of their experiments shut down the reactor. X-rays would not disclose the exact contents of a package, or whether any liquids were present. The Canadians were still relying on good American engineering for their final assurance.

11 Hewlett and Duncan, *Atomic Shield*, 481

12 Errington interview, 31 Jan. 1987

13 Curtis Nelson interview, Washington, 17 April 1985; Hewlett and Duncan, *Atomic Shield*, 523–4, 531–2

14 Hewlett and Duncan, *Atomic Shield*, 377–9

15 Lorne McConnell interview, Toronto, 17 Sept. 1986

16 Eric Perryman interview, Chalk River, 26 Sept. 1985

17 D.G. Breckon and J.H. Collins, *Control and Safety in the Operation of the NRX and NRU Reactors*, AECL-1486, April 1962, 4–5; current problems were outlined by Gilbert James in an AECL special report, NEI-12, 'Troubles and Problems in the Operation of NRX Reactor,' 16 Nov. 1951.

18 See Hurst, *Development of NRX*, 3–4, 6.

19 Cook, 'Birthpangs,' 174

20 Keys's report on the incident is quoted in ibid., 204; see also CRNL Records, 2181 Industrial Accidents –Bldg 224 Explosion, vol. 1, Preliminary Report on the explosion of the effluent evaporator, 13 Dec. 1950, Russell to Laurence, 14 Dec. 1950; 'Report by Board of Inquiry into Explosion of Effluent Evaporator in Building 224 Atomic Energy Project, December 13 1950'; this board, meeting on 15 December, established that ammonium nitrate had caused the explosion. See also F. Prosser to W.R. Livingston, 19 Dec. 1950, and Keys to Cockcroft, 2 Feb. 1951.

21 See Breckon and Collins, *Control and Safety in the Operation of the NRX and NRU Reactors*.

22 I am grateful to Dr Gordon Stewart of Deep River for the clarification: phone interview, 21 April 1986.
23 See Gowing, *Britain and Atomic Energy*, 284.
24 See Bothwell, *Eldorado*, 222–3.
25 Public Archives of Ontario, RG 10, box 2, file 'Port Hope, 1945–1947,' F.M.R. Bulmer and R.G. Elson to Dr Cunningham, 15 Oct. 1945
26 CRNL Records, 6000 NRX-Overall, vol. 1, Laurence to Cockcroft, 15 July 1944; see also Cook, 'Birthpangs,' 119.
27 Donald Hurst interview, Ottawa, 4 Feb. 1986; Eggleston, *Canada's Nuclear Story*, 254–5
28 Ibid., 5000 Montreal Labs – Reactor Concepts, vol. 2, Report of a visit to 'x' by C.H. Jackson, 11 Sept. 1944
29 Cook, 'Birthpangs,' 119
30 CRNL Records, 6001 NRX Reactor Incident 1952 Overall, vol. 1, Gilbert to Gray, 3 Oct. 1956
31 The above relies on a variety of sources, of which the most important is W.B. Lewis's report, dated 9 January 1953, DR-29, CRNL Records 6001 NRX incident 1952 overall, vol. 1; Les Cook interview (phone), 25 May 1987; Donald Hurst interview, Ottawa, 27 May 1987; Eggleston's account, in *Canada's Nuclear Story*, 215ff, is freely based on Lewis's. Les Cook's version is found in his memoir, 'Birthpangs,' 116–24. There is also a concise version in Patterson, *Nuclear Power*, 120–1, and a vivid reconstruction in John G. Fuller, *We Almost Lost Detroit* (New York 1975), 11–18.
32 CRNL Records, 6001 NRX Reactor Incident 1952 Overall, vol. 1, Gilbert to Gray, 3 Oct. 1956; Donald Campbell interview, Toronto, 13 May 1985
33 Polmar and Allen, *Rickover*, 621. Personnel, civilian and military, were sent from three AEC sites.
34 See Eggleston, *Canada's Nuclear Story*, 225–6, relying on Lorne Gray's 1953 account.
35 Quoted in ibid., 227
36 Cited in ibid., 227
37 E.R. Siddall interview, Aug. 1987
38 CRNL Records, 1615/N13, NRU Development Committee, vol. 1, Notes of Sixth meeting, 10 and 11 Sept. 1951
39 Ibid., 6000 NRU-Overall, vol. 5, Beynon to Laurence, Lewis and Gray, 12 Feb. 1953
40 Ibid., vol. 1, Laurence to Gray, 13 June 1950; 1615/F1 Future Systems Group, vol. 1, 'Cost Estimates for the Construction and Operation of the Atomic Energy Reactor,' July 1950
41 Ibid, Lewis to Mackenzie, 24 Oct. 1950, and Lewis to Ian MacKay, 10 Nov. 1950. On security, see Joe Holliday, *Dale of the Mounted: Atomic Plot* (Toronto 1959).

42 CRNL Records, 6000 NRU-Overall, vol. 2, Memorandum of a meeting between Lorne Gray and officers of the C.D. Howe and Foundation companies, 30 Aug. 1951. The assertion about the 1951 estimates is contained in a table labelled 'NRU Cost estimate,' attached to the minutes of the executive committee of the AECL Board of Directors, 21 Jan. 1954.
43 Ibid., 1615/N13 NRU Development Committee, vol. 1, Notes from ninth meeting, 17 Dec. 1951; ibid., memorandum by Winnett Boyd, 30 Nov. 1951; 6000 NRU, vol. 3, memorandum by I.N. MacKay in response
44 Ibid., 6000 NRU, vol. 3, Lewis to Gray, 29 July 1952, Culpeper to Gray, 6 Aug. 1952, Lewis to Gray, 12 Aug. 1952; vol. 5, Lewis to Gray, 17 Dec. 1952; Culpeper to Gray, 22 Dec. 1952. On designers see file 1615/P14, Power Reactor Group, vol. 1, minutes of fifth meeting of the PRG, 14 July 1954, for Hatfield's remarks.
45 Figures derive from the table 'NRU Cost Estimate' attached to the AECL executive committee minutes of 21 Jan. 1954.
46 See F.M. Sayers to W.J. Bennett, 17 Jan. 1956, attachment 2 to AECL executive committee minutes, 24 Jan. 1956; executive committee minutes, 4 Oct., 14 Dec. 1956.
47 AECL executive committee minutes, 5 and 18 Sept. 1957
48 See George Laurence, *The NRU Reactor*, AECL-525 (Chalk River 1957).
49 Ibid.
50 See AECL, *Annual Report 1958–59*, 11.
51 The first politicians actually to visit Chalk River were American. The trip was an effort to soothe their fears about the security of the US atomic secret.
52 Lewis, 'The Gleam in the Eye of the Atomic Scientist,' 15 Nov. 1949, DL-6, AECL Records, Chalk River, SDDO
53 See *Canada Year Book, 1947* (Ottawa 1947), 472–3 and 892. Oil had gradually replaced all other commodities as Canada's single most costly import.
54 *Canada Year Book, 1950* (Ottawa 1950), 563

CHAPTER 5: NUCLEAR POWER DECISIONS

1 See Bothwell and Kilbourn, *Howe*, chapters 15 and 16.
2 AECL, *Rapport Annuel, 1952–53*, Cédule II. Parliamentary votes netted the company $7.4 million; sales and rentals came to $418,000.
3 See the treatment in Sims, *Atomic Energy Control Board*, 42.
4 CRNL Records, file 1615/P14, Power Reactor Group, vol. 1, minutes of the fifth meeting of the Power Reactor Group. Lewis listed the greater chance of being financed as reason number one for the group's existence.
5 CRNL Records, Lewis office files, miscellaneous, C. Secord, 'Heat and Power from Nuclear Energy,' September 1946, Lewis, 'Possible Future

Projects,' 14 Oct. 1946; Lewis to Clayton, L.G. Cook, Hurst, Tupper, Sargent, and Watson, 29 Nov. 1946; 'Re: Heat and Power from Nuclear Energy'

6 Put another way, and in terms more commonly used today, a metric tonne of coal (a long ton in 1946) produces 29,000 megajoules (MJ); a tonne of oil yields 42,000 MJ; but 1 kilogram of natural uranium, one-thousandth of the quantity of oil or coal, produces 580,000 MJ. Figures taken from Janet Ramage, *Energy: A Guidebook* (Oxford 1983), 331.

7 CRNL, SDDO, DR-1, 'World Possibilities for the Development and Use of Atomic Power,' March 1947

8 Ibid., minutes of meetings to discuss Secord report, 6 and 9 Dec. 1946

9 Ibid., SDDO, DR-2, 'The Future Atomic Energy Pile,' 8 Sept. 1947. The branch heads at Chalk River believed Lewis's program to be too ambitious, something Lewis admitted in a letter to Mackenzie on 9 Sept. 1947: CRNL Records, file 6000 NRU-Overall, vol. 1. He recommended merging Chalk River's effort with the UK's to compensate for the laboratory's deficiencies. See also ibid., file 1615 F1, Future Systems Group vol. 1, correspondence from April 1948, for alternative ideas, such as plutonium cores (Laurence) or beryllium moderators (Hurst).

10 Ibid., file 1615 R4, vol. 1, 'First Report of the Reactor Development Panel to the Future Systems Committee,' 20 May 1949

11 Ibid., SDDO, Lewis, Director's Lecture (DL-5), 'Whither Atomic Energy Research?' 20 Oct. 1949

12 Ibid., file 6000 NRU-Overall, vol. 1, Laurence to Lewis, 30 March 1950. See also Laurence's later, and consistent, comment in file 5001 Candu Reactor Concepts, vol. 1, Laurence to Lewis and Mackenzie, 22 Oct. 1953.

13 CRNL Records, SDDO, DR-18, 'An Atomic Power Proposal,' 27 August 1951

14 Ibid., file 1800 UK/Overall, vol. 1, Hinton to Lewis, 28 Feb. 1952. See also Gowing, *Independence and Deterrence*, vol. 2, 298–9. Gowing comments that 'no attempts at formal integration were made, primarily because the [British] feared that the Canadians would be embarrassed in their relations with the Americans if they were too closely tied to Britain.'

15 CRNL Records, file 1615/F1, Future Systems Group, vol. 1, Minutes of Nuclear Physics and Reactor Engineering Group Meeting, 2 July 1952. The other items on the shopping list were a linear accelerator for studying particle physics, and a high temperature reactor consuming uranium-233.

16 Ibid., Minutes of Metallurgy and Chemical Processing Group Meeting, 25 July 1952. Lewis presided; in attendance were Les Cook, Runnalls, Hatfield, and Gilbert, among others.

17 Ibid., Minutes of the first meeting of the Fast Neutron Reactor Panel, 29 July 1952

18 Mackenzie diary, 12 Aug. 1952

19 Ibid., 3 Sept. 1952

20 Ontario Hydro Archives (OHA), OR 102.4, 'Major Development Program – Nuclear – General,' Osborne Mitchell, secretary, OHEPC, to Hearn, 1 Nov. 1946, drawing attention to reports emanating from the United Nations Atomic Energy Commission in New York.
21 Ibid., Saunders to O. Holden, 16 March 1950; Dobson to O. Holden, 11 May 1950; M.S. Oldacre, director of research, Utilities Research Commission, Chicago, to Dobson, 4 April 1950; R.C. Jacobsen, assistant research physicist, HEPC, to Dobson, 10 April 1951; Dobson's report to the executive committee was made on 22 May 1950: Dobson to G.D. Floyd, 31 Aug. 1955.
22 Ibid., Dobson to Holden, 14 Sept. 1951; Williamson, Holden's assistant, to Saunders, 7 Jan. 1952. The meeting occurred on 11 Jan. 1952.
23 AECL Records (Sheridan Park), Robert Belfield interview with Richard Hearn, 22 March 1982
24 On Saunders, see ibid., Mackenzie to Saunders, 3 Sept. 1952; CRNL Records, file 6000 Reactors-NPD Overall, vol. 1, Laurence to Hearn, 7 Aug. 1952, and Hearn to Laurence, 11 Aug. 1952.
25 CRNL Records, file 1615/F1, Future Systems Group, vol. 1, minutes of a meeting of the Nuclear Physics and Reactor Engineering Group, 19 Aug. 1952
26 CRNL Records, Lewis office files, miscellaneous, MacKay to Lewis, 19 Aug. 1952. Lewis minuted opposite MacKay's comments on the undesirability of 'universal' multipurpose reactors, 'What is this? Dogma?'
27 CRNL Records, file 1615/F1, Future Systems Group, vol. 1, Lewis, 'Summarized Recommendations for Presentation to Board of Directors Meeting 30 September 1952.' This is presumably RD-1, mentioned in the board minutes.
28 Board of Directors minutes, 29–30 Sept. 1952 and 5 Dec. 1952. A full complement of the board, except for Bennett, was present for these meetings. The major conference was at Harwell, 7–8 Nov. 1952, with Mackenzie, Lewis, and MacKay in attendance from Canada, and Lord Cherwell, General Sir F. Morgan, Cockcroft, and Hinton from the UK. Unfortunately the minutes of the meeting are unknown on the Canadian side and unavailable on the British: PRO, AB16/751.
29 McConnell interview
30 William Murray, *Historical Highlights: A Chronology Covering the Period 1900 to 1950* (Toronto 1956), 23. Despite the title, this commemorative pamphlet composed for Hydro's Golden Anniversary in 1956 listed events down to 1955. According to OHA, file OR 102.4, 'Major Development Program – Nuclear – General,' [Hearn] to Mackenzie, 14 Jan. 1953, D.O. Gregory and V.W. Ruskin were to be sent from Toronto; later, the names were changed to W.E. Cooper and Allan Ray Clink.

31 AECL Records (Ottawa), file 103-A-2, Advisory Committee on Atomic Power, Lewis to Guy Jarvis, 12 Aug. 1953; memorandum on symposium, 3 Sept. 1953. CRNL Records, SDDO, DL-12, 'Atomic Power Progress and Prospects,' Sept. 1953
32 AB16/751, Mackenzie to General Sir F. Morgan, 16 Oct. 1953; Morgan to Sir K. Hague, 19 Oct. 1953; see also extract from the minutes of the UK Atomic Energy Authority, 11 Oct. 1954. At that meeting Sir Edwin Plowden, AEA chairman, referred to 'earlier proposals that the United Kingdom Authority should arrange for the design and supply of a heavy water power reactor to be installed in Canada by the Ontario Hydro Electric Board.'
33 W.J. Bennett interview, Toronto, 31 July 1986: 'first I ever heard of it.' Board of Directors minutes, 17 Dec. 1953, by-law no. 5, authorizing the appointment of an executive committee of three members to be elected by the board.
34 AECL Records (Ottawa), file 101-4, Organization at Chalk River, Bennett to Howe, 24 Dec. 1953, V.W. Scully to Bennett, 28 Dec. 1953, Bennett to Scully, 5 Jan. 1954
35 CRNL Records, Lewis office files, miscellaneous, minutes of meeting, 6 Nov. 1953. In attendance were Bennett, Mackenzie, Lewis, Gray, and Hearn; AB19/13, Hinton to Plowden, 20 Sept. 1954, enclosing copy of a letter to himself from a British industrialist, possibly Sir Claude Gibb.
36 Board of Directors minutes, 17 Dec. 1953
37 AB16/1687, Talk with Dr W.B. Lewis, 6 May 1954: present were Sir E. Plowden, Sir J. Cockcroft, Sir C. Hinton, Sir W. Penney, and N.S. Forward.
38 Ibid., Sir C. Hinton, 'The Ontario Hydro Reactor Programme,' 12 May 1954; Atomic Energy Executive, 'The Ontario Hydro Reactor Programme,' 20 May 1954
39 Ibid., notes by Plowden of a talk with W.J. Bennett, 30 June 1954; Plowden to Hinton, 16 Sept. 1954
40 Bennett interview (phone), 15 Sept. 1987
41 AB16/1687, extract of minutes from AEX (54), 15th meeting held on 23 Sept. 1954; AB19/13, Hinton to Gibb, 2 March 1955; Hinton to Gibb, 1 April 1955
42 Bothwell, *Eldorado*, 336; the shape of the Canadian power program was discussed speculatively at an Anglo-Canadian technical conference in October 1953: AB19/85.
43 AB16/751, Bateman to J.G. Bower, 22 Dec. 1953
44 Bennett interview, 31 July 1986; minutes of the Board of Directors, 17 Dec. 1953 and 1 April 1954

45 CRNL Records, Lewis office files, notes of a meeting with Plowden, Cockcroft, Hinton, Penney, and N.S. Forward, 6 May 1954
46 AECL records, file 105-6-1 Reactors NPD, vol. 1, Harold Smith, 25 March 1957; AB16/1687, AEX (54) 152, 28 Sept. 1954
47 John Foster, 'The First Twenty-Five Years,' speech to the Canadian Nuclear Association, June 1985; OHA, Dobson, 'Report no. 1,' to Hearn, 12 Feb. 1954; CRNL Records, file 1615/P14, Power Reactor Group, vol. 1, minutes of meeting of Power Reactor Group, 3 March 1954. On the AECL side those in attendance included Lewis, Hatfield, Laurence, MacKay, Beynon, Lorne McConnell, Arthur Ward, and Donald Hurst.
48 AB16/1687, meeting of 12 May 1954
49 The divergence becomes clear in ibid., minutes of fifth meeting, 14 July 1954; see also AB19/13, Hinton to Plowden, 20 Sept. 1954.
50 AB19/13, Hinton to Plowden, 20 Sept. 1954
51 CRNL, file 1615/P14, vol. 1, minutes of the third meeting of the PRG, 25 May 1954
52 CRNL, SDDO, Limited Report Series, NPG 1, 1 July 1954
53 Ibid., minutes of the fifth meeting, 14 July 1954. My italics. According to John Foster (interview, 1 Aug. 1986) the fact of agreement on a heavy water system was a sufficient catalyst for the very rapid development that now followed.
54 CRNL Records, Lewis office files, miscellaneous, Lewis, memorandum for meeting held on 26–27 July on the nuclear power development program; notes of meeting, 26 July 1954
55 On the meetings at Harwell see AB16/255, minutes of meeting, 28 Sept. 1954: The Canadian team was made up of Bennett, Hearn, Gaherty, Lewis, and MacKay; the British consisted of Plowden, Cockcroft, and Hinton among others. For the Canadian meetings see Board of Directors minutes, meeting held on 4 and continued on 30 November 1954.
56 See especially Lewis, 'Development and Preliminary Design Study Programme for the Two Large Power Reactors,' DM-28, 21 April 1955. This memorandum was written well after both the board approval of the study and design of one large reactor, and a month after the federal cabinet had affirmed the same objective.
57 AECL Records (Ottawa), file 105-6-1, Reactors, NPD, vol. 1, Bennett to Lewis and Gray, 29 March 1955, and Bennett to Gaherty, 29 March 1955.
58 CRNL Records, file 1615/P14, vol. 1, minutes of PRG, 31 Jan. 1955
59 Board of Directors minutes, 17 Feb. 1955, containing detailed summaries of the bids; AECL Records (Ottawa), file 105-6-1, Reactors, NPD, vol. 1, H.M. Turner, president, CGE, to Bennett, 31 Jan. 1955; Bennett to Turner, 1 Feb. 1955; Turner to Bennett, 15 Feb. 1955; Bennett to Howe, 11 March 1955

60 Board of Directors minutes, 17 Feb., 10 March 1955; W.J. Bennett interview, Toronto, 31 July 1986

61 PCO Records, cabinet conclusions (RG 2, series 16), 23 March 1955. The Colombo Plan considerations occurred on 30 March 1955.

CHAPTER 6: A COMPETITIVE WORLD

1 See the very extensive correspondence among American officials and agencies reproduced in FRUS, 1952–1954, vol. 2, part 2 (Washington 1984).

2 Ibid., 1280, 'Second Report of the ad hoc committee on armaments and American policy (as amended by the NSC Planning Board) on disclosure of atomic information to allied countries,' 4 Feb. 1953, and altered thereafter. It was enclosed in a report to the US National Security Council by its executive secretary, 4 Dec. 1953.

3 On Eisenhower, see Stephen D. Ambrose, *Eisenhower*, vol. 2 (New York 1984), 94–5; FRUS, 1952–1954, vol. 2, part 2, 1289–90.

4 Text of the message is in Williams and Cantelon, eds., *American Atom*, 104–11. Lewis Strauss had explained to the British that the message was intended 'to make a gesture' and to put the Russians at a disadvantage; on that basis Cherwell and Churchill grudgingly accepted it: FO 371/105717, Note on the Conversations between Admiral Strauss and Lord Cherwell, 4 Dec. 1953.

5 See Bothwell, *Eldorado*, 380; on the AEA amendments, see *FRUS, 1952–1954*, vol. 2, part 2, 1360–1. According to Gerard Smith, memorandum to file, 19 Aug. 1954, ibid., 1503, Toronto was considered as a possible home for a future International Atomic Energy Authority.

6 See Gerard C. Smith to Lewis Strauss, 13 Oct. 1954, ibid., 1531–2.

7 The basic US policy was spelled out in NSC 5431/1, 13 Aug. 1954, and approved by the president on that date. As for the conference, Bennett advised against UN sponsorship, a view that others in the US government shared. As of August 1954 the US had not yet decided as to whether the proposed conference should be sponsored by the tripartite powers, or by the UN or by the US alone: see ibid., 1491, 1517.

8 Ibid., 1540–2, Acting Secretary of State to Henry Cabot Lodge, US ambassador to the UN, 20 Oct. 1954. Lodge, who was not noted for excessive intelligence, had opposed the UN conference idea.

9 Goldschmidt, *Atomic Complex*, 258; Pringle and Spigelman, *Nuclear Barons*, 165–6

10 Pringle and Spigelman, *Nuclear Barons*, 166; Goldschmidt, *Atomic Complex*, 258–9

11 Goldschmidt, *Atomic Complex*, 260–1
12 AECL, *Annual Report, 1955–56* (Ottawa 1956), 7
13 Pringle and Spigelman, *Nuclear Barons*, 118–19
14 AB19/20, J.A. Dixon to UK Liaison Office, Chalk River, 25 June 1956
15 CRNL Records, Lewis office files, Hinton to Lewis, 19 May 1955; AECL Records (Ottawa), file 105-3, Large Reactor Program, vol. 1, Lewis to Bennett, 26 Sept. 1955, Bennett to Lewis, 25 Oct. 1955; Fenning interview
16 AB19/22, Hinton to W.L. Owen, 5 Dec. 1956. Hinton added that 'the slight tension between Canada and the U.K. which has arisen from the Middle East crisis,' along with a reduction in American prices for enriched uranium, 'makes it almost impossible that the Canadians will wish to come in with us.'
17 Bennett interview, Toronto, 31 July 1986
18 Bennett to author, April 1987. In the 1970s the French government approached the government of Quebec with a proposal for an enrichment plant using power from the Baie James project.
19 Ibid.; AECL Records, file 105-17, USAEC-AECL Co-operation, Bennett to Fields, 25 Jan. 1956, and Fields to Bennett, 14 Feb. 1956
20 Board of Directors minutes, 7–8 Nov. 1955, appendix A: Lewis, 'The Chalk River Applied Development and Operations Programme 1955–60'
21 As if to emphasize Lewis's point two senior members of Commercial Products resigned in the fall of 1955; the senior chemist, L.G. Cook, also responded to the call of opportunity and left for American General Electric in Schenectady early in 1956. On the cost of NPD, see the minutes of the executive committee, 24 Jan. 1956.
22 See PCO Records, cabinet conclusions, 1 March 1956, agreeing to the NPD accord with Ontario Hydro: PC 1956–350.
23 See Board of Directors minutes, 8 March 1956. For Howe's views on private enterprise and atomic development ('we would hope that commercial interests would take over and carry on') see House of Commons *Debates*, 15 July 1955, 6213.
24 House of Commons, Special Committee on Research, *Minutes*, 5 June 1956, 241
25 Ibid.
26 Ibid., 5 June 1956, appendix I, W.B. Lewis, 'Why Heavy Water?' 250–1
27 OHA, file 1953–4, Questions in the Legislature, Number 20, Mr. Joe Salsberg (Labour-Progressive), which produced the answer, 'All coal purchased has been produced in the United States.'
28 SDDO, NPG-5, Harold Smith and John Foster, 'Report No. 1 on Preliminary Design of a Small Nuclear Electric Power Plant,' 31 Jan. 1955
29 Department of Trade and Commerce, Economics Department, *The*

Economics of Atomic Power Reactors (Ottawa 1955). Combining operating expenses and fixed charges, including interest, depreciation, and taxes, the Economics Department arrived at the following ranges, in mills per kilowatt-hour: coal, oil, or gas: 3.29–5.70; hydro: 0.96–3.21; atomic: 4.74–10.31.

30 AECL Records (Ottawa), file 103-A-2, Advisory Committee on Atomic Power, Bennett to Howe, 8 Feb. 1955, citing a conversation with Jack Fuller of Shawinigan

31 John Davis, *Canadian Energy Prospects* (Ottawa, March 1957), 246–50

32 CRNL Records, 6000 NPD-Overall, vol. 2, minutes of sixth meeting of the NPD Co-ordinating Committee, 25 June 1956

33 Hearn and Bennett interview, Queenston, July 1984; CRNL Records, file 6000 NPD-Overall, vol. 2, Clive Kennedy to David Keys, 11 Sept. 1956

34 See Hewlett and Duncan, *Nuclear Navy*, especially 288–90. Most of the production data on zirconium was declassified in 1955, to encourage new suppliers. On the original NPD design see AECL-356, Civilian Atomic Power Department, Canadian General Electric, Peterborough, 'Canada's First Nuclear Power Station,' mimeograph, 1956; Perryman interview.

35 Hewlett and Duncan, *Nuclear Navy*, 246

36 CRNL Records, file 1615-P14, Power Reactor Group, vol. 1, minutes of meeting of the Power Reactor Group, 28 April 1955

37 Foster, 'First Twenty-Five Years,' 14

38 Harold Smith, 'Atomic Energy in Industry,' *Electrical Digest*, June 1956, 100, 102. This article was first read as a paper at the National Industrial Conference Board in New York in October 1955.

39 Foster, 'First Twenty-Five Years,' 16

40 Smith, 'Atomic Energy in Industry,' 99; Perryman interview; CRNL Records, file 6000 NPD-Overall, vol. 2, Coates to Tilbe, 23 Aug. 1956

41 CRNL Records, file 5000 Reactor Concepts, Overall, vol. 1, Lewis to Laurence, 3 May 1956

42 Ibid., file 6000 NPD-Overall, vol. 2, minutes of ninth meeting of the NPD Coordinating Committee, 6 Dec. 1956; ibid., E. Siddall to all members of the NPD Technical Committee

43 Ibid., memorandum on containment, with attached letter, by John Foster, 28 Dec. 1956

44 Ibid., file 6000 NPD-Overall, vol. 3, minutes of the 10th meeting of the NPD Co-ordinating Committee, 1 Feb. 1957; minutes of the Executive Committee of the Board, 12 Feb. 1957, appendix 1, Gray and Lewis, 'Memorandum on NPD of 6 February 1957'; AECL Records (Ottawa), file 105-6-1, Gray to McRae, 6 Feb. 1957

45 CRNL Records, file 6000 NPD-Overall, vol. 3, Lewis to MacKay, 13 March 1957

46 Ibid., draft of DM-43, 26 March 1957
47 Executive Committee minutes, 27 March 1957
48 Ibid.
49 Board of Directors minutes, 11 July 1957
50 AECL Records (Ottawa), file 105-6-1 Reactors NPD, vol. 1, Smith to Bennett, 25 March 1957
51 Ibid., Bennett to Howe, 8 April 1957; Bennett to V.W. Scully, 17 April 1957
52 Ibid., press release, 13 April 1957; Bennett to Howe, 15 April 1957
53 Ibid., Bennett to Scully, 17 April 1957
54 On Churchill, see Bothwell, *Eldorado*, 414–15.
55 AB19/13, Hinton to Plowden, 20 Sept. 1954, quoting Gibb
56 The decision was made on 12 July 1957, but not recorded in the minutes until Bennett's state paper was ready for discussion and approval by the board: Minutes of the Board of Directors, 11 Oct. 1957.
57 AECL Records, file 105-3 Large Reactor Programme (CANDU), vol. 2, Bennett to Churchill, 12 Sept. 1957, enclosing 'Canada's Programme of Nuclear Power Development'
58 Ibid., Bennett to Duncan, 13 Sept. and 3 Oct. 1957
59 Ibid., Bennett to Bryce, 26 Nov. 1957; Privy Council Records, cabinet conclusions, 16 Jan. 1958
60 AECL Records, file 105-3, Lorne Gray, 'Canada's Programme of Nuclear Power Development,' 19 March 1959; this memorandum was approved by cabinet on 21 April 1959.
61 Executive committee minutes, 9 Nov. 1960
62 AECL, *Annual Report, 1961–62* (Ottawa 1962), 30. Gray repeatedly mentioned in interviews his dissatisfaction with CGE's costing.
63 Sims, *Atomic Energy Control Board*, 121–2
64 Interview with J. Herbert Smith, ex-president of CGE, Toronto, 30 May 1986
65 Bennett to author, April 1987
66 Gray interview, 16 April 1986
67 Goldschmidt, *Atomic Complex*, 269
68 Gray interview, 16 April 1986

CHAPTER 7: THE FLIGHT OF THE ARROW

1 John Diefenbaker, quoted in J.L. Granatstein, *Canada 1957–1967* (Toronto 1986), 22. See also Jon B. McLin, *Canada's Changing Defense Policy, 1957–1963* (Baltimore 1967), 61–84.
2 See Robert Bothwell, Ian Drummond, and John English, *Canada since 1945* (Toronto 1981), 9–14.

3 Bennett and Gray interview
4 Ibid.
5 Interview with Lorne Gray and R.B. Bryce, Finance Department, Ottawa, 13 Nov. 1985
6 Granatstein, *Canada 1957–1967*, 29. Diefenbaker's style of government put a premium on two things: frequent meetings of the cabinet (164 in 1958) and control by the prime minister. Bryce, whose principal responsibility was ensuring the smooth functioning of the cabinet, enjoyed an enhanced importance as a result.
7 Lorne Gray interview, Café Henry Burger, 13 Nov. 1985; Donald Watson interview, Ottawa, 13 Nov. 1985
8 See *Canada Year Book, 1961* for the first figure; by 1974 the federal government had stopped publishing the numbers in its *Yearbook* for reasons which are either unfathomable or disreputable; see *Historical Statistics of Canada*, 2nd ed. (Ottawa 1983), table Y211.
9 Herbert Smith interview; Board of Directors minutes, 10 Nov. 1958, referring to a forthcoming meeting with CGE on 15 Dec. 1958; executive committee minutes, 20 Feb. 1959, with Gray's recommendation that a contract be let to CGE for an organic cooled reactor at Chalk River?
10 Ian MacKay interview, Sidney, BC, 16 June 1986; Lorne Gray interview, Deep River, 16 April 1986
11 On estimates, see AECL Records, file 105-3, Large Reactor Programme, CANDU, vol. 2, Bennett to R.B. Bryce, 5 Nov. 1957: Recent American experience, Bennett wrote, confirmed his view that development should be undertaken first; otherwise 'the final cost, including development, obviously makes the original estimates quite unrealistic and destroys one's confidence in the ability of any nuclear power group to make a reasonably accurate estimate of final costs.'
12 Ibid., file 105-6-1, Reactors NPD, vol. 2, Gray to Duncan, 15 July 1957; Sayers to Duncan, 18 July 1957
13 Executive committee minutes, 9 Nov. 1960
14 John Foster interview, Toronto, 1 Aug. 1986; Bill Bennett interview, Toronto, 31 July 1986
15 AECL Records, file 105-6-1, vol. 2, Duncan to Bennett, 16 July 1957
16 OHA, file 102.4, Nuclear Power Feasibility Study, Smith to Duncan, 2 July 1957, 'Suggested Outline Approach to Large Nuclear Plant Development,' with attached memo, 'Preliminary Suggestions for a Joint AECL-Hydro Undertaking to Develop and Construct a Full-Scale Nuclear Power Plant'
17 AECL Records, file 105-3, Large Reactor Programme, CANDU, vol. 2, NA, 'The Canadian Nuclear Programme,' 5 Nov. 1958

18 Ibid., Smith, 'A Consideration of Possible Modifications to the Canadian Nuclear Power Programme,' 10 Nov. 1958; John Foster interview, Toronto, 1 Aug. 1986
19 Board of Directors minutes, 10 Nov. 1958
20 We have two sets of notes of this interview: Duncan's are in OH, HO F 28, Policy – Nuclear Power Development, no. 1 100.02; Gray's are in AECL Records, file 105-3, Large Reactor Programme, vol. 2.
21 Churchill's attitude was discussed by Gray in the executive committee minutes, 20 Feb. 1959. Gray confirmed this in an interview in Toronto, January 1987.
22 Letter reproduced in appendix 2 to the board of Directors' minutes of 13 March 1959
23 Cabinet conclusions, 19 May 1959
24 AECL Records, file number illegible, Gray to Churchill, 26 May 1959
25 Cabinet conclusions, 16 June 1959. In the interval AECL's ordinary capital budget had been approved on 4 June.
26 Ian MacKay interview, Sidney, BC, 16 June 1986. CRNL Records, file 5001 Candu Reactor Concepts/OCR/Overall, vol. 1, Gray to Brig. F.C. Wallace. On organic coolants, see W.M. Campbell, A.W. Boyd, D.H. Charlesworth, R.F.S. Robertson, and A. Sawatzky, *Development of Organic-Liquid Coolants* (AECL-2015), a paper presented to the Third U.N. International Conference on the Peaceful Uses of Atomic Energy, Geneva, Sept. 1964.
27 Board of Directors minutes, 10 Nov. 1958
28 CRNL Records, file 5001 Candu Reactor Concepts/OCR/Overall, vol. 1, Gray to Brig. Wallace, 4 Feb. 1959
29 Executive committee minutes, 20 Feb. 1959; Board of Directors minutes, 13 March 1959; executive committee minutes, 4 May 1959, referring to a letter of intent sent to CGE on 27 April
30 Executive committee minutes, 21 Dec. 1959
31 Ibid., 4 May 1959
32 Board of Directors minutes, 25 May 1959: Grinyer was to handle operations and services at Chalk River, as well as NPPD in Toronto.
33 AECL Records, file 101-13, Incorporation and Organization of AECL-Whiteshell Nuclear Laboratories Establishment, vol. 1, Gray to Churchill, draft, 14 July 1959; Gray, Memorandum to Cabinet, 27 Aug. 1959, appendix I to board minutes, 11 Sept. 1959. The latter document is much the same as the former.
34 Board of Directors minutes, Deep River, 8 Nov. 1959
35 Ibid., 20 Dec. 1961, with attached memorandum to cabinet, 21 Dec. 1961, 'Organic Test Reactor, Whiteshell Nuclear Research Establishment'
36 Ibid., 19 April and 17 June 1960; see WRNE Records, file A8505, T.W.

Morison to Gray, 1 March 1960, outlining necessary arrangements concerning the setting up of a new local government district.

37 Board of Directors minutes, 8 Sept. 1960, where the capital costs of the Whiteshell project, exclusive of the reactor, were set at $835,000; the name was considered at an executive committee meeting on 6 April 1961; the CMHC program was approved at the board meeting on 29 May 1961.

38 CRNL Records, 5001 Candu Reactor Concepts/OCR/Overall, vol. 2, Sayers to Lewis, 19 Aug. 1960. Sayers was a non-professional engineer who had worked in TRE with Lewis during the war. Lewis brought him over to handle dealings with suppliers during the building of NRU and later NPD: Foster interview, 1 Aug. 1986.

39 Gray interview, Toronto, 9 Sept. 1986

40 Board of Directors minutes, 8 Sept. 1960; Foster interview, 1 Aug. 1986

41 Cabinet conclusions, 25 Oct. 1960; executive committee minutes, 9 Nov. 1960

42 Executive committee minutes, 6 April 1961

43 Board of Directors minutes, 17 Sept. 1961

44 These reasons were laid down at an executive committee meeting on 31 Aug. 1961 by Grinyer. He added that helping the CGE team [to survive?] was another factor.

45 Board of Directors minutes, 20 Dec. 1961, with appendix, submission to cabinet, 21 Dec. 1961; see D.G. Turner, *The WR-1 Reactor: A General Description* (Whiteshell 1974), 1.

46 The executive committee toured Peterborough on 12 July 1961, where they were 'particularly impressed with the CGE fuel manufacturing line and the NPD fuelling machine.' Meeting in Toronto that afternoon, they heard Gray's remarks about CGE's future with special interest.

47 CRNL Records, 5001, Candu Reactor Concepts/OCR/Overall, vol. 14, NA, 'Status of Canada's CANDU-OCR Program,' 17 Nov. 1971

48 AECL, *Annual Report, 1966–1967* (Ottawa 1967), 29–31

49 Ted Thexton interview, Ottawa, 7 May 1987; Lorne Gray interview, Deep River, April 1985

50 See Alan Green, *Immigration and the Postwar Canadian Economy* (Toronto 1976), especially chapter six.

51 See CP Annual Report, 1958–9, 8; Errington interview, 31 Jan. 1987

52 Bennett to author, 18 Feb. 1987

53 Board of Directors minutes, 8 Aug., 10 Nov. 1958, 13 March 1959; executive committee minutes, 16 Jan. 1959

54 D. Watson interview, 8 Jan. 1987; R. Errington interview, 2 Feb. 1987

55 Figures derived from *Annual Reports*, 1957–64

56 AECL Records, C-103-P-1, vol. 1, minutes of the first meeting of PRDPEC, 23 March 1962

57 Strangely, no complete file of the SMC appears to exist anywhere in AECL's records.

CHAPTER 8: THE BURDEN OF PROOF: DOUGLAS POINT

1 OH Library, A brief presented to the Royal Commission on Canada's Economic Prospects by Richard L. Hearn, January 1956
2 See A.K. McDougall, *John P. Robarts: His Life and Government* (Toronto 1986), 41, 49.
3 Phil Stratton interview, Toronto, 3 Sept. 1986
4 John Foster interviews, Toronto, 1 and 25 Aug. 1986. The attached staff from other companies were paid by their 'home' companies. As time passed it was increasingly difficult to discern what advantage these companies would derive from the relationship, and such personnel were in any case acquiring a closer connection with AECL than with their points of origin.
5 Ibid., 1 Aug. 1986
6 Ibid., 25 Aug. 1986
7 Figures are derived from AECL, *Annual Reports*, 1961–9; 1961 was selected as a base because figures for earlier years are not available.
8 Foster interview, 25 Aug. 1986; AECL Records, file 105-3, Large Reactor Program – CANDU, vol. 3, Notes by Foster on CANDU organization, 3 April 1959
9 The organization chart on which this is based is dated 10 Sept. 1963: AECL Records, chart for 'Power Projects, Toronto.'
10 CRNL Records, 5001 CANDU Reactor Concepts Overall, vol. 2, Harold Smith to Lorne Gray, 17 Feb. 1959; AECL Records, Smith to executive committee, 12 March 1959; McConnell interview
11 McConnell interview; AECL Records, 105-3, Large Reactor Program, vol. 3, memorandum by Harold Smith, 26 June 1959
12 AECL Records, 105-3, Large Reactor Programme, vol. 3, Smith memorandum on CANDU siting, 26 June 1959; Foster interview, 25 Aug. 1986
13 Economic data taken from the Department of Energy, Mines and Resources' *National Atlas of Canada*, 4th ed. (Ottawa 1974), 249–50
14 AECL, CANDU Operations, *Power Projections*, June 1984, special edition: 'The Douglas Point Story,' 9
15 Quoted in ibid., 10
16 See F. Kenneth Hare and Morley K. Thomas, *Climate Canada*, 2nd ed. (Toronto 1979), 107.
17 Foster interview, 25 Aug. 1986
18 Ibid.; AECL executive committee minutes, 16 Jan. 1962
19 Morison interview; Nat Stetson interview, Aiken, , 22 April 1985; Julius Rubin interview, Washington, DC, 18 April 1985

20 Morison interview
21 See remarks by I.L. Wilson, quoted in Eggleston, *Canada's Nuclear Story*, 334.
22 The Board of Directors accepted a computer for CANDU on Gray's recommendation that it was 'a significant feature in the application of automation.' The computer cost $400,000. Minutes, 17 Sept. 1961
23 AECL Records, 105-3 CANDU, vol. 6, NA, Foster , 'Notes on Canadian Manufacture of Certain Items for Douglas Point Generating Station,' nd but probably 1961 or 1962
24 Stratton interview. See AECL Board of Directors minutes, 17 Feb. 1961, where it was noted that the twenty contracts for components so far made were all in line with the estimates.
25 Stratton interview; Board of Directors minutes, 29 May 1964, appendix C, 'Douglas Point Project: Notes for Status Report,' 28 May 1964; Bill Morison interview; CRNL Records, 6000 Douglas Point Overall, vol. 2, Report of a Discussion by S.G. Horton, 3 Nov. 1967
26 AECL, *Annual Report 1962–63* (Ottawa 1963), 7–8; Board of Directors minutes, 13 Dec. 1962; appendix D, Board of Directors minutes, 29 May 1964
27 AECL Records, 105-3 Douglas Point – CANDU, vol. 5, Foster to Gray, 8 May 1961
28 McConnell interview. See Board of Directors minutes, 14 Dec. 1965, where Foster reported that the variation between the original estimates and the total cost of the plant was virtually nil.
29 CRNL Records, 6000 Douglas Point Overall, vol. 2, News release, 15 Nov. 1966
30 McConnell interview
31 *Canadian Annual Review for 1960* (Toronto 1961), 289–90; AECL Records, file 'Winnett Boyd,' containing Boyd's 1959 speech and Gray's response
32 Ernest Siddall interview, Toronto, 18 March 1987

CHAPTER 9: PICKERING

1 McConnell interview
2 Figures derive from the *Canada Year Book, 1967* (Ottawa 1967), 640–61.
3 AECL executive committee minutes, 18 March 1963; Finance Department Records, memorandum by Lorne Gray, 16 July 1963, appendix I, Gray to Strike, 11 June 1963
4 AECL Records, 105-22, Foster to H. Smith and J.L. Gray, 17 Feb. 1964. Foster suggested a joint crown corporation to handle the reactor.
5 AECL Board of Directors minutes, 5 June 1963

6 Executive committee minutes, 17 July 1963; AECL Records, 105-22, Gray, draft memorandum to cabinet, 12 Aug. 1963. Finance Department Records, file 1202-15-6, Reisman to Bryce, 19 July 1963, with minutes by Bryce. See also PCO Records, cabinet conclusions, 12 Sept. 1963.
7 Finance Department Records, file 1202-15-6, contain the federal side of the deal.
8 PCO Records, cabinet conclusions, 12 Nov. 1963; cabinet document 362–63
9 Executive committee minutes, 25 Feb. 1964
10 AECL Records, 103-S-8 Senior Management Committee, minutes of the SMC, Whiteshell, 2 Dec. 1963. For Montreal Engineering, see Board of Directors minutes, 7 June 1965.
11 AECL Records, 103-S-8, SMC meeting, 2 Dec. 1963
12 Ibid., appendix A to minutes of the SMC, 13 Jan. 1964
13 AECL Board of Directors minutes, 14 Jan. 1964
14 Finance Records, file 1202-15-6, Bryce to Walter Gordon, 11 March 1964
15 Ibid., 'Negotiations of the financing of nuclear power plant in Ontario,' second meeting of the working group, 15 May 1964, S.A. Grenville to Bryce, 29 July 1964, conveying the approval of the arrangements by the cabinet committee on Finance and Economic Policy on 24 July; Ross Strike to Drury, 12 Aug. 1964
16 AECL executive committee minutes, 25 Feb., 4 Oct. 1964
17 PCO Records, cabinet conclusions, 19 Aug. 1964; AECL executive committee minutes, 20 Aug. 1964
18 Executive committee minutes, 27 July, 20 Aug. 1964
19 AECL Records, 105-22 Advanced Water Cooled Power Reactors – Pickering, vol. 1, H. Smith to Gray, 16 Oct. 1963. Smith urged that Douglas Point was not competitive on economic grounds with southern Ontario sites (Fairport and Nanticoke) for a large reactor.
20 AECL Records, C 103 P-1, PRDPEC, vol. 1, Lewis, 'Report on Power Reactor Development Evaluation,' 14 March 1963
21 See ibid. for PRDPEC's 'Summary Report' (2nd draft), conveyed to J.L. Gray on 17 April 1963.
22 See AECL Records, 105-22 Pickering vol. 1, Lewis to Gray, Foster, and Haywood, 21 Jan. 1964; Foster to Smith and Gray, 27 Jan. 1964; Foster to Lewis, 21 Feb. 1964; Foster to Haywood, 16 March 1964; Bill Morison interview.
23 Executive committee minutes, 29 Jan., 25 Feb., 9 April 1964
24 Ibid., Haywood to Gray, 20 April 1964
25 John Foster, 'Some Materials Aspects in the Canadian Nuclear Power Program,' a paper presented before the Toronto Chapter of the American Society of Metals, 1 Dec. 1964. I am grateful to George Laurence for drawing this paper to my attention.

26 See Foster, 'Twenty-Five Years,' 28–9; Foster interview, 25 Aug. 1986; Bill Morison interview.
27 Morison interview; Sims, *Atomic Energy Control Board*, 123–4
28 Morison interview
29 S.F. Borins and Lee Brown, *Investments in Failure* (Toronto 1986), 8–9; see also Hewlett and Duncan, *Atomic Shield*, 428–30, 524, 531.
30 See AECL Records, 103-A-3 Advisory Panel on Atomic Energy, memorandum to cabinet, 'Construction of a Nuclear Power Station in the Province of Quebec,' 11 July 1966: appendix IV, item 3(d) listed both the elimination of the possibility of expensive losses of heavy water and the 'potential health hazard from the tritium [another hydrogen isotope] produced in the irradiated heavy water.'
31 Ibid.
32 Borins and Brown, *Investments in Failure*, 10
33 Board of Directors minutes, 29 Sept. 1962; executive committee minutes, 5 Nov. 1962
34 AECL Records, 103-A-3, vol. 1, Gray to Advisory Panel, 11 Dec. 1962, enclosing memorandum. Philip Mathias, *Forced Growth* (Toronto 1971), cites a lowering of the US price from $28.50 to $24.50 (US), or $26 (Cdn).
35 Ibid. See also External Affairs Records, 14002-1-2. The paper assumed, among other things, 300 tons of firm Canadian requirements, with probable exports to India and to Great Britain; the future Pickering it listed as possible, along with further exports to Britain, Japan, Sweden, and so forth.
36 PCO Records, cabinet conclusions, 14 Feb. 1963
37 See Roy E. George, *The Life and Times of Industrial Estates Limited* (Halifax 1974), chapter 1, and Borins and Brown, *Investment in Failure*, 10; interview with Hon. Donald S. Macdonald, Toronto, 20 May 1987.
38 Board of Directors minutes, 5 June 1963
39 Gray interview, by phone, 7 Oct. 1986; Gray Papers, 'Discussions with the Minister,' 7 June, 22 July, 13 Aug., 18 Sept. 1963
40 PAC, L.B. Pearson Papers, vol. 39, file 154.2NS Stanfield, Stanfield to Pearson, 14 Aug. 1963, Mary Macdonald to MacEachen, 12 Sept. 1963, Stanfield to Pearson, 30 Sept. 1963
41 Mathias, *Forced Growth*, 107; Hellyer had become a Progressive Conservative MP by the time he made this statement: quoted in George, *Industrial Estates*, 79n; Pickersgill interview, 14 Oct. 1986; Borins and Brown, *Investments in Failure*, 11; Finance Department Records, vol. 5365, file 3949/G541, part 2, Simon Reisman to R.B. Bryce, 29 Jan. 1964, E.A. Oestreicher to Bryce, 30 Jan. 1964, quoting Gordon: 'We will try to get 50% where we can.'

42 Board of Directors minutes, 22 Sept. 1963
43 Gray interview, 7 Oct. 1986; the cabinet committee on economic policy, trade, and employment, meeting on 4 November 1963, made the crucial decision to locate in Nova Scotia: Finance Department Records, vol. 5365, file 3979/G541-3, part 2.
44 Gray Papers, 'Discussions with the minister,' 13 Aug. 1963; Finance Department Records, vol. 5365, file 3979/G541-3, part 2, Davis to Pearson, 18 Nov. 1963
45 George observes (*Industrial Estates*, 80) that Spevack was indubitably the world expert in heavy water technology and that the rival bidders never developed their schemes elsewhere despite offers of support from other provincial governments.
46 *Canadian Annual Review for 1963*, 122. The executive committee minutes for 1 December 1963 state that 'No decision has been made by the Government with respect to the proposals submitted for the construction of a heavy water plant.' By the time the minutes were written up Drury had made his announcement, and the minutes took note of the fact.
47 Mathias, *Forced Growth*, 110
48 Board of Directors minutes, 14 Jan. and 29 May 1964; Mathias, *Forced Growth*, 111
49 Mathias, *Forced Growth*, 112
50 Gray Papers, 'Discussions with the Minister,' 5 March 1964; Board of Directors minutes, 29 May 1964. Pakistan was also asking for a guarantee of heavy water for a proposed CGE reactor: executive committee minutes, 9 April 1964.
51 PCO Records, cabinet conclusions, 22 Oct. and 17 Dec. 1964. Put another way, regional programs were designed, in the words of a sympathetic analyst, 'to narrow the income gap between have and have-not regions through providing more jobs; that is, through self-help and not welfare or mass migration out of the region': A. Careless, *Initiative and Response* (Montreal 1977), 86.
52 Confidential source
53 Mathias, *Forced Growth*, 114
54 PCO Records, cabinet conclusions, 23 Feb. 1965
55 Ibid., 26 July, 11 Aug. and 1 Sept. 1965
56 Pearson Papers, vol. 39, file 154.2NS Stanfield, Stanfield to Pearson, 20 Dec. 1965, and O.G. Stoner to Pearson, 14 Jan. 1966, conveying a draft reply to Stanfield; executive committee minutes, 22 Nov. 1965, 9 Feb., 8 March 1966
57 Gray Papers, 'Discussions with the Minister,' 30 June 1965
58 Executive committee minutes, 5 Aug. 1965; Board of Directors minutes, 14 Dec. 1965

59 Board of Directors minutes, 14 Dec. 1965
60 Executive committee minutes, 8 March 1966; AECL Records, 103-A-3, memorandum to cabinet, 'Revival of Proposal for Further Production of Heavy Water from Glace Bay Plant,' 8 Sept. 1966
61 AECL Records, 103-A-3, vol. 6, minutes of the AEAP, 10 Nov. 1966; executive committee minutes, 9 Feb. 1966
62 Executive committee minutes, 7 July 1966: the deuterium content of Cape Breton water was 20 per cent higher than that of water in Alberta, CGE's second choice.
63 Quoted in Mathias, *Forced Growth*, 116
64 Board of Directors minutes, 11 Aug. 1967, 15 Jan., 29 Oct. 1968; executive committee minutes, 25 Nov. 1968, 24 March 1969
65 Executive committee minutes, 25 Nov. 1968
66 See *The Canadian Annual Review for 1968* (Toronto 1969), 152; *The Canadian Annual Review for 1970* (Toronto 1971), 235–6; Mathias, *Forced Growth*, 117–21.
67 J. Herbert Smith interview; Nat Stetson interview
68 Executive committee minutes, 20 March 1967, 7 March 1968; Board of Directors minutes, 11 Aug., 23 Oct. 1967, 15 Jan. 1968
69 Figures are derived from the *Canada Year Book, 1967*, chapter 15.
70 See Philippe Faucher and Johanne Bergeron, *Hydro-Québec: La société de l'heure de pointe* (Montréal 1986), 62–3.
71 Board of Directors minutes, 1 Feb. 1967. They replaced Hearn and Stewart.
72 See Dale Thomson, *Jean Lesage and the Quiet Revolution* (Toronto 1984), chapter 12 passim, and especially 273, 278; Faucher et Bergeron, *Hydro-Québec*, 62.
73 Thomson, *Lesage*, 288–9. See also the analysis of the Lesage, Johnson, and Bertrand governments in Pierre O'Neill and Jacques Benjamin, *Les mandarins du pouvoir: L'exercice du pouvoir au Québec de Jean Lesage et René Lévesque* (Montréal 1978).
74 Executive committee minutes, 27 July, 20 Aug. 1964; Board of Directors minutes, 18 Oct. 1964; Faucher et Bergeron, *Hydro-Québec*, 59 (my translation)
75 Executive committee minutes, 20 Aug. 1964. The committee concluded that there should be 'minimum innovations' in a 250-megawatt station for Hydro-Québec, but did not yet exclude CGE's proposals.
76 Executive committee minutes, 10 Dec. 1964
77 Gray interview, Toronto, 1 Oct. 1986; executive committee minutes, 14 May 1964: 'There was general agreement with the recommended policy [by Haywood] regarding the limitation, termination and redirection of

effort to minimize expenditures on organic cooled reactor systems projects.'

78 Gray interview, Deep River, 16 April 1986

79 AECL Records, 103-P-5, vol. 1, G.Y. Lougheed, 'Request from Quebec Premier re Reactor CANDU-BLW-250,' 27 April 1965 (memorandum to the Privy Council Committee on Scientific and Industrial Research)

80 Ibid.; see also memorandum to cabinet (#224/65), 30 April 1965, from the Committee on SIR, covering the same points as the Lougheed memorandum, and enclosing a draft letter from Pearson to Lesage.

81 PCO Records, cabinet conclusions, 3 May 1965

82 Executive committee minutes, 4 May 1965

83 Thomson, *Lesage*, 288–9; executive committee minutes, 8 March, 1 May 1966

84 Board of Directors minutes, 31 May 1966

85 AECL Records, 103-A-3, Memorandum to cabinet, 'Construction of a Nuclear Power Station in the Province of Quebec,' 11 July 1966

86 Executive committee minutes, 12 Aug. 1966

87 Board of Directors minutes, 23 Oct. 1967

88 AECL, 1969–1970 Annual Report (Ottawa 1970), 7; Board of Directors minutes, 7 April 1970

89 Figures taken from Statistics Canada, *Canadian Statistical Review: Historical Summary 1970* (Ottawa 1972), 25. Federal expenditure on gross capital formation in 1959 was $363 million (an unusually high figure) and in 1969 $486 million. The trend for federal capital expenditures is very much downwards after the mid-1950s and the room in the federal budget for austerity fiddlings proportionately restricted.

90 Tom Kent, 'Regional Development in a Federal Setting,' in O.J. Firestone, *Regional Economic Development* (Ottawa 1974), 126. See also Peter C. Newman, *The Distemper of Our Times* (Toronto 1968), 8–9; Nicole Morgan, *Implosion: An Analysis of the Growth of the Federal Civil Service in Canada (1945–1985)* (Ottawa 1986).

91 See the interesting discussion in Borins and Brown, *Investments in Failure*, 3–4.

92 Ibid., 132

93 George, *Industrial Estates*, chapter 9

CHAPTER 10: EXPERIENCE ABROAD

1 Walter Patterson, *Nuclear Power*, 2nd ed. (London 1983), 48–9

2 See ibid., 57–60, for a description; on Westinghouse, see Pringle and Spigelman, *Nuclear Barons*, 160–2.

3 Pringle and Spigelman, *Nuclear Barons*, 163–4
4 Goldschmidt, *Atomic Complex*, 306–9
5 Pearson's Scarborough speech of 12 January 1963, announcing that the Liberal party would not reject a commitment to station nuclear weapons in Canada, made mention of Canada's preceding activities in the nuclear field.
6 See L.B. Pearson, *Mike: The Memoirs of the Rt. Hon. L.B. Pearson*, vol. 2 (Toronto 1973), chapter 5.
7 See John Holmes, *The Shaping of Peace: Canada and the Search for World Order*, vol. 2 (Toronto 1982), 201–2; Escott Reid, *Envoy to Nehru* (Delhi 1981), 10. A splendid example of Canadian fellow-feeling is George Britnell, 'Under-Developed Countries in the World Economy,' *Canadian Journal of Economics and Political Science* 23 (Nov. 1957), 453: 'A Canadian may be in a better position than either an Englishman or an American ... to appreciate the complex problems of under-developed countries.'
8 An extreme example of anti-Americanism was the Indian defence minister, Krishna Menon, a favourite of Nehru's: see Michael Brecher, *India and World Politics: Krishna Menon's View of the World* (London 1968), 300–4.
9 Reid, *Envoy to Nehru*, 257; Goldschmidt, *Atomic Complex*, 258; Pringle and Spigelman, *Nuclear Barons*, 175–6
10 Ara Mooradian interviw, Ottawa, 28 Nov. 1985: Lewis was a devout Anglican, and his Third World concerns were very much part of his religious faith. On Bhabha and the thorium cycle, and fast breeders, see David Hart, *Nuclear Power in India, A Comparative Analysis* (London 1983), 32–5.
11 W.J. Bennett, 'The Reactor for India,' note to the author, April 1987; DEA file 11038-1-40, Bennett to A.E. Ritchie, 21 March 1955
12 DEA file 11038-1-40, Léger to Pearson, 21 March 1955
13 PCO Records, cabinet conclusions, 30 March 1955; Howe Papers, vol. 10, file s-8-2/10, Howe to Pearson, 15 June 1955
14 PCO Records, series 16, cabinet conclusions, 11 July 1955
15 Howe Papers, vol. 10, file s-8-2/18, Nehru to St Laurent, 4 Aug. 1955, Pearson to St Laurent, 4 Aug. 1955, and Bennett to Howe, 4 Aug. 1955; CRNL Records, 6000/CIR Overall, vol. 1, Lewis to Bhabha, 5 April 1955; DEA file 11038-1-40, Nehru to St Laurent, 2 Sept. 1955
16 See D.K. Mishra, *Five Eminent Scientists: Their Lives and Work* (Delhi 1976), 161–3; R.P. Kulkarni and V. Sarma, *Homi Bhabha: Father of Nuclear Science in India* (Bombay 1969), 23–6.
17 See DEA file 50219-AC-40, Léger to Pearson, 1 Dec. 1955, enclosing draft memorandum for cabinet, 'Proposed International Atomic Energy Agency.'
18 Ibid., A.F.W. Plumptre to Léger, 20 Oct. 1955

19 Goldschmidt, *Atomic Complex*, 185
20 See DEA file 14001-2-1, Wershof to Léger, 30 June 1958; W.B. Lewis to Wershof, 19 Nov. 1958.
21 Foster interview, 25 Aug. 1986; Gray interview, 1 Oct. 1986
22 Bennett, 'The Reactor for India,' April 1987
23 See Paul Martin, *A Very Public Life*, vol. 2: *So Many Worlds* (Toronto 1985), 284: Martin claims that while Nehru and Shastri did not want a bomb, Bhabha (whom he knew at the UN) certainly did.
24 DEA file 11038-1-13-40, Bennett to L. Couillard, 28 Jan. 1958; Kulkarni and Sarma, *Homi Bhabha*, 23–5
25 Board of Directors minutes, 10 Nov. 1958, 11 Sept. 1959
26 Ibid., 8 Nov. 1959, 29 Jan. 1960
27 Mishra, *Five Eminent Scientists*, 163
28 Pringle and Spigelman, *Nuclear Barons*, 375; Gray interview, February 1987; H.N. Sethna to Lorne Gray, 8 Dec. 1986. The full title of CIRUS, according to Dr Sethna, was 'Canada India Reactor Uranium System'; he presided at the meeting in December 1965 when the name was adopted.
29 Hart, *Nuclear Power*, 36–8
30 AECL Records, 103-A-3, vol. 1, Aide Mémoire, unsigned, 2 Dec. 1960; DEA file 14003-J2-3, G. Stoner to A.E. Ritchie, 7 Sept. 1960
31 DEA file 14003-J2-3, J.L. Gray, 'Notes on meetings and discussions with Dr. H.J. Bhabha, 3rd–10th November 1960'; executive committee minutes, 9 Nov. 1960, 6 April 1961. A CANDU-type reactor would have to be as close to the Douglas Point design as possible, since NPPD had no resources to spare for any new project.
32 Hart, *Nuclear Power*, 43
33 PCO Records, cabinet conclusions, 14 Sept. 1960, for instructions to the Canadian delegation to IAEA. It was hoped that the approval of IAEA standards would relieve the government of Canada of the onerous and unpopular duty of enforcing its own safeguards on customers. Board of Directors minutes, 17 Feb. 1961
34 See DEA file 14003-J2-3, D. Watson, AECL, to Gordon Churchill, 6 June 1961.
35 Ibid., Gray, trip report, 14 Feb. 1961; Norman Robertson to Howard Green, ca 20 April 1961; Green to Robertson, 27 June 1961, Gray to Bhabha, 29 June 1961; executive committee minutes, 6 April, 12 July 1961
36 Executive committee minutes, 16 Jan., 1 March 1962; DEA file 14003-J2-3/4, Gray to Bryce, 2 March 1962, Bhabha to Gray, 16 March 1962. The matter was discussed at Mme Berger's on 22 March.
37 Executive committee minutes, 27 March, 8 May 1962
38 DEA file 14003-J2-3, memorandum to the minister (Paul Martin), 2 May 1963

39 Hart, *Nuclear Power*, 40–4
40 Executive committee minutes, 29 Aug. 1962
41 Board of Directors minutes, 13 Dec. 1962; AECL Records 103-A-3, Gray to Plumptre, 28 Sept. 1962
42 Executive committee minutes, 5 Nov. 1962; see also Ramsey interview on the several grades of plutonium and its appropriate uses.
43 PCO Records, cabinet conclusions, 24 Jan. 1963
44 Quoted in Pringle and Spigelman, *Nuclear Barons*, 377
45 Executive committee minutes, 18 March 1963
46 Ibid., 15 May, 18 March 1963
47 Ibid., 17 July 1963; Board of Directors minutes, 22 Sept. 1963, 14 Jan. 1964. See Charlotte Girard, *Canada in World Affairs* (Toronto 1980), 240–1.
48 See E.L.M. Burns, *A Seat at the Table: The Struggle for Disarmament* (Toronto 1972), 165–8, 189. The Moscow Treaty was signed on 5 August 1963: *Canadian Annual Review for 1963* (Toronto 1964), 328–9.
49 Goldschmidt, *Atomic Complex*, 388–9
50 Executive committee minutes, 9 April 1964; Board of Directors minutes, 29 May 1964, appendix dated 28 May 1964
51 Board of Directors minutes, 19 Feb. 1965
52 PCO Records, cabinet conclusions, 1 June 1965; cabinet document 288–65, 'Uranium Policy,' 28 May 1965, and revision, #301–65, 2 June 1965
53 Hart, *Nuclear Power*, 46
54 AECL Records, 103-A-3, vol. 5, memorandum to cabinet, 'Export financing: Rajasthan Atomic Power Project II: Section 21A Special credits – India,' 17 May 1966
55 Executive committee minutes, 12 Sept. 1966
56 Burns, *Seat at the Table*, 215–16
57 Joseph Nye, 'Maintaining a Nonproliferation Regime,' in George H. Quester, ed., *Nuclear Proliferation: Breaking the Chain* (Madison, Wis 1981), 17–18; Pringle and Spigelman, *Nuclear Barons*, 301–7; Pringle and Spigelman leave the impression that the treaty was signed in 1970, when that was in fact the year in which it came into force. Goldschmidt's account, in *Atomic Complex*, 190–200, is worthwhile.
58 AECL Records, 103-A-3, vols 5 and 6, Atomic Energy Advisory Panel, 19 July, 10 Nov. 1966
59 H. Clayton interview, Deep River, Sept. 1985; Board of Directors minutes, 17 Feb. 1961
60 CRNL, W.B. Lewis office files, misc., Kenneth Taylor, deputy minister of finance, to Gray, 7 Jan. 1960, with enclosures; T.W. Morison to Gray, 25 Jan. 1960; minutes of three meetings to discuss Canadian government

policy, 28–29 Jan. 1960. Present were representatives from the CBC, NRC, NFB, Post Office, and AECL, as well as Sir Savile Garner, British high commissioner.

61 CRNL, Lewis office files, Lewis to B.F.J. Schonland, 4 Sept. 1957, and Lewis to Sir James Chadwick, 14 April 1958; Sir John Hill interview, London, 15 April 1987

62 See L. Hannah, *Engineers, Managers and Politicians: The First Fifteen Years of Nationalised Electricity Supply in Great Britain* (London 1982), 178–81.

63 Ibid., 186–90

64 Eric Booth interview, London, 9 April 1987; Sir John Hill interview; AB19/23, Hinton to Hearn, 17 April 1957; on Hinton's perception of 'the two schools of thought' in the AEA, see AB19/13, Hinton to Plowden, 28 March 1955.

65 Patterson, *Nuclear Power*, 51; Hannah, *Engineers, Managers and Politicians*, 238–40, 244; AECL executive committee minutes, 9 Nov. 1960

66 AECL Records, 105-A-3, vol. 1, Rough notes of Gray's trip to Japan and India, 26 Dec. 1960–10 Feb. 1961. Gray's last stop was London, where considerable, and urgent, interest was shown by Dr H.S. Arms of English Electric.

67 Gray Papers, 'Discussions with the Minister,' 13 Feb. 1961; Churchill, Gray recorded, 'supports anything we can do in foreign aid or for foreign contracts.' A draft memorandum to the cabinet was attached to his next meeting, 21 March 1961. See also PCO Records, cabinet conclusions, 28 April, 3 May 1961.

68 AECL Records, 103-E-1, vol. 1, Gray to executive committee members, 31 July 1961; Hannah, *Engineers, Managers and Politicians*, 242

69 Sir Chistopher Hinton, 'Nuclear Power,' *Three Banks Review* 52 (December 1961), 14

70 Hannah, *Engineers, Managers and Politicians*, 242

71 Ibid., 241–2

72 Executive committee minutes, 16 Jan. 1962; Gray Papers, 'Discussions with the Minister,' Hinton to Gray, 11 Jan. 1962, attached to 'Items for Discussion with the Minister,' 16 Jan. 1962. Churchill gave his consent to discussions with the British, and for some leeway in the size of the fee.

73 Gray Papers, meeting with the minister, 26 July 1962; Gray to Advisory Panel, 31 Aug. 1962

74 Executive committee minutes, 5 Nov. 1962

75 Ibid., 8 Nov. 1962

76 AECL Records, 103-A-3, vol. 1, Gray to Advisory Panel, 21 Nov. 1962. When the panel met at Madame Burger's on 23 November it agreed to Gray's proposed policy and recommended it to the cabinet.

77 Gray Papers, 'Discussions with the Minister,' 22 March 1963
78 Board of Directors minutes, 22 Sept. 1963
79 Executive committee minutes, 27 Oct. 1963
80 Patterson, *Nuclear Power*, 66, 165–6; Roger Williams, *The Nuclear Power Decisions: British Policies, 1953–78* (London 1980), 20, 29
81 Williams, *Nuclear Power Decisions*, 94–5
82 Patterson, *Nuclear Power*, 166: seawater was again the culprit, at a cost of £50 million.
83 Booth interview, Hill interview
84 Confidential interviews
85 Gray interview, Sept. 1986; Lord Sherfield interview, London, 14 April 1987; Hill interview
86 The question of CGE's dependence on AECL was discussed at the AECL executive committee meetings of 6 April and 12 July 1961, the latter meeting featuring a tour of CGE's facilities. The committee was 'particularly impressed' by CGE's fuelling and fuelling machine manufacturing lines. See also AECL Records, 103-S, Senior Management Committee, 9 April 1964.
87 See Appendix 'A' to Board of Directors minutes, 14 Jan. 1964: 'AECL-A Nuclear Engineering Consultant.'
88 Board of Directors minutes, 14 Dec. 1965
89 See Douglas Ross, *In the Interests of Peace: Canada and Vietnam, 1954–73* (Toronto 1984), 232, 252–4. Ross argues that an effective harmony of viewpoint existed down to 1964.
90 Board of Directors minutes, 17 Feb. 1961
91 Executive committee minutes, 8 May 1962; Board of Directors minutes, 13 Dec. 1962
92 Executive committee minutes, 22 Sept. 1963; advance purchasing arose again in 1964, for the same reason: ibid., 25 June 1964; AECL Records, 103-A-3, vol. 2, Trip Report by Lorne Gray, 9–22 Dec. 1963, with stop in Karachi on 18 Dec.; Advisory Panel minutes, 31 Jan. 1964
93 Executive committee minutes, 9 April, 25 June 1964; Board of Directors minutes, 18 Oct. 1964
94 Executive committee minutes, 20 Aug., 4 Oct., 10 Dec. 1964, 5 Aug. 1965
95 AECL Records, 103-P-1, Lewis to Gray, 5 Feb. 1965
96 Ibid., 103-A-2, vol. 1, Haywood to Gray, 8 Oct. 1964, expressing disappointment with the current state of the Canadian nuclear industry, with insufficient contributions from the private sector; Board of Directors minutes, 19 Feb. 1965
97 AECL Records, 105-19, vol. 2, Alan Wyatt to John Foster, 16 March 1965
98 Gray Papers, 'Discussions with the Minister,' 16 March 1965
99 AECL Records, 105-19, vol. 1, Watson to Foster and Haywood, 20 July 1965

100 Ibid., 105–19, vol. 2, minutes of a meeting of the Nuclear Reactor Export Promotion Committee, 1 March 1966
101 AECL Records, file 105-19, vol. 2, minutes of the Nuclear Power Reactor Export Promotion Committee, 1 March 1966
102 Ibid., 103-A-3, vol. 5, J.L. Olsen to Gray, 22 Sept. 1966, with copy to Denis Harvey; 'Organization of the Nuclear Power Industry for Export,' 7 Nov. 1966, with critical comments by Les Haywood: Haywood to Greenwood, 3 Nov. 1966
103 Ibid., draft memorandum to the cabinet, 'Exploitation of Canadian Nuclear Power Generation,' 28 Sept. 1965
104 Ibid., Gray to Robertson, 11 Oct. 1966; executive committee minutes, 21 Nov. 1966
105 Ibid., Advisory Panel minutes, 10 Nov. 1966; executive committee minutes, 21 Nov. 1966; on the meeting with Pépin, Gray interview by phone, 4 Nov. 1986
106 J.H. Smith interview, Toronto, 30 May 1986. Smith explained, 'My engineering background [had] kept my interest in it. Otherwise we would have washed it up in 1959.' See executive committee minutes, 20 March 1967; AECL Records, Gray to Pépin enclosing a draft memorandum to cabinet, 3 May 1968.
107 Executive committee minutes, 7 March 1968
108 AECL Records, 105-19, Gray to Pépin, 3 May 1968, with enclosure, 'Canadian Nuclear Power Generating Systems for Export,' May 1968
109 Gray Papers, 'Discussions with the Minister,' 17 July 1968: J.J. Greene
110 Executive committee, 22 Oct. 1969
111 MacKay interview
112 The misgivings emerged in a meeting of the SMC, 8 Sept. 1965: 'Mr. Haywood queried (and Dr. Lewis agreed), whether ... the Company should not leave foreign markets to be exploited by Canadian consulting engineering firms.'
113 AECL Records, file 103-S-8, vol. 1, Senior Management Committee, 15 May 1964
114 Goldschmidt, *Atomic Complex*, 336–7
115 Ibid., 319
116 Pringle and Spigelman, *Nuclear Barons*, 284–5

CHAPTER 11: BIG SCIENCE

1 See Statistics Canada, *Perspectives Canada III* (Ottawa 1980), 259–74. Other figures are taken from *Canada Year Book, 1973* (Ottawa 1973).
2 Gray interview, Toronto, 5 Nov. 1986

3 On federal budgetary policy, see *Canadian Annual Review for 1968* (Toronto 1969); AECL figures are from the *1968–69 Annual Report* (Ottawa 1969); Morgan, *Implosion*, 61.

4 *1967–1968 Annual Report* (Ottawa 1968), 23; see G. Bruce Doern, *Science and Politics in Canada* (Montreal 1972), 101.

5 See B.T. Eiduson, *Scientists: Their Psychological World* (New York 1962), 165.

6 Doern, *Science and Politics*, 113n; SMC committee minutes, 11 July 1966: 'Mr. Gray expressed concern regarding the fact that virtually no summer students apply for permanent employment ...' See also L.R. Haywood, 'The Role of AECL Laboratories,' in *The Twelfth AECL Symposium on Atomic Power on 19 April, 1968 at Ottawa* (AECL-3067), 41. NRC was the next most important Canadian destination; thirty-five man-years left the country from Chalk River every year.

7 The Treasury Board remark is in the SMC minutes, 29 June 1967.

8 Gray and Lewis constituted an ad hoc committee on research at Chalk River, which met under C.J. Mackenzie's chairmanship in April 1962: Board of Directors minutes, 31 May 1962, appendix 'A'; ibid., 24 Sept. 1962.

9 SMC minutes, 15 May 1964; Board of Directors minutes, 29 May 1964, appendix 'A': W.B. Lewis, 'High Flux Neutron Facility,' 28 May 1964; Doern, *Science and Politics*, 104

10 Board of Directors minutes, 29 May 1964, appendix 'A'; L.G. Elliott, W. Bennett Lewis, and A.G. Ward, 'ING: A Vehicle to New Frontiers,' *Science Forum*, Feb. 1968, 3–7

11 Elliott et al. predicted (7) that the annual volume of nuclear business in Canada by 1980 would be $1 billion; AECL Records, file 103-s-8, Lewis to A.H.M. Laidlaw, 27 Oct. 1964, protesting the failure to record his views in the SMC minutes for 6 Oct. 1964. He repeated the 12 per cent figure, and his view of the role of high technology in 'the national economy' in the SMC meeting of 8 Sept. 1965. Gray's opinion is recorded in Finance Department Records, vol. 5361, file 3950-03/w582, Gray to K.W. Taylor, 14 Oct. 1960.

12 Executive committee minutes, 9 April, 14 May 1964

13 Ibid. At the executive committee on 25 June 1964 Lewis estimated costs as $100,000 in 1964–5, $750,000 in 1965–6, and $1 million in 1966–7. Thereafter he predicted it would cost between $7 million and $10 million a year. He wondered whether money currently assigned to the BWR could not be diverted to ING, provided the Ontario government would fund the former project; and he predicted that ING would accrue great commercial value by producing isotopes.

14 Board of Directors minutes, 18 Oct. 1964, appendix 'H,' Lewis, 'AECL Estimates for 1965–66,' 17 Oct. 1964

15 Ibid.; *1965–66 Annual Report* (Ottawa 1966), 44
16 Gray Papers, 'Discussions with the Minister,' 17 Dec. 1964
17 Doern, *Science and Politics*, 114
18 Ibid., 114–16; Board of Directors minutes, 7 June 1965
19 Executive committee minutes, 7 July 1966
20 AECL Records, 110-5-12, vol. 1, G. Davidson to Gray, 8 Aug. 1966, and Gray to Davidson, 19 Aug. 1966
21 Executive committee minutes, 12 Aug. 1966
22 Ibid., 12 Sept. 1966
23 Board of Directors minutes, 12 Oct. 1966
24 Executive committee minutes, 21 Nov. 1966; Doern, *Science and Politics*, 116–17
25 Doern, *Science and Politics*, 117; SMC minutes, 7 Feb. 1967
26 Board of Directors minutes, 18 April 1967; *1966–67 Annual Report*, 39
27 CRNL Records, W.B. Lewis office files, Lewis to Thode, 29 Dec. 1967
28 Board of Directors minutes, 15 Jan. 1968. Gray already knew that the project would be cancelled: AECL Records, 110-5-14, vol. 1, Gray to Reisman, 29 Aug. 1968; executive committee minutes, 3 Oct. 1968; Gray interview, 5 Nov. 1986.
29 AECL Records, 105-19, Gray to Pépin, 3 May 1968, with enclosure, a draft memorandum for the cabinet, 'Canadian Nuclear Power Generating Systems for Export,' May 1968
30 Ibid., 103-A-3, Gray to Bryce, Harvey, Reisman, and Gordon Robertson, 4 June 1968, with agenda for their next lunch at Madame Burger's
31 Ibid., C-103-P-1, PRDPEC meeting #32, 12 Jan. 1968, appendix A, G.M. James to R.F. Wright, 10 Jan. 1968; McConnell interview; Foster interview, 25 Aug. 1986
32 Board of Directors minutes, 29 Oct. 1968, 14 Feb. 1969; executive committee minutes, 25 Nov. 1968
33 Executive committee minutes, 22 Aug. 1968, 24 March, 13 May 1969
34 Board of Directors minutes, 14 Feb. 1969
35 Ibid., 12 June 1969; see also AECL Records, C-103-P-1, PRDPEC #32, appendix A, James report, 10 Jan. 1968.
36 Board of Directors minutes, 22 July, 14 Aug. 1969; AECL Records, 105–22, vol. 2, Gray to Gathercole, 31 July 1969
37 AECL Records, 105-22, vol. 2, Gray to Harold Smith, 10 Sept. 1969; McConnell interview
38 AECL Records, C-103-P-1, PRDPEC #34, 6 June 1968; executive committee minutes, 12 Aug. 1970
39 McConnell interview; Morison interview

40 See *Canada Year Book 1973* (Ottawa 1973), 385–6; *1969–70 Annual Report* (Ottawa 1970), 10–11.
41 Foster, 'The First Twenty-Five Years,' 38–9; Stetson interview; AECL Records, 103-P-5, Watson to Gray, 20 Nov. 1969
42 AECL Records, 105-22, vol. 2, Wardell to O'Sullivan, 20 Nov. 1969, encl. 'Capital Cost of Units 1–4, Pickering Generating Station, October 1969 Forecast'
43 Ibid., 103-H-2, Foster to Smith and Gray, 16 Feb. 1970, Haywood to Gray, 3 Dec. 1970, and Haywood to Gray, 9 Dec. 1970
44 Ibid., Public Affairs files, Speeches – Harold A. Smith, F.H. Krenz to Smith, 15 Dec. 1969, and Smith to Krenz, 4 Feb. 1970
45 Ibid., 105-22, 'Pickering Progress,' weekly reports issued from the project director's office, 25 Feb.–28 Sept. 1971
46 Board of Directors minutes, 7 June 1971
47 AECL, *Annual Report, 1955–56* (Ottawa 1956), 8–9, 17; 1956–57 (Ottawa 1957), 15; 1957–58 (Ottawa 1958), 15. Executive committee minutes, 25 April 1956; board minutes, 2 Nov. 1956, 11 July 1957. Commercial operations did *not* make a profit in 1956–7 or 1957–8, partly because of the high premium on the Canadian dollar; at the same time its R&D grant from the rest of the company lapsed. Annual Reports stressed that Commercial Products operated independently (in a financial sense) of other divisions.
48 Board of Directors minutes, 7–8 Nov. 1955; AECL, *Annual Report, 1958–59* (Ottawa 1959), 24
49 Board of Directors minutes, 8 Aug., 10 Nov. 1958, 13 March 1959; executive committee minutes, 16 Jan. 1959
50 Board of Directors minutes, 17 June, 8 Sept. 1960; AECL, *Annual Report, 1960–61* (Ottawa 1961), 21–3, 31
51 Board of Directors minutes, 17 Feb. 1961; see photo of the Mobile Cobalt-60 Irradiator truck, *Annual Report, 1961–62* (Ottawa 1962), 18.
52 *1963–64 Annual Report* (Ottawa 1964), 17–19; executive committee minutes, 27 July 1964
53 Executive committee minutes, 25 Feb. 1964, 5 Aug. 1965; Board of Directors minutes, 29 May 1964, 19 Feb. 1965
54 Executive committee minutes, 4 May, 5 Aug. 1965
55 Ibid., 22 Nov. 1965, 1 and 31 May 1965; SMC, 13 Dec. 1965: Gordon, the principal objector, had retired.
56 SMC, 13 Dec. 1965
57 Executive committee minutes, 7 July, 12 Aug. 1966
58 Ibid., 12 Sept. 1966; *1967–68 Annual Report* (Ottawa 1968), 29
59 C.H. Hetherington, 'History of Commercial Products, October 1, 1946–August 16, 1974,' RCC files; Errington interview

60 Board of Directors minutes, 18 April 1968; AECL, *1968–69 Annual Report* (Ottawa 1969), 43; 1969–70 (Ottawa 1970), 43
61 Board of Directors minutes, 13 Feb. 1969
62 Ibid., 11 Dec. 1969

CHAPTER 12: EPILOGUE; OR THE PRICE OF SUCCESS

1 The definition is found in volume 2 of the *Report* of the Royal Commission on the Economic Union and Development Prospects for Canada (Ottawa 1986), 133.
2 Finance Department Records, vol. 5360, file 3950-02, NA, 'Nuclear Programme Study,' 5 Oct. 1973. The cabinet had mandated the study on 15 July 1973.
3 See ibid., vol. 5360, file 3950-02, 'Brief Outline of Potential Nuclear Power Developments in Canada,' probably late 1973 or early 1974.
4 Spencer Weart, 'The Heyday of Myth and Cliché,' in Len Ackland and Steven McGuire, *Assessing the Nuclear Age* (Chicago 1986), 88–9
5 Finance Department Records, vol. 5361, file 3950/03/C436, E.L. Fytche, 'Visit to Chalk River, May 9, 1973, Informal Notes,' 10 May 1973
6 Board of Directors minutes, 4 April 1974
7 Minutes of the Public Accounts Committee, 25 Jan. 1977, 9:14
8 Ibid.
9 Appendix PA-3 to the 12 Dec. 1976 minutes of the Public Accounts Committee
10 Ibid., 9 Dec. 1976, 5:32; 25 Jan. 1977, 9:27; 3 Feb. 1977, 12:6; Donald Macdonald interview
11 Pringle and Spigelman, *Nuclear Barons*, 374–5; see the thoughtful discussion in Hart, *Nuclear Power*, 56–9.
12 Pringle and Spigelman, *Nuclear Barons*, 375
13 Moher, 'The Policies of Supplier Nations,' 89
14 *Canadian Annual Review of Politics and Public Affairs for 1974* (Toronto 1975), 329–30; Ivan Head, Trudeau's foreign policy adviser, and a specialist in Third-World matters, represented the prime minister's concerns: Head interview, Toronto, 31 Aug. 1987.
15 Ibid., 331
16 *Canadian Annual Review of Politics and Public Affairs, 1975* (Toronto 1976), 270
17 Board of Directors minutes, 27 Feb. 1973; Gray interview, Deep River, 27 March 1985
18 Executive committee minutes, 5 Oct. 1973; Board of Directors minutes, 27 Feb., 7 Nov. 1973
19 Public Accounts Committee minutes, 25 Jan. 1977, 9:27

20 Hill interview; Macdonald interview
21 Board of Directors minutes, 23 Sept. 1977
22 Public Accounts Committee minutes, 3 Feb. 1977, 12:6
23 Richard D. French and R. van Loon, *How Ottawa Decides: Planning and Industrial Policy-making, 1968–1984* (Toronto 1984), 40
24 Boyer, *Bomb's Early Light*, 358
25 Macdonald interview
26 *Canadian Annual Review of Politics and Public Affairs, 1977* (Toronto 1979), 5–6. *The Financial Post* alleged that the Argentinian agent's fee had gone in a bribe to a Peronist minister; the minister denied the charge and shortly afterwards died. Nothing has ever been proven.
27 See Ivan Bernier and Andrée Lajoie, *Regulations, Crown Corporations and Administrative Tribunals* (Toronto 1985), 63–7, for a discussion of the comprehensive audit and its fate.
28 Public Accounts Committee, minutes, 10 May 1977, 32:22
29 The advantage was estimated at between 1.5 and 2 mills per kilowatt-hour in late 1977.
30 See Ray Silver, *Fallout from Chernobyl* (Toronto 1987).
31 See opinion poll results reported in the *Globe and Mail*, 16 June 1987.

Note on Sources

Primary Sources

Most of the documentation for this book came from the records of Atomic Energy of Canada Limited. At the time of writing, much if not most of this material was in the process of transferral to the National Archives of Canada (formerly the Public Archives of Canada, or PAC). Those files held at head office in Ottawa are designated AECL Records in the notes; those from Chalk River are listed as CRNL. There are, besides, references to AECL's many publications, some of which, being classified or proprietary, are held in the 'Special Documents Depository' at Chalk River; all AECL publications are listed by their AECL number in the notes, and are not individually cited in the bibliography.

Other manuscript sources consulted were:

Ottawa

Privy Council Office Records
Department of External Affairs Records
Eldorado Nuclear Limited Records
Department of Finance Records (PAC)
National Research Council Records (PAC)
Atomic Energy Control Board Records (PAC)
Mackenzie King Papers (PAC)
C.D. Howe Papers (PAC)
C.J. Mackenzie Papers (PAC)
L.B. Pearson Papers (PAC)

Washington

Manhattan District Records (National Archives)

London

Atomic Energy Authority Records (AB)
Cabinet Records (CAB)

Summit, NJ

Les Cook Papers (privately held)

Deep River, Ont.

Lorne Gray Papers (privately held)
George Laurence Papers (privately held)

Interviews

Colin Amphlett, 29 May 1985
Gilbert Bartholomew, 16 April 1986
W.J. Bennett, 27 March, 10 April, 16 July, 23 December 1985, 5 June, 15 and 28 November 1986, 22 April 1987
R.B. Bryce, 13 November 1985
Donald Campbell, 10 April, 13 May 1985
Ross Campbell, 7 May 1987
Derek Chase, 7 January 1987
Henry Clayton, 16 April 1986
Harry Collins, 15 April 1986
Sir William Cook, 7 June 1985
Gene Critoph, 20 October 1986
Jack Davis, 16 June 1986, 25 September 1987
James Donnelly, 9 June 1987
Roy Errington, 31 January, 1–2 February 1987
Eileen and Fred Fenning, 6 June 1985
John Foster, 14 December 1984, 1 and 24 August, 27 October 1986
Grant Gay, 6 January 1987
Bertrand Goldschmidt, 10 June 1985, 5 June 1986
Lorne Gray, 10 December 1984, 27 March, 25 September, 13 November 1985, 16 April, 9 September 1986
Jules Guéron, 10 June 1986
John Guthrie, 15 March 1985
Geoff Hanna, 28 August, 25 September 1985, 14 April 1986
Gordon Hatfield, 26 March 1986
Les Haywood, 25 September 1985
Sir John Hill, 15 April 1987
Donald Hurst, 4 February 1986
Gib James, 27 August 1985
Hon. Donald S. Macdonald, 20 May 1987
Ian MacKay, 16 June 1986
C.J. Mackenzie, 29 March 1976, 17 August 1977
Lorne McConnell, 17 September 1986
John Melvin, 26 September 1985
George Mercer, 18 April 1985
Ara Mooradian, 17 December 1985, 20 October 1986
W. Morison, 15 September 1986
Colonel Curtis Nelson, 17 April 1985
General Kenneth D. Nichols, 18 April 1985
E.C.W. Perryman, 26 September 1985
Hon. J.W. Pickersgil, 14 October 1986
George Pon, 20 October 1986
Maurice Pryce, 6 December 1985
Robert Ramsey, 19 April 1985
Gordon Robertson, 21 October 1986
Julius Rubin, 18 April 1985
B.W. Sargent, 26 October 1985
Hon. Mitchell Sharp, 4 February 1986
Lord Sherfield, 15 April 1987
T. Shoyama, 24 September 1987
E. Siddall, 18 March 1987
Ted Smale, 5 January 1987
Harold Smith, 21 June, 14 December 1984

J. Herbert Smith, 30 May 1986
Roger Smith, 12 March 1985
Nat Stetson, 22 April 1985
Gordon Stewart, 27 August, 25 September 1985, 15 April 1986
G. Stoner, 25 November 1986
Ted Thexton, 7 May 1987
H.G. Thode, 10 October 1985
Arthur Ward, 24 July 1985
Donald Watson, 13 November 1985, 8 January 1987
Leo Yaffe, 16 July 1985, 25 March 1986

Bibliography

Ashley, C.A., and R.G.H. Smails, *Canadian Crown Corporations*, Toronto 1965

Babin, Ronald, *The Nuclear Power Game*, Montreal 1985

Bernier, Ivan, and Andrée Lajoie, *Regulations, Crown Corporations and Administrative Tribunals*, Toronto 1985

Berton, Pierre, 'The Mayor of Atom Village,' *Maclean's Magazine*, 1 Dec. 1947, 9, 71–3

Borins, S.F., 'World War II Crown Corporations: Their Functions and Their Fate,' in J.R.S. Pritchard, ed., *Crown Corporations in Canada*, Toronto 1983

Borins, S.F., and Lee Brown, *Investments in Failure: Five Government Corporations That Cost the Canadian Taxpayer Billions*, Toronto 1986

Bothwell, Robert, *Eldorado: Canada's National Uranium Company*, Toronto 1948

Bothwell, Robert, Ian Drummond, and J.R. English, *Canada since 1945*, Toronto 1981

Bothwell, Robert, and J.L. Granatstein, 'Canada and the Wartime Negotiations over Civil Aviation,' *International History Review*, 2, 1980, 585–601

Bothwell, Robert, and J.L. Granatstein, eds., *The Gouzenko Transcripts*, Ottawa 1982

Bothwell, Robert, and William Kilbourn, *C.D. Howe: A Biography*, Toronto 1979

Boyer, Paul, *By the Bomb's Early Light: American Thought and Culture at the Dawn of the Atomic Age*, New York 1985

Boyle, Andrew, *The Fourth Man*, New York 1979

Brecher, Michael, *India and World Politics: Krishna Menon's View of the World*, London 1968

Britnell, George, 'Under-developed Countries in the World Economy,' *Canadian Journal of Economics and Political Science*, 23, Nov. 1957, 453–66

Brown, Anthony Cave, and Charles B. MacDonald, *The Secret History of the Atomic Bomb*, New York 1977

Burns, E.L.M., *A Seat at the Table: The Struggle for Disarmament*, Toronto 1972

Canada Year Book, Ottawa 1937–

Canadian Annual Review, Toronto 1961–

Careless, Anthony, *Initiative and Response*, Montreal and London 1977

Clark, Ronald W., *The Birth of the Bomb*, London 1961

Davis, John, *Canadian Energy Prospects*, Ottawa 1957
Dawson, R.M., *The Government of Canada*, 4th ed., Toronto 1963
Doern, G. Bruce, *Government Intervention in the Canadian Nuclear Industry*, Ottawa 1980
Doern, G. Bruce, *Science and Politics in Canada*, Montreal 1972
Doern, G. Bruce, and R.W. Morrison, *Canadian Nuclear Policies*, Montreal 1980
Eggleston, Wilfrid, *Canada's Nuclear Story*, Toronto 1965
Eggleston, Wilfrid, *National Research in Canada: The* NRC, 1916–1966, Toronto 1978
Eiduson, B.T., *Scientists: Their Psychological World*, New York 1962
Energy, Mines and Resources, Department of, *The National Atlas of Canada*, 4th ed., Ottawa 1974
External Affairs, Department of, *Documents on Canadian External Relations*, vol. 9, ed. by John Hilliker, Ottawa 1980
Faucher, Philippe, and Johane Bergeron, *Hydro-Québec: La société de l'heure de pointe*, Montreal 1986
Feynman, R.P., *Surely You're Joking, Mr. Feynman*, London 1985
Financial Post Survey of Markets and Business Year Book, Toronto 1944–60
Finch, Ron, *Exporting Danger: A History of the Canadian Nuclear Energy Export Program*, Montreal 1986
Finnegan, Joan, *Some of the Stories I Told You Were True*, Ottawa 1981
French, Richard D., and R. van Loon, *How Ottawa Decides: Planning and Industrial Policy-Making, 1968–1984*, Toronto 1984
Fuller, John G., *We Almost Lost Detroit*, New York 1975
George, Roy E., *The Life and Times of Industrial Estates Limited*, Halifax 1974
Girard, Charlotte, *Canada in World Affairs, 1963–65*, Toronto 1980
Goldschmidt, Bertrand, *The Atomic Complex: A Worldwide Political History of Nuclear Energy*, La Grange Park, Ill, 1982
Goldschmidt, Bertrand, 'Hans Halban (1908–1964),' *Nuclear Physics* 79 (1966), 1–11
Goldschmidt, Bertrand, *Pionniers de l'atome*, Paris 1987
Goldschmidt, Bertrand, 'Les premiers milligrammes de plutonium,' *La Recherche*, 131, mars 1982
Gowing, Margaret, *Britain and Atomic Energy, 1939–1945*, London 1964
Gowing, Margaret, and Lorna Arnold, *Independence and Deterrence: Britain and Atomic Energy, 1945–1952*, 2 vols., London 1974
Granatstein, J.L., *Canada 1957–1967*, Toronto 1986
Green, Alan, *Immigration and the Postwar Canadian Economy*, Toronto 1976
Guéron, Jules, 'Lew Kowarski et le développement de l'énergie nucléaire,' in J.B. Adams, ed., *Lew Kowarski 1907–1979*, np nd
Hannah, L., *Engineers, Managers and Politicians: The First Fifteen Years of Nationalised Electricity Supply in Great Britain*, London 1982
Hare, F. Kenneth, and Morley K.

Thomas, *Climate Canada*, 2nd ed., Toronto 1979

Hart, David, *Nuclear Power in India, a Comparative Analysis*, London 1983

Hartcup, G., and T. Allibone, *Cockcroft and the Atom*, Bristol 1984

Hayes, F. Ronald, *The Chaining of Prometheus: Evolution of a Power Structure for Canadian Science*, Toronto 1973

Heeney, A.D.P., *The Things That Are Caesar's: The Memoirs of a Canadian Public Servant*, Toronto 1972

Hendry, John, ed., *Cambridge Physics in the Thirties*, Bristol 1984

Hewlett, Richard, and Oscar Anderson, *The New World: A History of the United States Atomic Energy Commission*, vol. 1, Washington 1972

Hewlett, Richard, and Francis Duncan, *A History of the United States Atomic Energy Commission: Atomic Shield, 1947–1952*, Washington 1972

Hewlett, Richard, and Francis Duncan, *Nuclear Navy, 1946–62*, Chicago 1974

Hinton, Christopher, 'Nuclear Power,' *Three Banks Review*, 52, December 1961

A History of Deep River, Deep River 1970

Holliday, Joe, *Dale of the Mounted: Atomic Plot*, Toronto 1959

Holmes, John W., *The Shaping of Peace: Canada and the Search for World Order*, vol. 2, Toronto 1982

Ignatieff, George, *The Making of a Peacemonger*, Toronto 1985

Jackson, C.H., 'The Chalk River Atomic Energy Project,' *Engineering Journal*, June 1947, 265–6

Jones, Vincent C., *Manhattan: The Army and the Atomic Bomb*, Washington 1985

Judd, James, *Ploot: A Novel*, Ottawa 1978

Kennedy, Clyde C., *The Upper Ottawa Valley: A Glimpse of History*, Pembroke 1970

Kent, Tom, 'Regional Development in a Federal Setting,' in O.J. Firestone, ed., *Regional Economic Development*, Ottawa 1974

Kinsey, Freda, 'Life at Chalk River,' *Atomic Scientists Journal*, 3(1), Sept. 1953, 18–23

Kuhn, Thomas S., *The Structure of Scientific Revolutions*, 2nd ed., Chicago 1970

Kulkarni, R.P., and V. Sarma, *Homi Bhabha: Father of Nuclear Science in India*, Bombay 1969

Leacey, F.H., ed., *Historical Statistics of Canada*, 2nd ed., Ottawa 1983

Legget, Robert F., *Chalmers Jack Mackenzie*, a memoir prepared for the Royal Society of London, 1985

Lewis, W.B., 'The Development of Electrical Counting Methods in the Cavendish,' in J. Hendry, ed., *Cambridge Physics in the Thirties*, Bristol 1984

Martin, Paul, *A Very Public Life*, vol. 1, Ottawa 1983; vol. 2, Toronto 1985

Mathias, Philip, *Forced Growth*, Toronto 1971

McDougall, A.K., *John P. Robarts: His Life and Government*, Toronto 1986

McDougall, Walter A., ... *the Heavens and the Earth*, New York 1985
McKay, Paul, *Electric Empire: The Inside Story of Ontario Hydro*, Toronto 1983
McLin, Jon B., *Canada's Changing Defense Policy, 1957–1963*, Baltimore 1967
Mishra, D.K., *Five Eminent Scientists: Their Lives and Work*, Delhi 1976
Moher, Mark, 'The Policies of Supplier Nations,' in David DeWitt, ed., *Nuclear Non-Proliferation and Global Security*, London 1987
Morgan, Nicole, *Implosion: An Analysis of the Growth of the Federal Civil Service in Canada (1945–1985)*, Ottawa 1986
Moss, Norman, *Klaus Fuchs: The Man Who Stole the Atom Bomb*, London 1987
Murray, William, *Historical Highlights: A Chronology Covering the Period 1900 to 1950*, Toronto 1956
National Research Council of Canada Review, Ottawa 1948–52
Newman, Peter C., *The Distemper of Our Times*, Toronto 1968
Newman, Peter C., 'Deep River – almost the perfect place to live,' *Maclean's Magazine*, 13 Sept. 1958, 24–5
Nichols, K.D., *The Road to Trinity: A Personal Account of How America's Nuclear Policies Were Made*, New York 1987
Nicolson, Harold, *Public Faces*, Toronto nd, original edition 1932
Nye, Joseph, 'Maintaining a Non-Proliferation Regime,' in George H. Quester, ed., *Nuclear Proliferation: Breaking the Chain*, Madison, Wis, 1981
O'Neill, Pierre, and Jacques Benjamin, *Les mandarins du pouvoir: L'exercice du pouvoir au Québec de Jean Lesage à René Lévesque*, Montreal 1978
Patterson, Walter C., *Going Critical: An Unofficial History of British Nuclear Power*, London 1986
Patterson, Walter C., *Nuclear Power*, 2nd ed., London 1983
Pearson, Lester B., *Mike: The Memoirs of the Rt. Hon. L.B. Pearson*, vol. 2, Toronto 1973
Peierls, R., *Bird of Passage: Recollections of a Physicist*, Princeton 1985
Pickersgill, J.W., *The Mackenzie King Record*, vol. 1, Toronto 1961
Pickersgill, J.W., and D. Forster, *The Mackenzie King Record*, vol. 3, Toronto 1970
Polmar, Norman, and Thomas B. Allen, *Rickover: Controversy and Genius*, New York 1982
Porter, John, *The Vertical Mosaic: An Analysis of Social Class and Power in Canada*, Toronto 1965
Pringle, Peter, and James Spigelman, *The Nuclear Barons*, New York 1981
Ramage, Janet, *Energy: A Guidebook*, Oxford 1983
Reid, Escott, *Envoy to Nehru*, New Delhi 1981
Rhodes, Richard, *The Making of the Atomic Bomb*, New York 1986
Ross, Douglas, *In the Interests of Peace: Canada and Vietnam, 1954–73*, Toronto 1984
Royal Commission on the Economic Union and Development Pros-

pects for Canada, *Report*, Ottawa 1986
Silver, Ray, *Fallout from Chernobyl*, Toronto 1987
Simpson, John, *The Independent Nuclear State: The United States, Britain and the Military Atom*, 2nd ed. London 1986
Sims, Gordon, *A History of the Atomic Energy Control Board*, Ottawa 1981
Smith, Harold, 'Atomic Energy in Industry,' *Electrical Digest*, June 1956, 100–2
Spinks, John, *Two Blades of Grass: An Autobiography*, Saskatoon 1980
State, Department of, *Foreign Relations of the United States*, 1947, 1948, 1949, 1950, 1951, 1952–54, Washington 1973–84
Taylor, Graham D., and P. Sudnik, *DuPont and the International Chemical Industry*, Boston 1984
Thistle, Mel, ed., *The Mackenzie-McNaughton Wartime Letters*, Toronto 1975
Thomson, Dale, *Jean Lesage and the Quiet Revolution*, Toronto 1984
Villa, Brian, 'Alliance Politics and Atomic Collaboration,' in S. Aster, ed., *The Second World War as a National Experience*, Ottawa 1981
Weart, S.R., *Scientists in Power*, Cambridge, Mass 1979
Weart, S.R., 'The Heyday of Myth and Cliché,' in Len Ackland and Steven McGuire, eds., *Assessing the Nuclear Age*, Chicago 1986
Weart, S.R., and G.W. Szilard, eds., *Leo Szilard: His Version of the Facts: Selected Recollections and Correspondence*, Cambridge, Mass 1978
Williams, R.C., and Philip L. Cantelon, *The American Atom, 1939–1984*, Philadelphia 1984
Williams, Roger, *The Nuclear Power Decisions: British Policies, 1953–78*, London 1980
Wilson, David, *Rutherford: Simple Genius*, Cambridge, Mass 1983
Wyatt, Alan, *Electric Power: Challenges and Choices*, Toronto 1986

Index

accelerators 396
Advisory Panel on Atomic Energy: established 74–5; influence on Canadian atomic policy 75–6, 139, 251–2, 277, 319, 349, 378, 387–90
Akers, Wallace 15, 16, 26, 30, 32, 34, 50; character of 23
Alexander, Lord 141
Amphlett, Colin 30
Anderson, Sir John 13, 17, 23, 36, 43–4, 45, 53, 81; character of 38, 66, 67–8, 72
Appleton, Sir Edward 55, 80
Argentina 390, 407, 426–7, 432–3, 435
Arrow, *see* CF-105
atomic bomb 3–4; testing 13, 428; explained to Mackenzie King 16; refusal to pursue Canadian bomb 73–4; co-operation with UK, US 140–1, 349, 369; and disarmament 342
atomic energy: federal control over 77, 174; public impact 82; possibility of in Canada 111, 172–6; in cabinet 115, 141, 208–11, 242–3, 249, 260–2, 264, 266, 268, 306, 319–20, 336, 338, 379, 427; adoption of power program by AECL 202–5; private development plans 204–5, 206, 223–4; economics of 225–8, 278, 316; governments' influence over 278, 322ff, 332–9; political significance 339–40, 424, 433
Atomic Energy Authority (UK) 196–7, 245, 278, 371, 374–80
Atomic Energy Commission (US) 75, 104, 114, 138–44, 246, 273, 278, 318, 347, 391, 392
Atomic Energy Control Act 74ff; amended 182
Atomic Energy Control Board (AECD) 77, 115; and Chalk River estimates 123; and second research pile 126; control over Chalk River 138; and foundation of AECL 144–5; and safety 240, 244, 314–6; lack of influence over international policy 350
Atomic Energy of Canada Limited (*see also* Ontario Hydro)

– advisory committee on atomic power 223
– audit committee, 434
– Board of directors 145, 180–1, 189–90, 195, 205–7, 250, 276, 294, 307–8, 325, 338, 365, 410, 415–6, 418, 420–1, 430, 433–4, 436
– budget, 341, 395, 400–1, 402, 406
– competitive situation 348, 358, 361, 370, 376–7, 387–90
– engineering company 441–2
– engineering function: *see subhead below*, Nuclear Power Plant Division
– executive committee 195, 202, 234–5, 276, 303, 312–3, 318, 377, 379
– exports: *see subhead below*, sales and marketing
– finances 235
– foundation of company 137–45
– head office 275–8
– heavy-water program 316–32
– liaison office in London 371
– made independent 182
– Nuclear Power Plant Division (NPPD) 237, 239–40, 241, 253, 256–7, 268–9, 275, 283–6, 304–5, 306, 447–8; reactor construction by 285–6; expansion of 281–6, 303; as consulting engineer 306–7, 310, 407, 448; Sheridan site 309; reconstituted 441, 442
– nuclear power program: approved by board 204–6; approved by cabinet 208–11, 242–3, 260–2, 264, 334, 338; and Ontario Hydro 245, 274, 282, 308–9, 310
– Power Reactor Development Program Evaluation Committee (PRDPEC) 277–8, 311–12, 411
– radiochemical company 441–2
– reorganization 436–7
– sales and marketing effort 307–8, 348, 370, 377, 380–1, 407–8, 425, 428, 433, 435–6
– Senior Management Committee (SMC) 277–8, 415
– Whiteshell 259ff
'Atoms for Peace' 213–14, 350
Attlee, Clement 67–9, 72, 81
Auditor-General of Canada 435, 436
Auger, Pierre 25, 35, 36, 50, 51
Austin, Jack 394
Australia 425
Autopact 302–3

Babcock and Wilcox 231, 235
Bateman, George 68, 73, 74, 75, 114, 115, 143, 198
Beck, Sir Adam 281
Belgium 215–16, 348
Bennett, W.J. (Bill) 29; character of 208; early life 179–80; becomes president of Eldorado 76; on AECD 77, 115, 138; and C.D. Howe 179–81, 445–6; and isotopes 133–7; director of AECL 145; opposes Eldorado-AECL merger 146, 182; and Errington 146–7, 273–4; Eldorado experience 180–1, 197–8; and Board of Directors 180–1; and Lewis 181–2, 249–50; and Chalk River 182; and C.J. Mackenzie 194–5; reverses British procurement policy 194, 200, 371; and power reactor program 194, 200, 202ff; choice of prime contractor 206; and 'Atoms for Peace' 214, 215; and Geneva 214ff; and bureaucracy 349–50; negotiations with UK and US 220–2, 349; and the '5-year' program, 222–4; and Homi Bhabha 351–2; aid to India

352–3; and NPD 202–3, 206, 223, 227ff, 237; and Gordon Churchill 239; establishes NPPD 237, 239–40, 241; and Ontario Hydro 245; resigns 246; later career 438
Berkeley, University of California at 14
Berry, M.D. (Mel) 286
beryllium 127
Bettis laboratory 149, 230
Beynon, C.E. 167
Bhabha, Homi 215, 351–2, 354–6, 357, 359–60, 361, 362, 363–4, 365, 366
Boulton, B.K. 49
Bowden, Vivian 108
Boyd, Winnett 167, 297–9
Boyer, T.W. 111
Bragg, W.L. 55
Brazil 215
British American Oil Company (BA) 318
Bruce nuclear power plant 410, 433; heavy water at 412, 433
Bryce, R.B. 70, 142, 251–2, 308, 404, 432
bureaucracy 74–5, 75–6, 251ff, 301, 341, 360, 387–90, 432, 437
Burge, Ray 414
Burton, E.F. 82
Bush, Vannevar 14, 16, 24–5, 33, 42

Café Henry Burger (Madame Burger's) 252, 360
calandria 293
Calder Hall 203
Cambridge, University of 9, 14, 17, 26, 54, 55
Campbell, Donald 195
Campbell, Ross 425–6, 434, 435, 437, 438
Canadair 206
Canadian Curtiss-Wright 415–16
Canadian General Electric (CGE) 206, 222, 223–4, 228ff, 236–7, 253–6, 269–70, 307, 328, 329, 344, 381–91, 407
Canadian Industries Limited (CIL) 10, 23, 49, 57, 96, 116–17
Canadian Vickers 136, 206
Canadian Westinghouse 190, 264
CANDU: alternatives to 203; ancestry 186–9, 197–8; bi-directional fuelling 201 cabinet approval 260–2, 379; calandria 293; commissioning 286; coolant 200, 201; cost 203, 204, 261; design 199ff, 202, 408, 410; doubts 410; enriched fuel for 411; export prospects 307–8, 359–60, 362, 370, 407, 413–14, 425, 430; fuel rods 258; fuelling machines 294; importance of, to AECL 278; name 256; natural uranium fuel 187, 200, 224; Ontario Hydro and 199, 222–3; pressure vessels and pressure tubes 187, 206, 231–2; quality control 285, 294–5; size 201; symbolism 340; timing of 199, 201, 223–4, 257–8, 259–61, 286, 295; zirconium in 187. *See also* NPD; Douglas Point
Cape Breton Island 322, 324, 326, 329, 344, 432, 433, 449
Carmichael, Hugh 170, 188
Caron, Marcel 333
Carter, Jimmy 163
Cavell, Nik 352
Cavendish Laboratory 9, 30, 54, 107, 351
C.D. Howe Company 142, 167–8, 196, 205, 297
Central Electricity Generating Board: *see* United Kingdom

Central Mortgage and Housing Corporation 91–2, 265, 266
CF-100 148
CF-105 247–8, 264, 284, 447–8
Chadwick, James 9, 13, 46–9, 51–2, 61, 71, 80, 99
chain reaction 6; defined 11; divergent 10
Chalk River: administration of 117–22, 275–6, 306; and AECD 77; budget 65, 123, 126, 128, 399, 400–1, 406; chemistry division 66, 95, 99–100, 111, 155–6, 188; chosen as atomic site 58–60; Commercial Products and 275, 420; construction of 61ff; co-operation with United Kingdom 65–6, 78, 220–1, 278, 371; co-operation with United States, 78, 221, 278, 349; engineering at 195, 240, 306 (*see also* AECL main entry); Future Systems Group 128–9, 138–9, 185; Geneva conference 216–17; geography 83–4; health and safety 123, 126, 129–30, 225–6; Hydro-Québec and 336; industrial co-operation 420, 425; intellectual climate 79, 106, 110, 395–6, 398; language at 401; metallurgy 154, 188; morale 112, 183–4; Nuclear Physics group 188; Nuclear Power Plant Division and 237, 240; nuclear power studies at 183ff, 191, 195ff; Ontario Hydro at 191–2; perfectionist attitudes at 167–70, 205, 229; personnel 396; physics 111, 123, 125, 395ff; Pickering and 312–13; PRDPEC and 278; and regional research centres 122; relations with government 106, 183; reorganized 195; salaries 112; site description 95–6; waste management 126; ZED 395. *See also* NRV; NRX
Chernobyl 165–6, 443
Cherwell, Lord 42, 50
Chicago, University of 14, 32; *see also* Metallurgical Laboratory
China 359, 362, 366, 425, 428
Chrétien, Jean 432
Churchill, Gordon 238–9, 242–3, 246, 249, 250, 258–61, 301, 302, 342, 362, 377, 378, 380
Churchill, Winston 17, 36, 38, 43, 67, 213
Cipriani, André 126, 129–30, 158, 162, 225
CIRUS (or CIR) 357–8, 362, 366
cobalt-60 135–7, 272, 417–18; see also isotopes; therapy machines
Cockcroft, John D. 11, 26, 47–8, 52, 62; early life 54–6; character of 55, 56; appointed to Montreal laboratory 49, 55–6; arrival 50; choice of Chalk River 58–9; and ZEEP 60–1; and the end of the war 63; refuses to stay in Canada 66; appointed to Harwell 67, 70, 71; and NRX 72, 79, 81, 125, 157–8; and security 78; life at Deep River 86–7; departs 81; importance of 81–2, 444; later contacts 138, 220, 374; death 438
Collège de France 5–6, 53
Colombo Plan 209, 352, 361
Columbia University 14, 21
Combined Policy Committee (CPC) 45, 48, 52, 72, 74, 114–15, 213
Commercial Products Division (of Eldorado and AECL): access to reactors 415, 417; budget 415, 416–17; establishment of 133; expansion and markets 271–5, 341, 391, 415;

location 134, 135, 272, 273–4, 417; merged with NRC 135, 146; privatization suggested 415–16, 418, 420; renamed 441; therapy machines 135, 272ff, 415, 416–17, 419–20, 432, 441–2
Compton, Arthur 14–15, 33
Conant, James B. 14, 16, 33–4, 35, 40, 42, 54, 62
Consolidated Mining and Smelting Company (COMINCO) 11, 17, 31
Cook, Leslie (Les) 30, 97–8, 99, 155, 160, 162, 453n
Cook, W.H. 117–18
Cormier, Ernest 27–8
cosmic rays 125
Cripps, Sir Stafford 81
cross-section, defined 6
Curie, Marie 5, 29
Czechoslovakia 408

Davis, Jack 226–7, 324, 386, 438–9
Dean, Gordon 142
Deep River 84, 85, 86ff, 250, 405
Defence Industries Limited (DIL) 49, 57ff, 93, 96, 112ff, 116ff
Defence Research Board 151
Desbarats, H.J. 119
deuterium: *see* heavy water
Deuterium of Canada 318, 321ff, 342, 343–4
Deutsch, John 251, 349
Diefenbaker, John 238, 243, 247–8, 250, 252, 300, 319
Dobson, W.P. 190–1, 193, 200
Dominion Bridge 206, 293
Donnelly, James: early career 438; becomes president of AECL 438; reorganization of company 438–9; confidence in atomic energy 438–9
Douglas Point: site 286–9; output 286–7; economic impact 288; containment shell 290–1; and heavy-water scarcity 290, 317; calandria for 293; quality control at 294–5; operational 295–6, 407, 408–9, 410; significance 296–7, 299, 359. *See also* CANDU; Bruce
Drury, Charles Mills (Bud) 301, 305–6, 321, 323–5, 327, 394, 403, 412
Dulles, John Foster 215
Duncan, James S. 238, 240, 258, 259–60, 276, 281
Duplessis, Maurice 198
Du Pont de Nemours company 33–4, 39, 96, 153, 321
Dupuis, René 145, 189, 198
Dynamic Power 321, 327

economics, of nuclear power 224–8, 316
education 91–2, 93–4
Eggleston, Wilfrid 164
Eisenberg, Shaul 427, 434
Eisenhower, Dwight D. 213
Eldorado Gold Mines Limited 16, 24, 31, 37, 43; postwar history as Eldorado Mining & Refining 76, 77, 132–7; proposed merger with AECL 145–6; *see also* Commercial Products
election: 1957 238; 1958 243, 246; 1962 300; 1963 300, 320–1, 363; 1974 429
electricity, in Canada 174–5, 189–90, 333, 440–1
energy 174, 177–8, 224–5, 393, 422–3, 433, 441; *see also* atomic energy; electricity
Errington, Roy 133–7, 146, 272–4, 275, 391, 416, 417, 418–21, 431, 438
espionage 78–9

Estevan, Saskatchewan 321–2
Euratom 263–4, 348
Evergreen Area 96
Export Credits Insurance Corporation 387
External Affairs, Department of 69–70, 222, 251, 342, 387, 426; character of 350; and safeguards 356, 357–8, 359, 360, 362–3, 364, 366, 407, 429–30

Fermi, Enrico 32, 39–40
Fields, General Kenneth 221
Finance, Department of 251–2, 305ff, 320, 341, 361, 362, 422, 425
Finland 387, 389, 390
Fleming, Murray 142
food irradiation 417
Foster, John 199ff, 229, 275, 283, 285; and Douglas Point 288–95, 408, 409; and NPPD's needs 303, 306, 307; and Pickering 312; and India 361, 365; defends CANDU concept 410; becomes president 431–2; and Ontario Hydro 432; and ministers 431–2, 434, 437; later career 437, 438
Foundation Company 167–9, 357
Fowler, R.H. 11
France: early experiments in 4–9; postwar atomic program 51, 392, 408; safeguards 366; therapy machines 441–2
Fraser Brace contractors 59, 101
Frisch, Otto 81
Frost, Leslie 191, 229, 280–1
Fuchs, Klaus 78, 141
functionalism 70

Gaherty, Geoffrey (Geoff) 145, 189, 202, 207, 276, 283, 359–60, 364
Gandhi, Indira 424, 429
Gathercole, George 409–10
Gaulle, General Charles de 7, 48, 300
General Electric (US) 154, 245–6, 278, 347, 361–2, 423; *see also* Canadian General Electric
Geneva Conference (1955) 212–17, 354
Gentilly: origin 332–5; design 335–6, 337–8; construction 338–9, 414; negotiations with Quebec 335–7, 432, 441; heavy-water supply 410, 432; operations 428
Geoffrion, Claude 333
Gilbert, F.W. 268, 357
Gillespie, Alastair 428, 434
Ginns, D.W. 98
Glace Bay 322, 324, 329, 330–1, 344, 411
Golden, David 403
Goldschmidt, Bertrand 29–30, 35, 36–7, 40, 51–2, 72, 78, 88, 97–8, 99–101, 216, 217–18, 355, 364
Gordon, Andrew (Andy) 145, 267–8, 271, 276, 303, 312, 398–9
Gordon, Walter 145, 225–6, 309, 323
Gouzenko, Igor 78, 113
graphite 31, 37, 40, 41, 53
Gray, Lorne 87, 93, 109; early life 249; appointed to Chalk River 120; character of 121–2, 250, 380–1; management style 252–3, 275, 449–50; view of Chalk River 122, 396–7, 398; negotiates acquisition of Commercial Products 135, 146; plutonium contract and 142; and golf course 153; and NRX accident 163, 165, 166; made vice-president 195; and nuclear power program 202, 258–60, 261; and NPD 233ff; and CGE 253–6,

269–70, 384, 385; becomes president 246, 249; relations with Lewis 249–50, 275, 430, 449–50; relations with Gordon Churchill 250, 259–61, 270, 342, 377, 378, 380; and the bureaucracy 251ff, 301, 341, 360, 387–90, 432; private lunches 252–3, 320, 360; and cabinet 323–5, 327–8, 427–8; reverses Bennett's policy 253–6, 447; second thoughts on organic reactor 271, 430–1; and Commercial Products 274–5, 391, 415–21; on CANDU timing 286, 295; and heavy water 289–90, 316–32, 322–5, 327–8, 342, 406–7; and Winnett Boyd 298; and Douglas Point 414; and Drury 301, 305, 321, 323–5, 327–8, 386, 394; and Pickering 303ff, 310, 312– 3; and NPPD 253–6, 304–5, 386, 407; and Quebec 333–5, 337; and Indian project 357ff, 424; and UK 374ff; and Pépin 389, 394, 407; and Greene 390, 394; and ING 398–406; as salesman 407–8, 425–6; retirement 430, 431; before Public Accounts Committee 435–6; death 438
Green, Don 134
Green, Howard 342
Greene, Joe 390, 394
Grinyer, Charles E. 264, 265, 268, 275, 284, 303, 312, 420
Groves, General Leslie 25, 31–3, 36, 43, 46, 48–9, 54, 58, 62, 65, 68, 78, 80, 149, 444; approves Chalk River 59, 104, 114; disapproves of ZEEP 60–1; view of Montreal laboratory 98–9; visits Chalk River 103–4
Guéron, Jules 30, 35, 51, 61, 78, 97–8
Haddeland, George 201
Halban, Hans von 4–6, 8–12, 14–7, 35, 36, 39–40; character of 4–6, 52; and ICI, 9, 12; visits United States 14–15; in Canada 15–16; and C.J. Mackenzie 21–2, 24–5, 38–9, 41, 46, 64; quarrel with Kowarski 25–6; and Montreal project 28ff; and General Groves 31–3, 46–7, 51; dismissal of 47–9; trip to France (1944) 50–2; later career 52; importance of 53
Hanford, Washington 45, 49, 56, 96, 98
Hanna, Geoff 123
Harvey, Denis 388–9
Harwell 67, 70, 197, 265, 374
Hatfield, Gordon 120, 124
Hawker Siddeley 306
Haywood, Les 306, 312, 313, 336–7, 385, 404, 405, 412, 431
health: in Montreal 27; at Deep River 90; at Chalk River 123, 129–30, 156–9; radio-isotopes and 130ff
Hearn, Richard 145, 189–91, 195, 199, 202, 206, 229, 238, 276, 279–80, 303, 312, 371, 403, 438
heavy water: 7–8, 10, 31, 33, 40, 45, 53, 62, 97, 114, 220–1, 289–90; defined 317; production in Canada 305, 316–32, 409, 410, 428, 432; location of production 320–5; reflections on 343–5
Heeney, Arnold 70ff, 75–6, 349
Hellyer, Paul 322
Hill, Sir John 431
Hinks, Ted 125
Hinton, (Sir) Christopher 125, 138, 187–8, 197, 200, 220, 316, 374, 375–6, 379–80
Hiroshima, impact of 62–3

Hitler, Adolf 3, 4, 10
housing 86–8, 92–3, 109
Howe, Clarence Decatur (C.D.) 23, 27, 34–6, 37–8, 42, 43, 48, 51–2, 57, 84–5, 444; character of 43, 64, 178–9, 250; decides to set up Montreal laboratory 16–7; and the British 68–9, 71–4, 80, 115, 197, 221; and the Americans, 35–6, 37, 43; visits Montreal laboratory 49–50; opposes Chalk River site 58–9; visits Chalk River 72–3; and postwar atomic policy 73–4, 106, 114, 349; and Atomic Energy Control Act 75–6; and Eldorado 132; and therapy units, 138; decision to establish AECL, 139ff; view of AECD 144–5; and Treasury Board 145–6; and business 177, 433; and energy questions 177–8, 191, 210; and W.J. Bennett 179–81, 445–6; and Leslie Frost 191; approves private nuclear power development program 204–5, 479 note 23; presents program to cabinet 208–11; and Indian reactor 353; approves rising costs 222–3, 224; visits NPD 229; and NPD-2 237; defeated 238; fondness for super-projects 249; death 278, 437; and Robert Winters 386
Hurst, Donald 132, 159, 188
Hydro-Québec 198, 271, 301, 332–9
hydrogen, as moderator 7

Ilsley, J.L. 19
immigration 272, 455n
Imperial Chemical Industries (ICI) 9, 12, 15, 23, 26, 40; US mistrust of 23
Imperial Oil 321, 328
India 209, 215, 306, 307, 308, 317, 351–71, 383, 384, 424, 428–30
Industrial Estates Limited (IEL) 321, 323, 325–6, 328, 330, 343, 344
industrial policy or strategy 65, 148, 197, 199, 204–5, 206, 223–4, 247–8, 254–5, 308, 309, 320–1, 326, 399
Industry, Department of 301, 387, 394
Industry, Trade and Commerce, Department of 394
Inglis 206
Intense Neutron Generator (ING) 395–406
International Atomic Energy Agency (IAEA) 217–20, 241, 355–6, 359, 360, 364–5, 366, 374, 424
isotopes 66, 130ff; US production 131; manufacture of, at Chalk River 132–3; marketing 146–7, 442
Italimpianti 425, 426, 427
Italy 347, 425

Jackson, J.F. 26–7, 30–1
James, Gib 154, 408, 410
Japan 347, 425–6, 443
Johns, Harold 136
Johnson, Daniel 338
Joliot-Curie, Frédéric 5, 51, 52
Joliot-Curie, Irène 5

Kemmer, Nicholas 62
Kennedy, Clyde 163
Kent, Tom 341, 344
Keys, David 118–20, 163, 190, 195, 276
King, William Lyon Mackenzie 16, 42, 43–4, 69, 70, 72, 73, 75–6, 81
Kinsey, Freda 88, 94
Kirkbride, David 59
Kissinger, Henry 429

Korea, South 427, 435
Kowarski, Lew 5–7, 8–10, 25, 39–40, 50; abilities of 26; design of ZEEP 60–1
Kuhn, Thomas 106

lattice 96
Laugier, Henri 28
Laurence, George 15–16, 21, 24, 29, 41, 48, 50, 57, 62, 75, 76, 87, 128, 139, 158, 186, 188, 191, 202, 314, 438, 444
Léger, Jules 251, 349, 353
Legget, R.F. 20
Lesage, Jean 302, 333, 335–6, 338
Lessard, J.C. 277, 302, 333, 335, 338, 339
Lewis, Wilfrid Bennett 55, 87, 93, 104, 105, 106; appointed director of Chalk River 80ff; early life 107–8; beliefs and character 108–10, 200, 257, 351, 352, 369, 392, 444; and United Kingdom 108, 200, 371; view of Chalk River 112, 222, 276, 396–7, 398, 400–1, 405; and NRX 159–60; and NRU 126–7; and George Laurence 128, 186; and Cockcroft 151; conception of nuclear power supply 172–6, 184–9; prefers natural uranium concept to breeder 185–6; and British industry 196; and Ontario Hydro 199, 430; and Power Reactor Group 199–202; at Geneva 215–16; negotiates with United States 221; reactor program 222–3; and economics of nuclear power 225–8, 311–12; and redesign of NPD 230–1, 233ff, 236–7; relations with Gray 249–50, 275–6, 449; confidence in CANDU 257, 276, 312; and the organic reactor 263–4, 385, 430; autonomy of 250, 275–6; chairs PRDPEC 277–8, 311–12; and Winnett Boyd 298; and Pickering 311, 312–13; and Homi Bhabha 351; and ING 395–406; retirement 430, 431, 438; death 438; importance of 445
Lilienthal, David 138
Lummus 412

Macaulay, Robert 304, 306
Macdonald, Donald 428, 432, 433–4, 436
MacDonald, Malcolm 17, 24, 42, 51, 52, 80
MacEachen, Allan J. 322, 329, 429–30
MacKay, Ian 128–9, 188, 192, 199, 200, 202, 262, 270, 385
Mackenzie, Chalmers Jack (C.J.) 16; and C.D. Howe 19, 23, 144; character of 19–21, 24–5, 63–4, 116; view of Canadian science 21–2, 65; understanding of atomic research, 22, 65, 69; and the Americans 25, 33–6, 38, 42, 43, 46, 104; and the British 38, 65, 67–9, 71–4, 80–1, 100, 115, 138, 192, 194, 196, 371; and the Montreal project 27, 28–9, 30, 34–9, 41, 46–8, 49–50, 54; and Cockcroft 55–6, 64–5, 66; and Chalk River 58, 62, 100, 103, 105, 106, 115–6, 143–4, 162–3; and Canada's postwar atomic policy 71ff, 99; on AECD 77; and Deep River, 90, 92; and NRU 126–7, 139–44, 166–7; and creation of AECL 144–5; poor relations with Errington 146–7; and nuclear power proposal 189, 192, 202; contacts Ontario Hydro 189, 191;

retirement 179, 194; and Lorne Gray 249; and organic site 265; and Winnett Boyd 298–9; death 437–8
MacNabb, Gordon 434
Makins, Sir Roger (Lord Sherfield) 377
Manhattan Engineering District 25, 29, 43, 75
Manitoba 261, 265, 431
Massue, Huet 145, 189
Mathias, Philip 324
MAUD committee 11–12, 22
May, Alan Nunn 78–9, 98
Mayneord, W.V. 81, 129
McConnell, Lorne 295, 304, 408
McElligott, Father 86, 95
McGill University 27, 82, 86, 118, 122, 128–9
McGoey, May 134
McLagan, T.L. 136
McMaster University 29, 122, 244
McNaughton, General A.G.L. 19, 76, 77, 114, 128
McRae, Ian 233–4
Metallurgical Laboratory (Chicago) 29–30, 32, 36–7
Mexico 407, 425
Mitchell, J.F. 57, 131
moderator: defined 6; 7, 53; in CANDU 291; graphite and heavy water compared 297
Montreal 27–8, 94
Montreal Engineering Company 145, 189, 195–6, 199, 283, 285, 307, 359–60, 365
Montréal, université de 27–8, 85
Montreal laboratory: chemistry division 40–1, 97–8, 99–100; co-operation with Americans 33, 36–7, 39–40, 42, 45–6; decision on heavy-water reactor 48; engineering at 41, 61; future of 40–4, 46–7; future systems group 57; health and safety 57, 157–8; heavy water for 31; morale at 38, 41–2, 47–8; organization 24, 38; physics at 40–1, 97; personnel 29–30; technical physics division 56; theoretical physics division 39
Mooradian, Ara 327, 330, 411
Morison, W.G. (Bill) 290, 295, 315
Munro, R. Gordon 17

National Defence, Department of 138
National Energy Policy 423
National Health and Welfare, Department of 417
National Research Council 9–10, 11, 27; co-operation sought 13, 16–17; history of, 18; organization and procedures 18, 28–9; finances, salaries and budget 18, 23, 30, 38; and C.J. Mackenzie 19–21, 64; as agent for AECD 77–8; and espionage 79; and townsite administration 90; relations with Chalk River 117; Gray's view of 121; limitations of 137, 143, 146
nationalism, economic 279, 281, 301
Nehru, Jawaharlal 352, 354, 355
Nelson, Colonel Curtis 153
New Brunswick 423, 433, 439–40
New Democratic Party 270
Newell, R.E. 25, 41
Nichols, Colonel Kenneth 43
Nicolson, Harold 4
Nobel, Ontario 57
North Atlantic Treaty Organization (NATO) 148
'Northern Vision' 263–4
Norway 7

Nova Scotia 321ff, 328–9, 330, 341–2, 344–5
Nova Scotia Light and Power 207
NPD: ancestry 201–2; cost 202, 222–3, 227–8, 229, 233–4, 237, 240, 243; pressure vessel 206, 230, 231–2; prime contractor 206; approved by cabinet 208–10; construction of 228ff, 253; redesign of 229, 230–1; safety at 228, 232–3; criticism of CGE 236–7; fuelling machines 294; operating record 294
NRC 127–9, 139–40, 186; design and construction 167–70; budget, 166, 168, 222; 1958 accident 170; significance of 171–2, 235; impact on nuclear power program 201–2; isotopes 271; preferred by Indians 354
NRX 59–60, 66, 71, 72, 79, 81; site of 95; design 96ff; cost 101, 102–3, 202; fuel 102, 269; safety 102; operating procedures 104, 156; goes critical 104–5; importance in Chalk River 109; isotopes 271; leaks in 123–4; operating record 123–5, 154–5; high neutron flux 125, 139; estimated lifespan 126–7; in US submarine program 151ff; accident 154–66; design declassified 352; transmission to India 354
Nuclear Power Plant Division (NPPD): *see* Atomic Energy of Canada Limited

Oak Ridge 54, 56, 149; as model for Chalk River 59, 84, 96, 143, 150; isotope production 131, 273
Obninsk 217
Office of Scientific Research and Development (OSRD) 11, 14
Ontario: Conservative government of 191, 280–1, 302; prosperity of 279, 303; resentment of 209, 223, 259, 341, 342; energy requirements 249, 279; reluctance to import energy 279, 281; government and Hydro 281, 308
Ontario Hydro 145, 189, 223, 229, 443; as provincial utility 190, 276–7, 280–1, 288; central generating stations 282; and CGE 237–8, 253–6, 382, 385; corporate culture 280–2, 317–8; engineering 282, 285, 407; first mission to Chalk River 193; Generation and Development Plan 433; heavy-water needs 325, 329, 331–2, 339, 345, 409, 410–2; hesitations 409–14; impact on AECL 196, 199, 204, 237, 238, 253–6, 259–60, 276–7, 283, 304, 317–18; nuclear partnership with AECL 206–7, 283–4, 304, 406–14, 439–40; and building of NPD 228ff, 237; of Douglas Point 285ff; power needs 279, 282, 303; preference for heavy-water reactors 317–18; quality control 285; rate structure 430; steam generating 279–80, 282
Orenda 206
organic reactor (OCR, OTR) 254, 262–71, 342, 381, 430; and Andrew Gordon 267–8; cabinet scepticism 268; and NPPD 268–9; and Gentilly 335, 385
Ottawa 90, 134
Ottawa Group 359

Pakistan 353, 359, 363, 382–5, 389, 390–1, 408, 425, 428
Paneth, F.A. 25, 35

Parizeau, Jacques 334
patents 6, 9
Patterson, Walter 374, 379
Pearson, Lester B. (Mike) 70ff, 300ff, 336, 350, 353–4, 366, 393
Peirce, Commander C.B. 57
Pembroke 84, 85–6, 89–90
Pépin, Jean-Luc 389, 390, 394
Perón, Evita 426
Perón, Isabel 426
Perón, Juan 426
Perrin, Francis 6
Perrin, Michael 16–17, 138
Petawawa 84, 89
Peterborough 270, 381
Pickering: origins 303–5; Ontario-Ottawa agreement 308–9; site 310; size 304, 310–11, 339, 433; design 312–16; heavy-water needs 339, 409, 410, 411; construction 409–10; cost, 412; operations 415, 428, 443–4
Pickersgill, J.W. 322
piles: *see* reactors
Pinawa 266, 270
Pitfield, Michael 437
Placzek, George 25, 35, 50
Plowden, Sir Edwin (later Lord) 179, 197, 214, 372, 374
plutonium 10, 14, 22, 29–30, 32, 37, 56, 69, 96, 346, 358, 363; production in Canada 68, 95–6, 99–100, 111, 127, 139, 151, 226
Pon, George 277, 337
Pontecorvo, Bruno 30, 35, 78–9, 105, 125, 159
Port Hawkesbury 329, 382, 411, 414
Portal, Lord 138
Power, Bill 87
Power Reactor Development Program Evaluation Committee (PRDPEC) 277, 311–12
pressure tubes 231–2, 313–14
pressure vessels 187, 206, 230–1, 346, 372
Privy Council Office 69–70, 389, 433
Public Accounts Committee 428, 435–6

Quebec 198, 300, 325, 332–9, 393, 440–1
Quebec agreement 44–5, 48
Quebec Conference (1943) 43–4
Quebec Hydro-Electric Commission: *see* Hydro-Québec
Queen's University 29

Rabi, Isidore 215–16
radiation hazards 156–60
radiation standards 164
radium 133–4, 136
Rajasthan 361, 362; RAPP I 365, 366; RAPP II 365, 366
reactors 10, 14, 24, 31; advanced gas-cooled (AGR) 347, 379; boiling light water (BLW) 201–2, 246, 271, 309, 337, 347–8, 385, 387; boiling heavy water 379; breeder 57, 79, 127, 129, 185, 186, 187–8, 352; cooling systems 129; design, 40–1; fog-cooled 309, 311; graphite 31–2, 45, 56, 346; heterogeneous 7; homogeneous 7, 193; heavy water 33–4, 47, 48, 56, 61, 153, 310–11; magnox 203, 346, 374, 375; pressurized light-water reactors (PWR) 203, 246, 347. *See also* NRU; NRX; ZEEP, organic reactor; NPD; Douglas Point; Pickering; Gentilly; CIR; Rajasthan
refugees 10
regional development 301–2, 308,

320–1, 322–5, 326, 341–3, 398–9
Reid, Escott 353
Reisman, Simon 301, 302, 305, 308–9, 386, 388–9
religion 90–1, 93
Renfrew County 84, 90
Rickover, Admiral Hyman 149ff, 163, 230, 347
Ritchie, A.E. 352–3
Robarts, John 302
Robertson, Gordon 388–9, 404
Robertson, Norman 70, 139–40
Roblin, Duff 265
Rolphton: see NPD
Romania 408, 443
Roosevelt, Franklin D. 33, 44
Rutherford, Lord 24, 54–5, 107

safeguards, international 354–5, 357, 359, 360–1, 362, 364–5, 366, 424; bilateral 430
safety: see Atomic Energy Control Board
St Laurent, Louis 177, 247, 251, 350, 354
Sargent, B.W. 29, 40, 87, 105
Saskatchewan 77, 327
Saskatchewan, University of 19
Saunders, Robert 190, 206–7, 229, 281
Sauvé, Maurice 336
Savannah River project 153, 220, 289, 322, 331, 342, 412
Sayers, Frank 267
Science Council of Canada 404–5
Scull, B.P. 286
Scully, V.W. (Bill) 143–4
Seaboard Power 324
Seaborg, Glenn 29–30
security 28, 30, 37, 39, 51, 78–9, 82–3, 107, 141, 193, 216
Senior Management Committee 277–8
sex 87–8
Sharp, Mitchell 336, 429
Shastri, Lal Bahadur 356
Shawinigan Chemicals 201
Shawinigan Power 226, 332
Shippingport 217, 347
Shoyama, T.K. 434
Shrum, Gordon 276, 420
Siddall, E.R. (Ernie) 232–3, 290, 291
Smith, George Isaac (Ike) 331
Smith, Greville 116–17
Smith, Harold 195, 199, 202, 234, 236, 245, 256–7, 275, 276, 409, 413–14, 438, 446–7, 451; character of 282–3
Smith, Herbert 253, 389, 390
Smith, Ivan 135
Solandt, Omond 74, 151
Spevack, Jerome 316–18, 322ff, 342
sports 93
Stanfield, Robert L. 321, 329, 331
Steacie. E.W.R. 56, 64–5, 66, 99, 145, 265
Stephens, D.M. 276, 385
Stratton, Phil 285, 289, 293–4
Strauss, Lewis 131, 179, 213–15, 356
Strike, Ross 276, 304, 305, 385
Suffolk, Earl of 8
Sweden 216, 298, 391, 409
Szilard, Leo 4, 131

Taiwan 390, 408, 424–5
Test Ban Treaty, 364
therapy machines 135, 272ff, 415, 416–17, 432
Third World: as focus for Canadian policy 350ff, 368–9, 392
Thode, Harry 29, 403, 405

Thompson, Robert 325
Thomson, G.P. 17
Thomson, Lesslie 30, 38, 43
thorium 57, 66, 74, 100–1, 127, 127–8, 129, 352, 358
Thornton, R.L. 80
Three Mile Island 423–4
Toronto, University of 82, 120, 244
Trade and Commerce, Department of 318, 320, 385–90, 394, 407
Trail, BC 11, 31, 33
Treasury Board 18–19, 142, 146, 235, 250, 251, 320, 401, 402, 412
Trombay 357
Trudeau, Pierre Elliott 339, 393, 394, 424, 429, 430
Truman, Harry S 62, 69
Tupper, K.F. 81, 104, 120, 124
Turkey 407

Union of Soviet Socialist Republics (USSR) 78, 113, 213ff, 217, 364, 443
United Kingdom: and Canadian uranium 16–17, 31, 37, 220; radar in 10, 80–1; heavy-water research in 10, 12, 220, 298, 379; wartime atomic project 9–17; and postwar Canadian atomic project 65–6, 68–9, 80–1, 112–13, 25–6, 138, 151, 194, 196–7, 202; establishment of separate British project 67–8; radium sales to 134; early nuclear power projections 184; industry and Canadian power program 194, 196–7, 371, 372–3; nuclear power program, 203–4, 216, 220–1; competition in reactor design 221, 245, 278, 346–7, 373; immigration from 272, 372–3; market for CANDU 371; relations with Canada 371–2, 373; Central Electricity Generating Board (CEGB) 373–4. *See also* Atomic Energy Authority; magnox; Montreal laboratory; reactors
United Nations: Atomic Energy Commission (UNAEC) 72–3, 74, 75–6, 113, 128; Geneva Conference (1955) 212–17; Relief and Rehabilitation Agency (UNRRA) 134
United States: atomic submarine program 149–50; 'Atoms for Peace' plan 213–14; bilateral agreement with Canada 348, 349; competition in reactor design with 221, 408; heavy-water development 391; heavy-water supply to Canada 114, 151, 221–2, 289–90, 316, 318, 27, 329, 407; IAEA and 217–18; isotope production and imports 131, 272; liaison officer at Chalk River 152; organic reactor and 269; plutonium sales to, 139–42; relations with Canada 37–8, 52, 68, 113–14, 220, 290, 298, 300, 391–2; relations with the United Kingdom 14–15, 23, 32, 44, 113, 141, 213; wartime atomic project (Manhattan Project) 11, 12, 14–15, 25, 29, 56
uranium 132; fuel 96, 269; natural uranium as fuel 186–7, 216, 230, 269; oxide 229, 230; production in Canada 13, 16, 37, 43, 180, 197–8, 225; perceived scarcity of 14, 114; separation of uranium-235 13, 56, 129, 216; in Canada 13, 221; uranium-233 57, 66, 100–1, 127–8, 129, 188
Urey, Harold 39–40
Usmani, Dr 383–4

utilities 189–90, 206–7, 433

Vancouver 265
Varian Associates 420
Viet Nam 370, 424, 433
Volkoff, George 29

Wallace, F.C. 145, 403, 420
Ward, Arthur 125
Washington conference (1945) 72
Watson, Donald 125, 265, 275, 405
Watson, W.H. 104
Weart, Spencer 424
Wells, H.G. 4
Western Deuterium 321–2, 326–7
Westinghouse Corporation 149, 278, 347, 392, 408, 423
Whiteshell 265–71, 275, 276, 277, 381, 391–2, 400, 402, 410, 411, 430
Wigner, Eugene 98
Wilson, I.L. (Willy) 286
World Bank 360–1
Wrong, H. Hume 70ff, 141

xenon poisoning 98

Yaffe, Leo 30, 87, 89, 120

ZED 395
ZEEP 60, 95; and Kowarski 60–1; construction of 61–2; goes critical 62
Zimmerman, Adam 265
Zinn, Walter 79–80, 149, 221
zircaloy 150, 230
zirconium 150, 313